T0309568

INTRODUCTION TO ELECTROPHYSIOLOGICAL METHODS AND INSTRUMENTATION

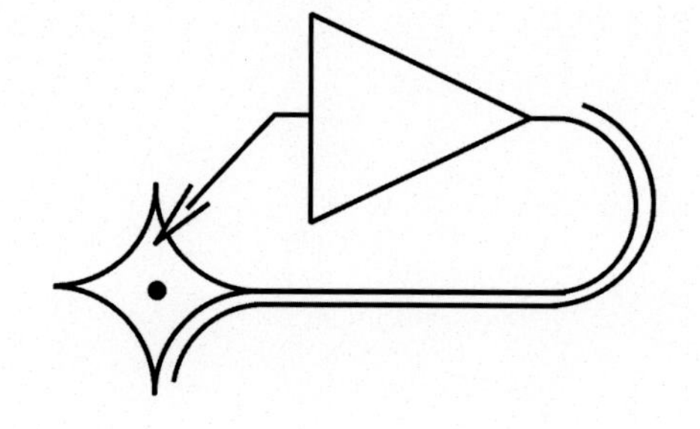

INTRODUCTION TO ELECTROPHYSIOLOGICAL METHODS AND INSTRUMENTATION

SECOND EDITION

FRANKLIN BRETSCHNEIDER
Utrecht University, Utrecht, The Netherlands

JAN DE WEILLE
University of Montpellier, Montpellier, France

Academic Press is an imprint of Elsevier
125 London Wall, London EC2Y 5AS, United Kingdom
525 B Street, Suite 1650, San Diego, CA 92101, United States
50 Hampshire Street, 5th Floor, Cambridge, MA 02139, United States
The Boulevard, Langford Lane, Kidlington, Oxford OX5 1GB, United Kingdom

Background image:
In vitro culture of neocortex cells from the rat brain. Labelling: Red: anti-neurofilament antibody for neurones. Green: anti-glial fibrillary acidic protein antibody for astrocytes. Blue: Hoechst 33342 for nuclei.

White graphs:
Front cover right: **Instrumentation**; A patch-clamp amplifier simplified to its iconic essence. Front cover left: **Recording**; Whole-cell currents from a moto-neuron of the spinal cord of a rat embryo in response to voltage-clamp pulses. Back cover: **Analysis**; A current to voltage plot using the data from the middle panel

Purple graphs:
Front cover: **Recording**; Synaptic currents recorded from rat neocortex neurones. Back cover: **Analysis**; The frequency spectrum of a long stretch of synaptic current data. The spectrum is then used for the detection of synaptic events (marked by the ticks in the right panel above the current trace).

Notices

Library of Congress Cataloging-in-Publication Data
A catalog record for this book is available from the Library of Congress

British Library Cataloguing-in-Publication Data
A catalogue record for this book is available from the British Library

ISBN: 978-0-12-814210-3

For information on all Academic Press publications visit our website at https://www.elsevier.com/books-and-journals

Publisher: Nikki Levy
Acquisition Editor: Natalie Farra
Editorial Project Manager: Kristi Anderson
Production Project Manager: Sujatha Thirugnana Sambandam
Cover Designer: Christian Bilbow

Typeset by TNQ Technologies

Contents

Preface

All living cells use electricity for some function or another. Even bacteria communicate with each other through electricity and have been shown to elicit electrical spikes of millisecond duration, uncannily akin spiking by our own neurones. Mitochondria, which are thought to be of bacterial origin, use electricity to fuel the phosphorylation of ADP into the energy carrier ATP. Perhaps viruses are the sole creatures that are electrically inert, even if some viruses contain DNA coding for ion channels. The sensitive plant or touch-me-not (*Mimosa pudica*) closes its leaflets upon mechanical stimulation. An action potential carried by potassium ions propagates slowly away from the site of stimulation causing other leaflets to close as well. The Venus flytrap, *Dionaea muscipula*, fires a couple of action potentials before closing the trap over an insect within a fraction of a second. Plants do not merely use electricity for exotic behavior but more generally rely on ion channels for volume regulation, most notably that of guard cells and signaling, involving photosynthesis and chloroplast movement. Electrophysiological signals are the fastest in living nature: The brain resolves time differences of less than 20 μs between the arrival of sound in the left and right ears in real time to determine the direction the sound is coming from. In addition to fast signaling, electrical processes are implied in the sensitive detection of weak signals from the environment. Certain fish bestowed with electroreceptors, sense organs for electricity, react to voltages as small as 1 μV across their body wall.

Ever since the archetypical protozoan ion channel came into being, evolution has engendered a wealth of ion channels with diverse functions, modes of operation, and selectivities. The human genome harbors hundreds of genes coding for (units of) ion channels of which more than 200 are destined for the plasma membrane alone. An equally impressive number of ion transporters add to the diversity of electrical signals in living cells. The study of all those channels and transporters in isolation or in their physiological context is a valid goal by itself and is much the realm of electrophysiology. Electrical signals can equally be used as tokens of health and disease, for diagnostic purposes as in clinical medicine. The electrocardiogram (ECG) and the electroencephalogram (EEG) are standard techniques used in that sense. Even if it might be conceivable to routinely operate equipment for diagnostic purposes without the slightest inkling of its inner workings as if it were a kitchen appliance, such lack of comprehension is certainly not to be recommended in the context of research. Whenever we try to measure something, we cannot prevent perturbing the object of observation to some extent. This is of course a general problem in science, but it appears most acute in the practice of electrophysiology, where it seems to be far much easier to record artifacts than the real events. The best way to dominate these problems is knowledge of the potential interactions between equipment and object under study. This is to justify "Instrumentation" in the title of the book. The "Methods" aspect of the book partly encompasses instrumentation, since selecting the right equipment for the job is part of a method to record particular events. Other methods apply to the subsequent analysis and presentation of the recorded data. "Introduction" alludes to our target group, which are graduate students who wish for a solid and up-to-date basis in electrophysiology. It is specifically destined for readers without a formal training in electronics, signal analysis, or electrochemistry and serves as a thorough, yet easy-to-digest introduction that should lead all the way up from a first recognition of principles to the understanding and the routine application of the various methods.

In the early days of electrophysiological recording, amplifiers and other tools were often built by the physiologists themselves. Nowadays, many types of instruments for recording, processing, and stimulation, versatile and almost perfect, can be delivered off the shelf. Despite the streamlined technology and the many computer algorithms available for filtering or postprocessing of the signals, all students of electrophysiology should gain proper insight in the working principles of their principal tools, specifically the vital stages like preamplifiers and electrodes that are connected to the living preparation under study. In planning experiments with the concomitant purchase of instruments, one has to know the possibilities to choose from and their consequences for the validity of the measurements. Since most of these instruments depend heavily on electronic circuitry, introductory electronics takes a significant part in this book. The chapters concerning it are preceded by a summary of the basics of electricity theory. An important source of artifacts in electrophysiological records concerns the chemistry of electrodes. Electrodes are not inert objects but exchange ions with the medium possibly creating an oxidative, reductive, or hyperosmotic environment. They may induce the formation of capacitive double layers that are a blessing in some situations and a nuisance in others. The description of electrode properties could be the subject of an entire volume of which our book resumes the principal aspects.

The second edition contains several updates that we will not enumerate at length here. We just mention a few. Notably, the chapter that overviews some of the classical techniques of recording electrophysiological data includes several updates since the first edition. Although electrophysiology is still the field of predilection of skilled personnel, several companies make efforts to shift it to a push-the-button exercise. Their efforts are possibly sparked by the huge interest that the pharmaceutical industry has in the development of automated electrophysiology due to the compulsory electrophysiology tests imposed by the FDA and the EMA. In the wake of these developments, several unrelated "on-a-chip" applications have appeared that we now discuss as well in the second edition. The chapter about optical methods in electrophysiology is new to the second edition too. Of course, every electrophysiologist has to know how to use a microscope. In addition, several fluorescent dyes and engineered proteins have become available with which cells can be manipulated and examined optically.

Acknowledgments

This book is the product of years of experience and cooperation with many colleagues and students. In the first place, we would like to thank our teacher and colleague Dr R.C. Peters, who laid the foundation of the first edition of this book and encouraged one of us (FB) to extend, improve, and publish it over the many years of our cooperation. We owe him many contributions and suggestions. Dr P.F.M. Teunis gave valuable comment and kindly provided the statistical data pertaining to gamma distributions. Many more people provided valuable comment on the first draft, among them Dr A.C. Laan, Mr W.J.G. Loos, Mr R.J. Loots, Mr A.A.C. Schönhage, Mr R. van Weerden, and several anonymous referees. We also acknowledge the smooth cooperation of Ms K. Anderson and Ms N. Farra and other people at Elsevier. Finally, we would like to thank all our students for explicit or implicit contributions and for their patience with the earlier versions of this book.

Summary

Introduction to Electrophysiological Methods and Instrumentation covers all topics of interest to electrophysiologists, neuroscientists, and neurophysiologists, from the reliable penetration of cells, the behavior and function of the equipment, to the mathematical tools available for analyzing data. It discusses the pros and cons of techniques and methods used in electrophysiology and how to avoid their pitfalls.

Particularly in an era where off-the-shelf solutions are available that in some cases reach high levels of automation requiring less and less intervention of the user, it is important for the electrophysiologist to understand how his or her equipment manages the acquisitions and analyses of low-voltage biological signals. *Introduction to Electrophysiological Methods and Instrumentation* addresses this need. The book presents the basics of the passive and active electronic components and circuitry used in apparatuses such as (voltage-clamp) amplifiers, addressing the strong points of modern semiconductors, and the limitations inherent to even the highest-tech equipment. It concisely describes the theoretical background of the biological phenomena. The book includes a very useful tutorial in electronics, which will introduce students and physiologists to the important basics of electronic engineering needed to understand the function of electrophysiological setups. The vast terrain of signal analysis is dealt with in a way that is valuable to both the uninitiated and the expert. For example, the utility of convolutions and (Fourier, Pascal) transformations in signal detection, conditioning, and analysis is presented both in an easy to grasp graphical form and in a more rigorous mathematical way.

Although the basics of electrophysiological techniques remain the principal purpose of this second edition, it now integrates several current developments. These developments are, among others, automated recording for high-throughput screening and multimodal recordings to correlate electrical activity with other physiological parameters that are often, but not exclusively, collected by optical means. A discussion on signal transducers and relevant aspects of microscopy has been included for this reason.

CHAPTER

1

Electrical Quantities and Their Relations

OUTLINE

Electrophysiology is the field of research on electrical processes in living creatures. This includes everything from current flow through wires to signals in our instruments to electrochemical processes, in living cells and organs, and in the electrodes we use to study them. So, although most of the vast field of electricity theory is outside the scope of this book, we need to deal with a handful of quantities that play important parts, such as charge, voltage (potential), current, resistance, impedance, and especially their complicated changes in time. We start with the main definitions and the relationships between these quantities.

ELECTRIC CHARGE, CURRENT, AND POTENTIAL

The basic quantity is the electric charge, buried in the atomic nucleus as what we call a positive electric charge and in the electrons surrounding it, which we call negative charge. The unit of electric charge (abbreviated Q) is the coulomb (abbreviated C), defined in (macroscopic) electric circuits in the 18th century.

The underlying fundamental constant, found much later (in 1909, by R. Millikan), is the elementary charge, the charge of one electron, which amounts to 1.6021×10^{-19} C. Since

Introduction to Electrophysiological Methods and Instrumentation, Second Edition
https://doi.org/10.1016/B978-0-12-814210-3.00001-3

this "quantum of electricity" is so small, most electric phenomena we will describe may be considered as continuous rather than discrete quantities.

By the nature of atoms, most substances, and indeed most materials in daily life, are neutral. Obviously, this does not mean that they have no charges at all, but that (1) the number of positive charges equals the number of negative charges and (2) the opposite charges are so close together that they are not noticeable on a macroscopic scale. This means that a number of substances can be "teased" to release electricity, e.g., by rubbing them together. This was indeed the way electricity was discovered in antiquity and was examined more systematically from the 18th century on. Many science museums are the proud owners of the large static electricity generators invented by among others van Marum and Wimshurst. These machines generated rather high voltages (around 50 kV), but at very low current strengths (1 μA), and so were not of much practical use.

Nowadays, most sources of electric energy are electrodynamic, made by rotating machines such as the generators in our power plants, in cars, and on bicycles.

These machines can produce almost any voltage and current needed, usually as alternating current (AC) that can be transformed into lower or higher voltages as desired. In addition, electrochemical processes, found originally by Galvani and Volta, are employed in the arrays of galvanic cells we call batteries and accumulators. Both forms of source deliver the electrical energy at lower voltages (say, 12 V) but allow far larger currents to be drawn (hundreds of amperes in the case of a car battery).

This brings us to the most important quantities to describe electrical phenomena: the unit of tension, the volt (V), and the unit of current, the ampere (A). Note that the correct spelling of the units is in lowercase letters when spelled out, but abbreviated as a single capital. In Anglo-Saxon countries, tension is often called "voltage." Both units have practical values, that is, it is perfectly normal to have circuits under a tension of 1 V or carrying 1 A in the lab or even at home. The definitions are derived from other fundamental physical quantities:

Charge (Q): 1 coulomb is defined as the charge of 6.2415×10^{18} electrons.
Current (I): 1 ampere is a current that transports 1 coulomb of charge per second.

An overview of electrical quantities, their units, and symbols is given in the Appendix.

The origin of the definition of tension, or potential difference, is a bit more intricate. The electrical forces that act on charges (or charged objects) depend not only on the field strength but also on the distance traveled. Thus, the electrical potential (abbreviated as U) is defined in terms of the amount of energy, or work (abbreviation W), involved in the movement of the charge from a certain point in the electric field to infinity (where the electrical forces are zero by definition). If one does not move to infinity, but from one point in the field to another point, less energy is involved. This is called the potential difference between the two points. Where to choose the two points will be a matter of practical, quantitative discussion.

Electrophysiologists measure what they call the membrane potential by sticking one electrode into a cell. Formally, then, the reference electrode should be placed at infinity, where the potential is defined to be zero. In practice, however, the potential difference between inside and just outside the cell is measured. For this purpose, the potential just outside the cell can be considered to be sufficiently close to zero. This is because the membrane has a resistance that is many orders of magnitude higher than the fluids inside and outside the cell.

Other circumstances, however, change this view radically: Many electrophysiological quantities are recorded entirely outside the cells, such as electrocardiogram, electroencephalogram, and signals from nerves and muscles. In this case the potential outside the cell cannot be considered zero! Instead, the potential difference between two extracellular points constitutes the whole signal. Nevertheless, the potential difference across the cell membrane is called potential in the long tradition of electrophysiology. The unit of tension, or potential difference, is the volt.

Tension (U): 1 volt is the tension between two points that causes 1 joule (J) of work (W) to be involved in carrying 1 coulomb of charge from one point to the other.

We use "involved" because the energy is either necessary for or liberated by the movement, depending on the direction.

RESISTANCE

The concept of resistance stems directly from these fundamental quantities: If a certain current flows through an object as a consequence of a tension applied to this object, it exhibits the phenomenon of resistance, which is defined as the ratio of voltage to current.

Resistance (R): one ohm (Ω) is one volt per ampere.

These relations are remembered better in the form of equations:

$$I = Q/t \quad \text{or ampere} = \text{coulomb/second:} \quad 1\,A = 1\,C/sec$$

$$U = W/Q \quad \text{or volt} = \text{joule/coulomb:} \quad 1\,V = 1\,J/C$$

$$R = U/I \quad \text{or ohm} = \text{volt/ampere:} \quad 1\Omega = 1V/A$$

The latter law is known as Ohm's Law and is familiar to all people who handle electrical processes. It is often seen in the two other forms, depending on which is the unknown quantity:

$$U = IR \quad \text{and} \quad I = U/R$$

This means that, knowing any two quantities, Ohm's Law gives you the third one. This is used very frequently. In electrophysiology, for instance, one needs to calculate electrode resistances from the voltage that develops when feeding a constant current through the electrode, membrane resistances from measured current values together with the clamping voltage, and so on.

Resistance is the property of an object, such as a micropipette or a cell membrane. Solids, such as copper, and fluids, such as water, also have resistance, but the value depends on the dimensions of the body or water column. The resistance per unit of matter is called specific resistance or resistivity. The dimension is Ωm ("ohm meter"). In electrochemistry, where the small unit system (cgs system) is still used frequently, the unit of resistivity is the Ωcm. As a guideline, fresh water has a resistivity of about 1 kΩcm (10 Ωm) and seawater about 25 Ωcm (0.25 Ωm). Obviously, metals are better conductors, i.e., they have far lower resistivity values: in the order of 10^{-5} Ωcm.

The dimension "ohm meter" may seem odd at first but is easily explained since the resistance is proportional to the length of a water column and inversely proportional to the cross section, which is width × height, or the square of the diameter. So, it is actually a simplification of $\Omega cm^2/cm$.

Other, related quantities we have mentioned already are power and energy, or work. The quantity energy (symbol W) has a unit called the joule (J). The related, often more interesting, quantity of energy per unit of time is called power (symbol P) and has the unit watt (W). So, the performance of loudspeakers, car motors, and stoves is expressed in W. The longer they are used, the more energy is spent (which must be paid), but power is the best characteristic. Electrical power depends on voltage, current, and, through Ohm's Law, resistance:

$$P = UI \quad \text{or} \quad P = I^2R \quad \text{or} \quad P = U^2/R; \quad 1W = 1VA \quad \text{or} \quad 1W = 1V^2/\Omega$$

and so on. Work is simply power times time:

$$W = Pt \quad \text{or} \quad W = I^2Rt$$

and so on. The latter equation is known as Joule's Law.

CAPACITANCE

The quantity to be discussed next is capacitance. This is the ability to store electric charge associated with a voltage. Now what is meant with "store"?

The phenomenon shows up, either wanted or not, when two conducting wires, or bodies in general, are brought close together. If one of the conductors carries a positive charge and the other one a negative charge, a (relatively high) voltage exists between the two. When brought closely together, however, the electric fields influence each other, thereby partially neutralizing the effect. (If equal positive and negative charges would coincide exactly, or have the same center of gravity, the net result would be zero charge or neutrality. This is why atoms, in general, are neutral.) In other words, by bringing two conductors together, the voltage decreases. Therefore, the charge is partially "hidden" or stored. The shorter the distance, and the larger the surface area, the more charge can be stored.

Note that this works only if the two conductors are separated by a very good insulator, such as a vacuum or dry air. Otherwise, a current would neutralize the charges. Other good insulators are glass, most ceramics, and plastics. Note also that charge storage is different from what we saw with resistors: A voltage exists across a resistor only as long as a current is flowing through it. The moment the current stops, the voltage will be zero. A capacitance behaves differently. This can be seen by comparing electric quantities with hydraulic ones. An amount of water is the analogue of an electric charge, a flow of water is the analogue of an electric current, and a water level corresponds to an electric voltage. Capacitance is an analogue of a vat or water butt. When water flows in a vat, the water level builds up slowly, depending on the total amount of water poured in. In a small vat, a certain water level is reached with a smaller amount of water than in a large vat. The larger vat is said to have a larger storage capacity.

In the same way, a capacitor is a vat for electric charge, and the word capacitance is derived directly from this analogy. The unit of capacitance (symbol C) is the farad (symbol F, after Faraday).

Capacitance (C): 1 farad is the storage capacity that causes a tension of 1 V to arise by transferring 1 coulomb of charge or

$$C = Q/V$$

Check the following derived equations:

$$C = It/V \quad \text{and} \quad C = t/R \quad \text{and} \quad Q = C\,V$$

The capacitance exhibited by two conductors depends on their distance and hence on the form of the objects. Wires, spheres, and irregular shapes have part of the surface area closer and part farther from the other conductor. So computing capacitances can be complicated, but for two parallel plates, it is easy:

$$C = \varepsilon_0 \varepsilon_r A/d \tag{1.1}$$

Here, A is the surface area ($l \times w$ for a rectangle, $\pi\, r^2$ for a circle), and d is the distance between the plates. Note that the substance between the plates, called the dielectricum, plays an important part. This property depends on the material used and is called dielectric constant, symbol ε. It consists of two parts: ε_0 is a physical constant called the absolute permittivity of free space, or absolute dielectric constant, and has a value of 8.854×10^{-12} F/m. The second part, ε_r, is called the relative permeability or relative dielectric constant, often dielectric constant for short, and is determined by the material between the plates. By definition, the vacuum has a dielectric constant of unity, air has a value only slightly higher (1.00058), but other insulators have more marked effects on the capacitance. The list below gives approximate values (the dielectric constant is temperature dependent and often also frequency dependent):

Glass	About 3.8…6.7
Mica	6.0
Paper	3.7
Polyethylene	2.35
PVC	About 4.5
China	About 6…8
Water	About 81

Apparently, the dielectric medium influences the electric field in the gap between the plates. The relatively high value of water is due to the dipole form of the water molecules, together with their mobility in liquid water. Thus, a large part of an applied electric field "disappears" in the reorientation of water molecules. But since even the purest water conducts electricity a bit, water is not suited as a dielectric.

MAGNETISM

Magnetic processes, like electrical ones, are known since antiquity. The intimate relations between the two were found much later, however: only about two centuries ago.

The main relation is the following: An electric current induces a magnetic field, but the converse is not true; a magnetic field does not induce an electric current, only *a changing magnetic field* generates an electric current.

Because of this relation, magnetic quantities are often expressed in electrical units. A useful measure of magnetic processes is the magnetic flux density (or magnetic inductance or inductance for short), which has a unit called tesla.

Inductance: 1 tesla (T) is the inductance that exerts a force (F) of 1 newton (N) at 1 meter (m) distance from a wire carrying a current of 1 A or

$$1\text{T} = 1\text{N}/(\text{Am}) \text{ ("newton per ampere meter")}$$

The term flux density can be seen by the following relation:

$$1\text{T} = 1\text{J}/(\text{Am}^2) \text{ ("joule per ampere meter squared")}$$

which means that, integrated over one square meter, a magnetic induction corresponds to an amount of energy.

SELF-INDUCTANCE

The interaction of electric current and magnetic field leads to the notion of self-inductance in the following way: a changing electric current induces a changing magnetic field, but a changing magnetic field induces a current again. Since this induced current is in the opposite direction, the net effect is that the current that flows is lower than it would be without self-inductance. The unit of self-inductance (symbol L) is henry (symbol H). Since self-inductance arises by changing current strength, the description of the process involves time and follows from the following relation:

$$U = -L\,dI/dt$$

In words, the induced voltage is proportional to the change of current strength dI/dt and the magnitude of the self-inductance L (and has the opposite polarity). Hence,

Self-inductance: 1 henry is the self-inductance that causes a tension of 1 volt to develop when the current strength changes (minus) 1 ampere per second.

Apparently, self-inductance is a fundamental property of any conductor carrying current. Nevertheless, self-inductance is strongest when a long wire is wound into a coil, also called an inductor or solenoid. By joining many turns of wire together, the magnetic fields caused by one and the same current are added. Adding a core of iron, ferrite, or any other ferromagnetic material enhances the magnetic induction still further. So, the magnetic properties of iron can be expressed in a quantity similar to the dielectric constant in storing electric charge. This is the relative magnetic permeability and has the symbol μ_r. The vacuum has a permeability of unity, and many substances have values close to 1. Materials that have a (slightly) smaller value, such as water, are called diamagnetic (μ_r about 0.99999); they are repelled by a magnetic field (like with the famous experiment in which a frog is hovering in a very strong magnetic field). Materials with a slightly higher permeability (about 1.01) are known as paramagnetic. In fact, all materials show diamagnetism to some degree, but in paramagnetic materials the latter property dominates the first. Far more conspicuous are the so-called ferromagnetic materials that have permeability values far higher than unity. The best known is iron, of course, but some iron oxides (ferrites), nickel, and chrome are also ferromagnetic. The table below shows approximate values for a few well-known materials:

Cast iron	600
Ferrite	1000
Pure iron	5000
Alnico	8000
Mu metal	20,000
Permalloy	10^5
Supermalloy	10^6

The highest values are from metal alloys specifically designed to have extremely high μ_r. Since μ_r depends on the magnetic field strength, the values in the table are approximate maximum values. Because of this behavior, ferromagnetic materials are used as core materials in electromagnets, transformers, etc. and to store information, such as on computer disks. Iron is suited only for low frequencies (50 Hz–50 kHz), whereas ferrites are useful for both low and higher frequencies (up to about 30 MHz). At still higher frequencies, self-inductance is getting so dominant that no core is needed. This is exploited in radio and television sets but does not play a role in electrophysiology. Note, however, that coils made to block (high-frequency) radio waves from an electrophysiological setup must use ferrites as core materials to be effective.

DIRECT AND ALTERNATING CURRENT; FREQUENCY

An electric current that flows in a certain direction and does never change direction is called a direct current, abbreviated DC (or dc). In a more strict sense, a DC is a current that is constant over time. The tension (or voltage) associated with a direct current is called a "DC voltage." Static electricity, such as that caused by the charge built into "sticky" photo albums, and the voltages of galvanic cells (batteries and accumulators) are examples of DC voltages. In science, the term is used in the more strict sense of an absolutely constant voltage. A battery or other power supply is said to deliver a DC voltage. For the batteries, this is only approximately true, since the actual voltage depends on the current drawn from it (the load), and in most types the voltage decreases slowly by exhaustion during use. Nevertheless, the notion of a DC is useful in contrast with an alternating current, or AC, which is a current changing direction all the time, usually on a regular basis. In many countries the mains voltage, for example, changes direction 50 times per second (in parts of the Americas and Japan 60). Strictly speaking, the mains voltage changes direction 100 (or 120) times per second: first from plus to minus, then from minus to plus again. The convention, however, is to count the number of repetitions, or cycles per second, called the frequency. The unit of "per second" is the hertz (Hz). 1 Hz = 1 cycle/s.

In electrophysiology, one distinguishes frequency from rate: A sinusoidal or other continuous waveform, such as a sound wave or a modulated light intensity, is said to have a frequency (in Hz), whereas a pulse train, such as a train of spikes (nerve impulses), is said to have a certain rate. This distinction is made for an important reason: If a frequency is changed, all parts of the process are accelerated or slowed down. A receiving apparatus, such as an amplifier, must be adjusted to allow the changed frequency to pass. Pulses, such as nerve "spikes," however, have the same shape, and hence the same speed the voltage is changing with, irrespective of their rate. This has consequences for the recording apparatus (amplifiers): see Chapter 3. To underline the difference, a separate unit called the adrian (after the pioneer electrophysiologist Lord Edgar D. Adrian) has been proposed but did not catch on among electrophysiologists. For rates, the unit is usually notated simply as "/sec," "sec^{-1}," or "sp/sec." The difference is illustrated in Fig. 1.1. The sinusoidal shape is considered to be the "basic" alternating current. There are several reasons to do this. In the first place, a sine wave arises in electric generators, such the bicycle dynamo, by rotating a magnet in a coil, or alternatively rotating a coil in a magnet.

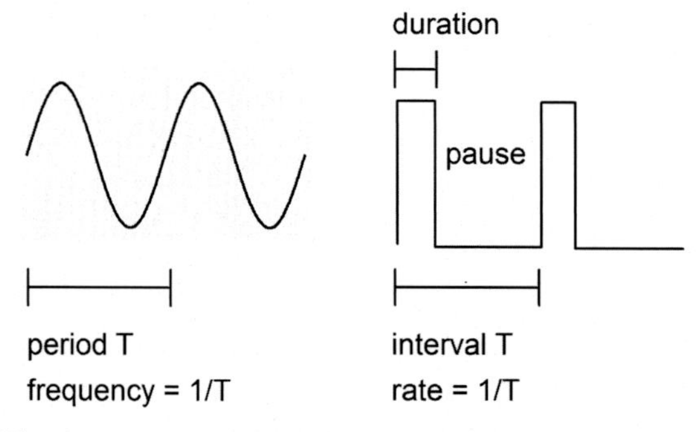

FIGURE 1.1 Parameters of (sine) frequency (left) and (pulse) rate (right).

Thus, most electric power distributed in society is made up of a sine wave at 50 Hz (in America 60 Hz). Secondly, it can be shown mathematically that, upon transforming a signal waveform to the frequency domain, a sine is the "building element" having only a single frequency. Other waveforms, such as square wave, sawtooth, or the tone of an organ pipe, can be considered as combinations of sine waves at different frequencies.

A square wave with a frequency of 100 Hz for instance has components at 300, 500, 700 Hz, and so on. These higher frequencies are always a multiple of the basic, or fundamental frequency, and are called harmonics, overtones, or partials. Note that the fundamental frequency, or fundamental for short, is also called the "first harmonic" (so the "first overtone" is the second harmonic).

Actually a sine wave should be called a "sinusoidally changing voltage" rather than a "wave," since the word wave means the spatial spreading of disturbances in a medium, whereas our sine "wave" is only present at the output terminal of an instrument. Nevertheless, it is a common habit to speak of sine and square waves. We prefer the term "signal."

A sine signal is a function of time, and so the voltage at any moment can be described by the following function:

$$U_t = U_{max}\sin(\omega t)$$

Here, U_t is the voltage at any moment t, U_{max} is the maximum voltage, usually called the amplitude, and ω is the angular frequency, i.e., the number of radians per second. The angular frequency is related to the frequency (f; the number of cycles per second) by:

$$\omega = 2\pi f$$

When connected to a resistor, an alternating voltage will give rise to an alternating current

$$I_t = I_{max}\sin(\omega t)$$

where I_t and I_{max} are the momentary current and the current amplitude, respectively. The relation of voltage and current are again determined by Ohm's law: $U_t = RI_t$ and $U_{max} = RI_{max}$.

REACTANCE

Things get more complicated when we apply alternating voltages and currents to parts that are not simply resistors. For example, we have seen earlier that applying a DC voltage to a capacitor will lead to charging, but once the capacitor is charged to the full input voltage, no current will flow: The two conductors are separated by an insulator, allowing no current to flow between them. When one applies an AC voltage to a capacitor, however, a current (an alternating current) will flow and will keep flowing as long as the AC voltage is applied. Because the input voltage changes polarity many times per second, the capacitance will be charged positively, then negatively, then positively again, and so on. Thus, despite the insulator, a current seems to flow through it.

Note that, although no charge can flow through the insulating layer, the AC current flow through a capacitance is very real. It is not difficult to understand that the magnitude of the current that flows will depend on the capacitance value: A large capacitor stores a larger charge at a given voltage and will sustain a larger current when reverse-charged many times per second. Thus, a 1 pF (pF) capacitor will support only a minute current, even when connected to the 230 V mains. In fact, this is the situation that arises when one picks a mains cable with the hand: The stray capacitance formed by the copper leads and the hand separated by the plastic insulation amounts to one or a few pF. The resulting current, about 300 nA, is harmless, even imperceptible. To the contrary, a capacitor of, say, 100 μF would support a rather large current when connected to a high-voltage AC circuit so that it might be fatal to touch.

Although the mentioned currents resulting from stray capacitances are minute, they can be measured and are often picked up inadvertently by sensitive electronic devices, causing a 50 or 60 Hz interference known as hum. This is frequently a nuisance for electrophysiologists trying to measure microvolts in laboratory rooms, fitted with 230 V (or 120 V) mains cables, outlets, lighting, and the like. The best strategy is to keep mains cables far from sensitive instruments and especially as far as possible from the specimen in an electrophysiological setup. Alternatively, the specimen or the entire setup can be kept in a Faraday cage that will screen any static electromagnetic fields effectively.

Apart from the capacitance value, the current through a capacitance will depend on the frequency, i.e., on how often the charge is reversed. Thus we need a notion similar to resistance to describe the ability of a capacitance to sustain a current. This is called the reactance, or in this case capacitive reactance, and can be computed using the following equation:

$$I_{max} = CdU/dt = Cd(U_{max}\sin(\omega t))/dt$$

which can be converted to:

$$I_{max} = \omega C U_{max}\cos(\omega t)$$

The capacitive reactance (symbol X_C) follows from the ratio of voltage to current:

$$X_C = 1/(\omega C)$$

Like resistance, it is expressed in ohm. Simple calculations show that at 50 Hz, a 100 pF capacitance has a reactance of about 32 MΩ, one of 100 μF of 32 Ω.

Note that the capacitive reactance is inversely proportional to frequency, which explains why high-frequency radio waves, such as the ones from our cell phones, are often a nuisance when performing sensitive electrical measurements (see Chapter 3) and why even standing close to an electrophysiological setup may cause interference.

Estimating the capacitance of a person standing next to a mains outlet at 10 fF, the current flow would be 690 pA. Although this is minute (and harmless), it is nevertheless far more than the current through a single ion channel. Note that at 10 MHz, a 100 pF capacitance would have a reactance of only 160 Ω. Thus, a capacitor that blocks the mains frequency may pass radio-frequency interference.

By a similar argument, the way in which an inductance (solenoid) interacts with AC can be described. Because the counter voltage that develops in a coil depends on the change of current strength, the reactance of a self-inductance, or inductive reactance (X_L), is also frequency dependent:

$$X_L = \omega_L$$

Here, the reactance increases with frequency. Since an inductance does not block a direct current, it can be used to smooth the current from mains-operated power supplies (see Chapter 4).

In Chapter 2, we will describe practical applications of electricity theory and electric circuits.

CHAPTER

2

Electrical and Electronic Circuits; Measurements

OUTLINE

CURRENT AND VOLTAGE SOURCES

The theoretical entities, (ideal) voltage source and (ideal) current source, play important roles in electricity theory. A voltage source is defined as a component that maintains a constant voltage across its two terminals, irrespective of the current drawn from it. We will discuss later on to what extent practical voltage sources can approximate this ideal. In an analogous way, a current source is defined as a component that drives a certain current strength through any connected load, irrespective of the voltage that is needed to do so (and hence irrespective of the resistance of that load). Again, practical current sources will only approximate this ideal.

Introduction to Electrophysiological Methods and Instrumentation, Second Edition
https://doi.org/10.1016/B978-0-12-814210-3.00002-5

COMPONENTS, UNWANTED PROPERTIES

Most of the aforementioned electrical quantities can be put in the form of components, where one property has some well-defined value. A component sold specifically for the property of resistance is called a resistor. The usual construction is a small rod with two axial connecting wires. The rod is either made of a composite material having the desired resistance (composition resistors) or of an insulating ceramic material covered with a film of metal or carbon (metal film and carbon film resistors, respectively). It is important to note that the properties of carbon composition resistors are rather bad: They have a high temperature coefficient, have a high drift and aging, and are by far the noisiest type. Carbon film resistors are slightly better, but metal film is the best buy. Metal film resistors have a far lower temperature coefficient and are more precise and stable than the other two types of resistor. In professional instruments, almost all resistors are of the metal film type. A fourth type is the wire-wound resistor, in which a wire of a metal alloy, again with a low temperature coefficient, is wound on a ceramic body. Metal film resistors are used for most purposes; wire-wound resistors are useful from arbitrarily low values (0.1 Ω or less) up to about 1 kΩ and can be built for high power dissipation. However, a coiled wire might have too much inductance for high-frequency applications. Metal film resistors are made from about 100 Ω up to about 5 MΩ. For still higher values, as often used in electrophysiology, one has to rely on carbon film resistors and compensate for the larger errors involved, e.g., by selection of components and by calibration of the instruments.

The value of the resistor is printed on the rod body, usually as colored rings (see Fig. 2.1). This is done because printed numbers wear off very quickly, notably the decimal point. Therefore, in cases where the value is printed in numbers on the component's body, the decimal point is replaced by the multiplier, e.g., "5k6" is printed instead of "5.6 kΩ," "1n2" instead of "1.2 nF."

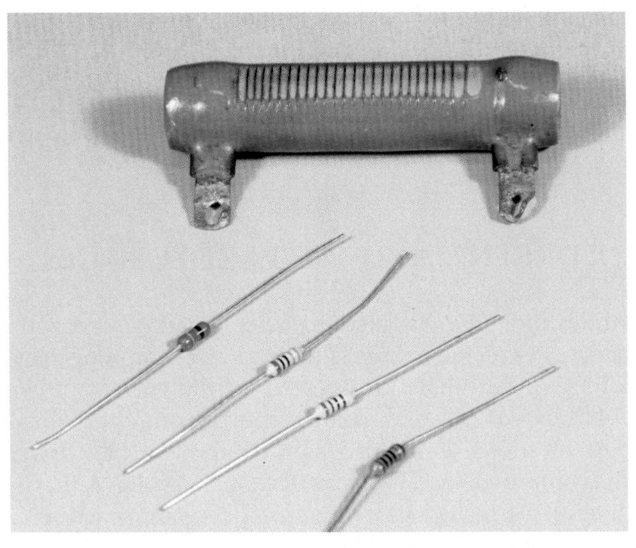

FIGURE 2.1 Wire-wound resistor (top) and metal film resistors (bottom).

The color code is as follows: carbon resistors have three rings. The first two stand for digits of the resistance value and the third one codes for a multiplication factor, i.e., the number of zeros following the two digits. The meaning of the colors is shown below; it is rather easy to remember: the 10 digits are formed by the colors of the spectrum, preceded by earth colors (black and brown), and followed by sky (i.e., cloud) colors (gray and white):

Black	0	or no zeros
Brown	1	or 0
Red	2	or 00
Orange	3	or 000
Yellow	4	or 0000
Green	5	or 00 000
Blue	6	or 000 000
Purple	7	or 0 000 000
Gray	8	
White	9	
Gold	–	1/10
Silver	–	1/100

Thus, a resistor with red, purple, yellow has a value of 270,000, or 270 kΩ, and a resistor with orange, white, gold has a value of 3.9 Ω (39*0.1).

Things can be more complicated, however. The abovementioned resistors have usually a fourth ring that codes for the tolerance, the accepted, one-sided error. Here, gold signifies 5%, and silver, 10% tolerance. Since metal film resistors are far more stable, they are usually made to better tolerances: usually 2% or 1%. Therefore, these resistors have five rings: three digit rings, a multiplier, and a tolerance ring, the latter being mostly red or brown.

A capacitor is a component that behaves as an almost pure capacitance. In principle, capacitors consist of two conducting plates, or electrodes, separated by a thin insulating layer. However, practical designs may differ widely, depending on the desired magnitude range, the necessary precision, and the voltages the insulator must be able to withstand (see Fig. 2.2). Capacitors can be obtained in the range from about 1 pF (10^{-12} F) up to about 10 mF (the abbreviation MFD, still printed occasionally on capacitors, means μF). The first capacitors, used to collect static electricity by 18th century pioneers, consisted of a glass jar, covered with aluminum on both sides. These "Leyden jars" could withstand thousands of volts. The smallest modern capacitors (1 pF–10 nF) consist of a mica sheet, silvered or aluminized on both sides. These capacitors are very stable and may be manufactured to narrow tolerances. Larger values (about 1 nF–5 μF) are obtained with a sandwich of two aluminum foils with a plastic foil in between. These are handy, cheap, and fairly stable. The values are less precise, however. Capacitors with still larger values would either grow to unwieldy sizes or need impractically thin plastic films. Therefore, a metal oxide coating on an aluminum foil is used as insulator instead. These capacitors consist of a metal

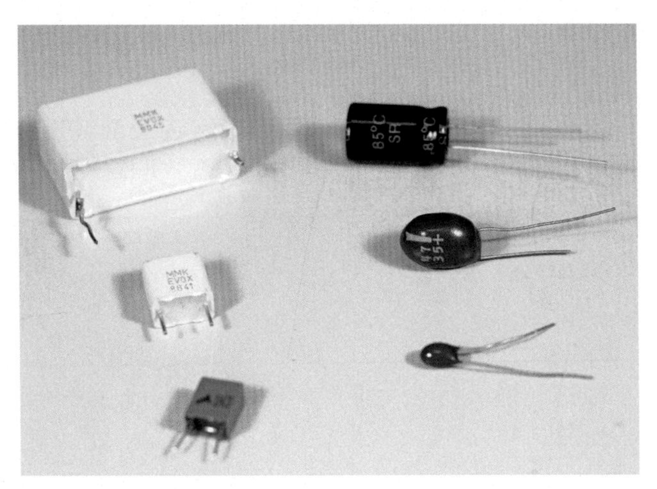

FIGURE 2.2 Capacitors. The ones in the left column have plastic as dielectric; the right column shows electrolytic capacitors (middle and lower of tantalum type).

can, filled with a conducting (yes, conducting) salt solution in which a coiled aluminum foil is suspended. Can plus solution (electrolyte) together form the first electrode, the aluminum foil is the second one. The insulator is merely a thin coating of aluminum oxide (Al_2O_3), made by forcing a current through the finished assembly. Because the insulator can be made very thin, these so-called electrolytic capacitors can be made up to very large values: over 10 mF (10,000 μF).

Unfortunately, this ingenious form of capacitor has a number of disadvantages. First, the thickness of the oxide layer cannot be made very precise so that rated values are only approximate. Factory tolerances might be stated as "$-10 + 50\%$." This means that a capacitor of nominal value 100 μF might have a true capacitance somewhere between 90 and 150 μF. To make it worse yet, the value may change during use. Secondly, the thin insulators cannot withstand very high voltages so that most high capacity "electrolytes" are intended for low-voltage circuits. A lucky circumstance is that the composition of the salt solution can be chosen to oxidize the aluminum and so to "repair" small holes automatically. The electrolyte (salt solution) serves further to warrant the contact between the metal electrode and the oxide layer, and so it is part of one of the conductors. In view of the dielectric constant, water would make an ideal dielectric to make capacitors with, if only it would not conduct. A further peculiarity of electrolytic capacitors is that they are polarized: The oxide-carrying foil must be the anode, i.e., must always be kept positive with respect to the other one.

A special type of electrolytic capacitor is the tantalum capacitor. Here, the anode is made of tantalum, and the dielectric is a thin layer of tantalum oxide (Ta_2O_5). The electrolyte is absorbed in a thin paper foil. These capacitors can be made relatively small, are more stable than conventional "electrolytes," and have slightly better tolerances. They are the preferred choice for most electronic devices, especially in the signal chain. Conventional electrolytes are used mainly in power supplies and in the cheaper forms of consumer electronics.

Still larger values, more than a farad, can be obtained by using an electrical double layer as the dielectric. This double layer between a metal and an electrolytic solution (see Chapter 5) is extremely thin: only a few tens of nanometers, which yields a very high capacitance with a relatively small surface area. However, because the dielectric is so thin, such a double-layer capacitor or supercapacitor can only be used at very low voltages.

UNWANTED PROPERTIES, IMPEDANCE

The different types of components described above can all be obtained commercially and are usually good-quality products. Nevertheless, it is important to note that resistors and capacitors do have unwanted properties, which are in part due to fundamental, hence unavoidable, laws. In short, resistors do have capacitance, and capacitors do have resistance. And both have a (usually very small) self-induction.

Fundamentally, the two ends of a resistor can be considered as two conductors that are fairly close together (mostly less than a centimeter). Therefore, any resistor will have a few pF of capacitance. This is hardly a problem with low resistances (less than about 10 kΩ) but becomes very prominent at higher values, especially in the GΩ resistors used in intracellular and patch-clamp amplifiers. In addition to resistors, all parts of an electronic circuit have unwanted stray capacitance with surrounding objects. Adjacent wires in a cable, or even adjacent lanes on a printed circuit board, have a small capacitance with one another, and any wire has a stray capacitance with its environment, such as the instrument case which is usually grounded.

Conversely, capacitors have at least two unwanted forms of resistance. The first one is the resistance of the dielectric, which should be infinitely high, but is often noticeable as a "leak." The best insulators, such as mica or glass (the latter one only at room temperature), have nearly ideal insulating properties, but capacitors with plastic insulators may show a noticeable leakage resistance that shows up in parallel with the capacitance. Although this is usually high, hundreds of megohms, its existence must be kept in mind and may play a role in electrophysiological measurements, where gigaohms are often involved. Capacitors will also show a fundamental series resistance, since the conductors—metal coatings or foils and connecting wires—have a small but nonzero resistance. Usually, the series resistance may be neglected but will become prominent with large foil-capacitors carrying relatively high currents. In addition to resistance, any coiled-up capacitor foil will exhibit self-inductance.

Practical self-inductances are made by winding long thin copper wires into a coil, or solenoid, often around an iron or ferrite core. This may yield values from less than 1 μH, useful in radio and TV receivers to tune to a specific broadcast program, up to tens of henries, used to reduce ripple in power supply circuits (see Chapter 4).

Because of the use of often long and thin copper wire to cram a large inductance value into a small volume, inductors tend to have a relatively high series resistance that can be neglected rarely, if at all. Current through a coil thus heats it, spending (or dissipating) energy according to Joule's law. Thus, coils are complex components and must in fact always be considered as a (perfect) self-inductance in series with a resistor.

This means that the effective "resistance" of a coil is composed of a "true" (ohmic) resistance caused by the wire and the reactance due to the inductive processes. The notion describing this "total AC resistance" is called impedance (symbol Z, unit Ω). Thus, for a practical inductor, the impedance is the sum of these two properties:

$$Z = R + XL \quad \text{or} \quad Z = R + \omega L$$

Since impedance is frequency dependent, any statement about impedances must include an explicit or implicit statement about the frequency pertaining to it. Often, this leads to conventions, such as in the case of loudspeakers, where the nominal impedance, such as the familiar 8 Ω, is to be taken at a frequency of 1 kHz. In practice, a loudspeaker coil has a resistance of about 3 Ω, and a reactance of 5 Ω at 1 kHz. Note that the reactance will be lower at low frequencies and higher at high frequencies.

The impedance of other components, such as a leaky capacitor, can be derived by a similar argument. Here, capacitance and leakage resistance are parallel rather than in series. Parallel and series circuits will be dealt with in Chapter 3.

Note that perfect or ideal capacitances and inductors do not dissipate energy, only resistances do. This is comparable to the mechanical equivalents of self-inductance, capacitance, and resistance in mechanical systems: mass, spring stiffness, and friction, respectively. A car that would have only mass (inevitable) and springs (mainly added) would bounce restlessly on an uneven road, since the bouncing energy would not be dissipated (in reality, a little energy is dissipated by air friction, but this is far from sufficient). Therefore, real cars have friction added, in the form of oil-filled dashpots, to dissipate most of the bouncing energy.

Inductors (coils, solenoids, see symbols in Fig. 2.7) are not used much explicitly in electrophysiology. An exception is the trick to wind a mains or signal cable onto a ferrite rod to reduce radio frequency interference. To be effective, this makeshift coil must be situated close to the wall of the Faraday cage that harbors one's setup. Practical forms of solenoids are shown in Fig. 2.3.

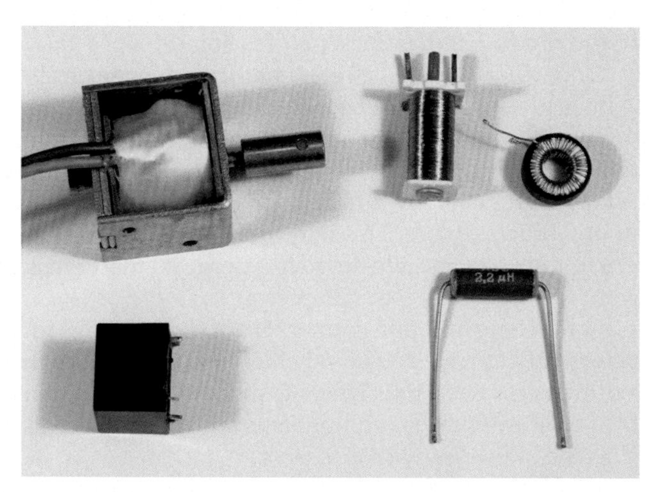

FIGURE 2.3 Practical coils (left top: solenoid, bottom: relay; right self-inductions).

Transformers, composed of two or more coils onto a single core, are much used in all sorts of technical setups (Fig. 2.4). The first coil, where an input AC voltage is applied, is the primary winding or primary for short. The second and any further coils where other voltages can be tapped are all called secondary windings. The mutual relations of electric currents and magnetic fields are exploited in the transformer for the benefit and flexibility of electric circuits. That this device is able to transform voltages needs no argument, but it is nevertheless useful to discuss the virtues as well as the limitations.

The most basic characteristics of a transformer are the turns ratio and the impedance ratio. Since the two solenoids share a common magnetic field, the turns ratio determines the ratio of input and output voltages. But a certain turns ratio can be reached in different ways: a primary of 10 turns, together with a secondary of 1 turn has the same ratio as the combination of 10,000 turns and 1000 turns. The difference lies in the impedances of the two windings. The impedance of a transformer winding is related to the square of the number of turns, and so the *impedance ratio is the square of the turns ratio*. Usually, the impedances must be matched to the voltages and the intended load.

In addition, the dimensions and the construction of a transformer determine how much energy (or rather power) may be converted without overheating it. Therefore, most transformers are specified to these three quantities, e.g., "230—12 V, 25 W." Note that the latter designation states the amount of power that can be transferred, not the power dissipated by the transformer. Since power is voltage times current, the power rating is occasionally stated as "25 VA." In addition to passing useful power on, a transformer also dissipates a bit of energy itself. This is caused partly by resistance of the copper wires and partly by currents generated in the core. This may amount to about 20% of the power transmitted. As an example, a transformer that delivers 24 W to a load (12 V $\times$ 2 A) might spend about 29 W from the source (230 V $\times$ 0.125 A).

It will be obvious that the mutual influence of electric current and magnetic field is limited to alternating current, that is, any DC applied to a transformer is not transmitted. Note that it does, however, cause dissipation, i.e., heating the device up. A second factor limiting the

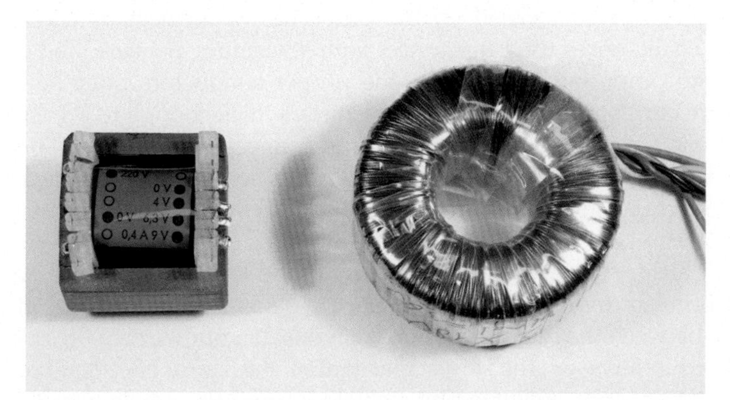

FIGURE 2.4 Practical transformers. A conventional, E-core type at left, and a more modern ring core type at right. The thin wires are connected to the primary winding (high voltage, low current), whereas the thick ones, to the secondary (lower voltage, stronger current). Ring core transformers are preferred because they produce less stray magnetic fields.

amount of DC that may be forced through a transformer is that a direct current magnetizes the core. This may cause the core to be saturated, preventing proper functioning. Transformers are used in most situations where a high voltage must not only be converted into a lower voltage but may also be used in the reverse situation: pushing a low voltage up (a step-up transformer). Note, however, that a transformer does not add energy to a signal and so cannot be used instead of an amplifier.

And, transformers have more bad habits than power loss. Even with AC, the impedances of the windings are frequency dependent and so are the magnetic properties of the core. Therefore, transformers can be used only in a limited frequency range. Even then, a transformer will almost always distort a signal. This is why transformers are hardly ever used to transform signals. They are indispensable, however, in power supply circuits transforming the mains voltage into either higher voltages needed to drive gas discharge tubes used in among others in flat TVs and computer screens and microscope illuminators or to lower voltages needed for most transistor circuits. LED backlight panels running on low voltages are taking over in flat LCD screens.

Apart from transforming voltages, transformers are also useful to insulate two current circuits electrically from each other. A well-known example is the "shaver outlet" found in bathrooms. This is a 1: 1 transformer (230 V in, 230 V out) that may save lives because it insulates the circuit connected to the shaver from the mains voltage, one of the conductors of which is grounded. The shaver circuit is called "floating," since none of the terminals is grounded. The effect is that unintentional contact with one "live" wire does not cause a current to flow through the human body to ground. By the same token, specially designed transformers are used to insulate medical instruments from the patient circuit (ECG electrodes).

In Chapter 3 we will discuss the virtues of and methods for letting an electrophysiological recording circuit "float" (differential recording), which is used to reduce interference.

CABLES

Special attention must be focused on the properties of connection wires, usually called cables. The form of the wires used to connect lamps, vacuum cleaners, and other household appliances is hardly important. Obviously, the copper conductors must be thick enough to carry the necessary currents, and the insulation must be thick enough to prevent shock hazard. Often a third wire is added to provide for a ground connection to the metal parts of the appliance. The same demands hold for cables used to power scientific instruments, but the handling and measuring of weak signals adds new and special demands on the connections used. All signal links must employ shielded cables. The simplest form, used occasionally, is to put one conducting "shield," usually a copper braid, around both signal wires. Although this resolves the hum problem by keeping out the electric fields emanating from mains cables, the stray capacitance between the two signal wires remains. If a number of wires carrying different signals are bundled in this way, the stray capacitances may cause one signal to "leak" to other wires in the bundle. It is not prevented by a shield around the whole bundle (Fig. 2.5).

This cross talk between the wires can be prevented by covering each wire with a separate shield. Since a cylindrical shield around a single wire is concentric or coaxial with it, these cables are called coaxial cables (coax for short). Interference by electrical fields, both from

FIGURE 2.5 Cables. Left: coaxial (Ethernet cable); middle: four wires in one shield (USB cable); right: two separately shielded wires (audio cable).

the mains and from radio waves, and cross talk are prevented to a large extent by using coaxial cables for the transport of weak electric signals. However, the stray capacitance between the wire and its shield remains and this will prove to be a nuisance in several electrophysiological recording situations. The commercially available coax cables have a capacitance of about 100 pF per meter of length. This is often no problem, but the stray capacitance becomes dominant in two situations: high-frequency (i.e., in radio engineering) and high-impedance, i.e., in many electrophysiological recording chains. In fact, the cable capacitance together with the resistance of a microelectrode forms a filter that will block higher-frequency components if not well designed. These filters will be dealt with later on. The most obvious solution is to keep cables short in critical parts of the setup.

A second solution is to use a preamplifier that does not need an input cable at all because it is situated directly behind the electrode holder. This construction is called a probe amplifier. A third solution consists of the use of special low-capacitance cable. This form of coax uses plastic foam instead of massive plastic between wire and shield. Because foam is half plastic, half air, the relative dielectric constant of a foam is between that of plastics (about 2–3) and that of air (1). These cables may have capacitances of about 30 pF/m. Often the use of the "probe" type configuration is mandatory as in intracellular and patch-clamp recording. The bottom line: users of electronic components must always be aware of the unwanted properties of a component (know the order of magnitude) and so estimate the effects before application of that component in a specific research setup.

CIRCUITS, SCHEMATICS, AND KIRCHHOFF'S LAWS

Electrical components are joined into electric circuits. The most primitive way to do so is simply to tie the connecting wires firmly together. Because the surfaces of wires tend to oxidize and thereby lose contact, other methods were invented to make connections that last permanently. Soldering, i.e., the connection of (copper) wires with an easily melting lead/tin alloy, is the method used most often. Mass production leads to the development

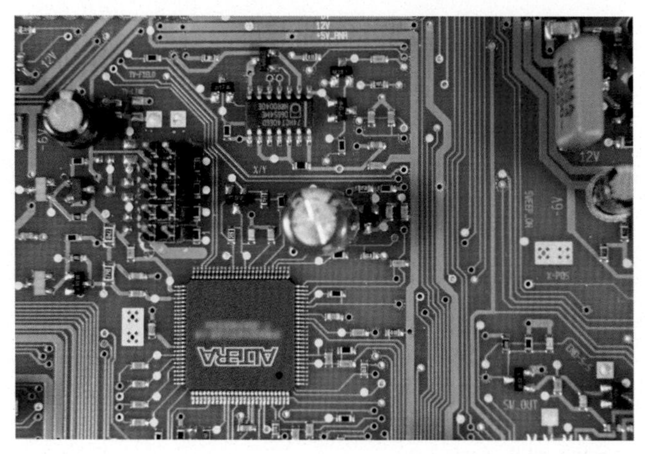

FIGURE 2.6 Part of a printed circuit board showing several components in addition to the printed copper lanes.

of the printed circuit board (PCB), where all wires necessary for a particular circuit are manufactured in one stroke, being etched in the form of lanes on a metalized, glass-fiber reinforced plastic sheet (Fig. 2.6).

The course of the connections between components on a PCB is often very long and sinuous, running around numerous other components. Therefore, it is almost impossible to analyze the function of a circuit by looking at the PCB. To explain the action of an electronic circuit, a schematic, graphical representation called a circuit diagram is drawn, in which specific symbols that indicate the different components are connected by clear, straight lines.

Different forms of symbols are used throughout the scientific and engineering practice, all more or less abstract representations of the form or function of the components they represent and unfortunately laid down in industrial standards in different countries. The symbols used in this book do not conform fully to one of these standards (although they are close to the main European use) but are chosen mainly for clarity. The symbols used for the components dealt with in this chapter are depicted in Fig. 2.7.

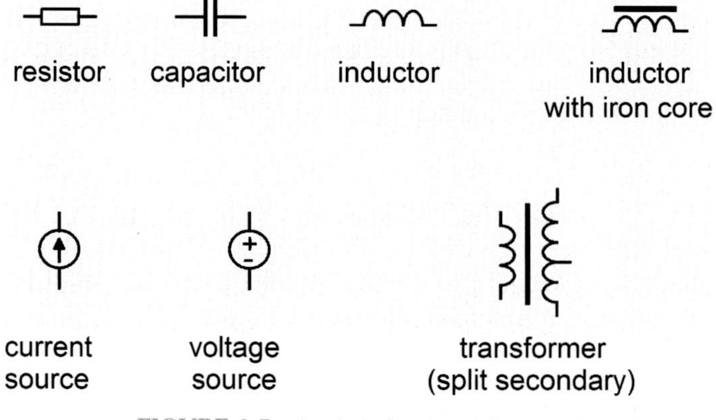

FIGURE 2.7 Symbols for circuit diagrams.

In addition, we will need symbols for switches, voltmeters, outlets, and plugs of both "genders," for electrodes used in electrophysiology, and for such things as earth and ground connections. Several accessory symbols are shown in Fig. 2.8. Note the distinction between case ground (case, frame, or all connected metal parts of an instrument or setup that may not only have contact with the ground but may also float) and ground (the same as before, but definitely tied to ground). Switches exist in a variety of forms that differ in the number of poles and the number of positions, hence SPST (single pole, single throw) for a single-pole on/off switch, and so on.

Note also that the (case) ground symbols suggest an "open end," where in reality all ground connections form a closed circuit. Finally, note the distinction between a wire crossing and a wire connection. The latter form reduces the risk of ambiguity. Further symbols will be introduced where necessary (Fig. 2.8).

An overview of schematic symbols is given in the Appendix.

Most symbols, such as resistors and capacitors, can be used in both the horizontal and vertical position. For some components, a specific orientation is recommended. A second convention is to let the main signal path point in the reading direction, i.e., from left to right. Thus, the primary winding of a transformer is drawn preferably as the one left of the core. A further useful standard is to draw the power supply line at the top of the diagram and the reference (ground) line at the bottom. In the case of dual power voltages, the positive one is at the top and the negative one is usually drawn below the ground (reference) level. Note that most symbols represent a single component, such as a resistor or capacitor, but some are "abbreviations" for more complex entities, such as the notions of a current source or voltage source. Symbols representing entire instruments, used frequently in so-called block diagrams, will be dealt with in Chapter 3.

Before discussing the properties of circuits, we need two fundamental laws pertaining to the behavior of electric quantities in circuits. Joining components together creates nodes where several components meet and loops through which electrical currents may flow. The two notions are illustrated in Fig. 2.9.

The two important laws are formulated first by Kirchhoff (around 1850 AD). The first one, Kirchhoff's node law, states that the sum of all currents entering a node must be zero. Thus, in the example of Fig. 2.9A, the sum I1 + I2 + I3 must be zero. Taking the directions of the currents into account, if I1 = 1 A and I2 = 3 A (toward the node), I3 must be −4 A, or 4 A away from the node.

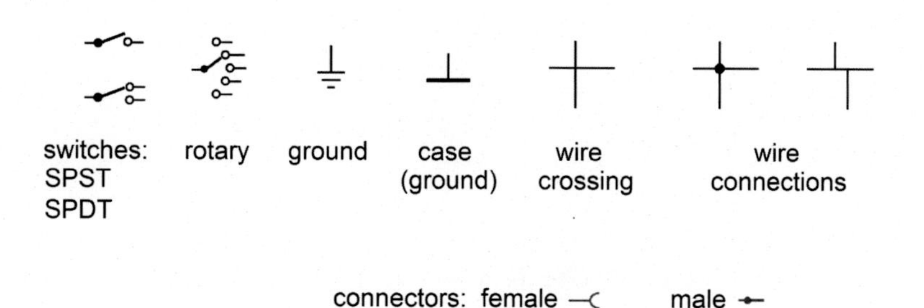

FIGURE 2.8 Accessory symbols.

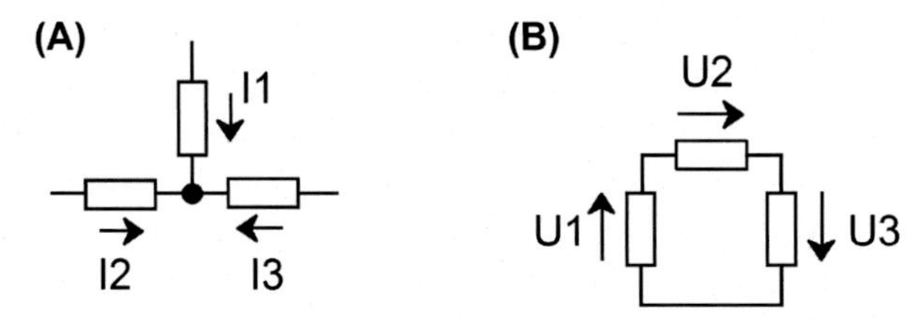

FIGURE 2.9 Node (A) and loop (B).

The second law is the loop law that states that the sum of all voltages in a loop must be zero. Thus, in Fig. 2.9B, if U1 is 1.5 V, U2 and U3 must add up to −1.5 V. Kirchhoff's laws follow directly from fundamental laws like the conservation of energy. Stated simply, neither voltages nor currents can get "lost" in a circuit. Kirchhoff's laws form a simple and powerful tool to analyze the processes in any electric circuit quantitatively.

COMPOSITIONS OF SIMILAR COMPONENTS; ATTENUATORS

The most simple and basic circuits are compositions of components of the same type, such as resistors, connected in parallel, in series, or in a combination thereof. In any case, a composition of two or more resistors connecting two points (nodes) can be replaced by a single resistor, the value of which can be computed using Ohm's and Kirchhoff's laws. Examples are given below. The simplest form is the connection of two resistors in series. The total resistance between the two end points is simply the sum of the two component resistances. This principle holds for any number of resistances in series: see Fig. 2.10.

$$R_{tot} = R_1 + R_2 + \cdots + R_n \quad \text{or} \quad R_{tot} = \sum R_i$$

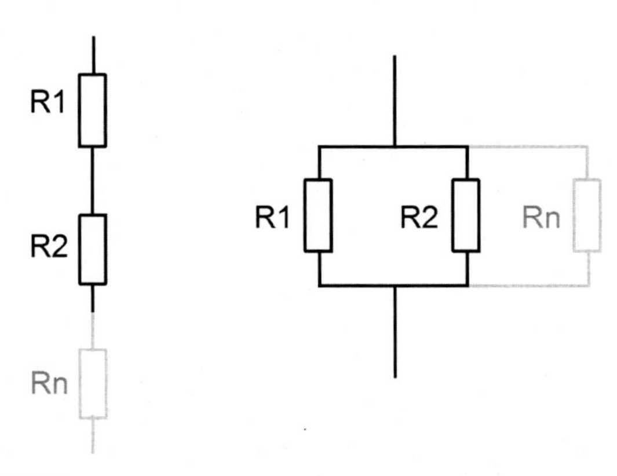

FIGURE 2.10 Resistances in series (left) and in parallel (right).

When connecting resistors in parallel, the total resistance between the two points is lower than any of the component resistors and is often formulated as follows:

$$1/R_{tot} = 1/R_1 + 1/R_2 \cdots + 1/R_n$$

In fact, it is easier to use a quantity called conductance (symbol g) here, which is the inverse of resistance.

$$g = 1/R$$

The notion of a conductance is used very often, especially in electrophysiology, such as in dealing with the amount of current through a membrane or passed by an ion channel.

The unit of conductance is the siemens (S). This unit, in fact the "inverse ohm" or Ω^{-1}, is often called the mho (ohm spelled backwards) in the United States. So, a resistance of 1 MΩ is the same as a conductance of 1 μS. Thus, the equation for parallel resistances reduces to the simple addition of the component conductances:

$$g_{tot} = g_1 + g_2 + \cdots + g_n \quad \text{or} \quad g_{tot} = \sum g_i$$

Similar arguments lead to the effect of combinations of other parts, such as capacitances. It is easy to see that connecting capacitors in parallel increases the total capacitance level of the liquid.

$$C_{tot} = \sum C_i$$

To the contrary, a chain of capacitors in series has a lower capacitance. This can be understood by noting that it amounts to increasing the thickness of the dielectric. The equation is analogous to the case of the parallel resistances:

$$1/C_{tot} = \sum 1/C_i$$

The inverse of capacitance has been used formerly but has never become popular. Guess how Americans used to call this unit ... (yes, the daraf; no kidding!).

With self-inductances, the case is similar to the connections of resistances: in series circuits, the self-inductances sum; in parallel circuits, the inverses of the self-inductances sum.

So far we dealt with multiple components connecting two nodes, but as argued these circuits may be reduced to a single component. Parallel and series circuits are used occasionally, for instance, to compose values that are not on the market. Another application is to spread the power dissipated over more than one component: In this way a 1-Ω resistor that can dissipate 10 W safely may be composed by connecting 10 resistors of 10 Ω, each rating 1 W, in parallel.

More often, however, resistors are combined to create different voltages, for example, to attenuate a signal that is too strong. A well-known example is the oscilloscope probe, intended to extend the range of input voltages by using resistors to divide the input voltage by a factor of (usually) 10. Two resistors in series, used as a voltage divider or attenuator circuit, are shown in Fig. 2.11.

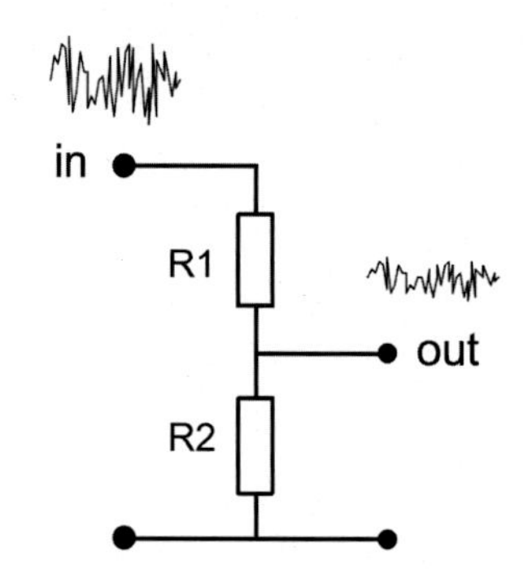

FIGURE 2.11 Voltage divider circuit.

Here, the circuit has three nodes: ground, input, and output. The output voltage as a function of the input voltage can be derived with the following equation.

$$U_{out} = U_{in} \times \frac{R_2}{R_1 + R_2}$$

The factor $R_2/(R_1+R_2)$ is called the attenuation factor.

Two capacitors or two self-inductances connected in the same way can be used as attenuators for alternating current. Often, a voltage needs to be attenuated in a variable way. The most familiar example is the sound volume control on a radio, TV, or audio set. Thus, we need two resistors with a variable attenuation factor. This is done by providing a (carbon, wire-wound, or metal film) resistor with a moveable tap. Such a component is called a potentiometer and is shown schematically in Fig. 2.12.

Since it is purely resistive, a potentiometer may be used to attenuate AC and DC voltages. Note, however, that stray capacitances may lurk around the corner. Practical forms of potmeters and switches are shown in Fig. 2.13.

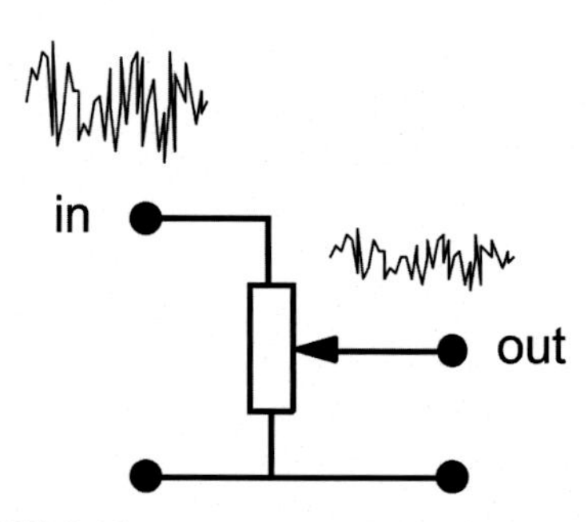

FIGURE 2.12 Potentiometer as a volume control.

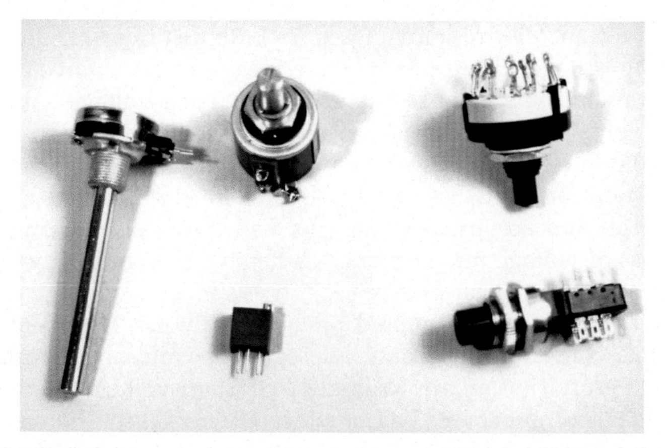

FIGURE 2.13 Practical potentiometers (left and center) and switches (right).

PRACTICAL VOLTAGE SOURCES AND CURRENT SOURCES

Apart from the useful attenuators described above, voltage division often occurs unintentionally. This is shown best when we consider practical voltage and current sources. The ideal of a voltage source maintaining a constant voltage across a load is approached to a fair extent by the well-known batteries and accumulators. However, short-circuiting a battery results in a drop in voltage rather than an infinite current.

In fact, the stated (nominal) voltage of 1.5 V of the familiar alkaline battery exists only when no current is drawn from it. The more current is drawn, the more the voltage decreases. This can be explained as a resistance in the voltage source.

Thus, a real voltage source can be considered as an ideal voltage source in series with a (small but noticeable) source resistance (R_{src}), also known as the internal resistance or the output resistance, since it is manifest at the output of the voltage source. Thus, a practical voltage source can be represented in the way shown in Fig. 2.14A. The source resistance is not an added component but an unavoidable property of any voltage source. The load resistance (R_{load}) is the representation of the lamp, motor, instrument, or whatever is connected to this voltage source. It is easy to see that source resistance and load resistance together form a voltage divider, each getting their share of the total voltage, which was historically called the electromotive force (it is not a force, however).

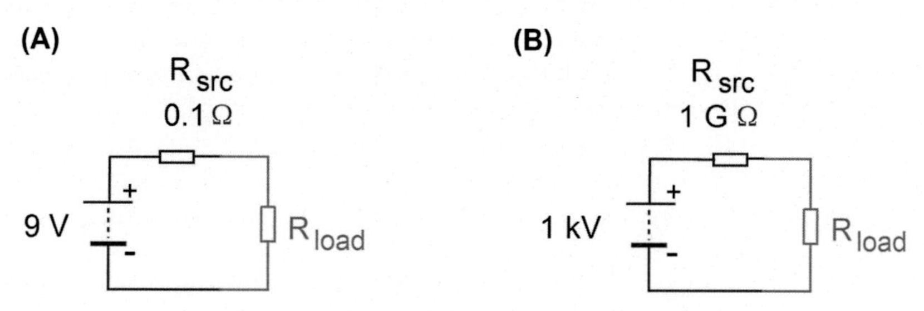

FIGURE 2.14 Practical voltage (A) and current (B) sources.

The symbol used for an "electromotive force" is E, to distinguish it from the "practical," "actual," or useable voltage U. Apparently an ideal voltage source has a zero source resistance, and the lower the source resistance, the better a voltage source behaves. As a rule of thumb, the voltage can be considered to be constant as long as the load resistance is high relative to the internal resistance. How much higher will depend on the precision wanted. Usually, a ratio of 1: 100 is acceptable, since this will introduce an error of only 1%. Although batteries and accumulators are not bad, the design of any instrument must take the inherent loss of voltage into account. Readers who wondered why a 3-V flashlight needs a 2.2-V bulb have the answer here.

Practical current sources are encountered rarely in daily life. A possible exception is the solar cell, which at a constant illumination yields a constant current, proportional to the amount of light falling on it (at an approximately constant voltage of 0.5 V). Contrary to a voltage source, an ideal current source has *infinite source resistance*. Therefore, a practical current source can be built using a high-voltage source in conjunction with a high series resistance, such as shown in Fig. 2.14 B. In this example, the 1 kV source combined with the 1 GΩ resistor delivers approximately 1 μA to the load irrespective of the load resistance, as long as the load resistance is small with respect to the applied source resistance. How much lower will again depend on the wanted precision. Note that in this case, the low source resistance of the 1 kV power supply is increased artificially by the 1GΩ resistor. Resistors of this order of magnitude are used among others in microelectrode preamplifiers to provide a test current for measuring the resistance of the glass capillary. Since capillary resistance in an experiment may exceed 100 MΩ, the errors are often not negligible and must be corrected by means of a table or a calculation.

VOLTAGE AND CURRENT MEASUREMENT

The reliable measurement of voltage and current faces the same problems related to internal resistances. The classical way to measure a voltage is by a galvanometer or moving-coil meter in which a thin-wired coil fitted with a pointer is suspended in a magnetic field. A current through this coil causes it to rotate around its pivot so that the pointer moves over a graduated scale. The deflection may be made to be very nearly linear with the current strength. Moving-coil meters may be made fairly sensitive, such as 100 μA for the maximum, full-scale deflection (fsd). Since the coils have resistances of about 1 kΩ, full-scale deflection is reached at a voltage of about 0.1 V. Today, the cheapest voltmeter for use in domestic electrical circuitry is still based on this moving-coil or microammeter (see Fig. 2.15, left).

Not surprisingly, however, this basic device is superseded by electronic ones, called DVMs (digital voltmeters; Fig. 2.15, right). The voltage to be measured is amplified electronically and usually converted into digital form, presented on an LCD display.

Unfortunately, the mode of operation and the limitations of such devices are hard to fathom. Therefore, we will digress briefly on voltage and current measurement with a moving-coil meter. DVMs may have better performances, but the principles to observe remain the same.

The functioning of a moving-coil voltmeter is not hard to understand. A voltage across its terminals causes a current to flow through the wire. The current causes a magnetic field to be

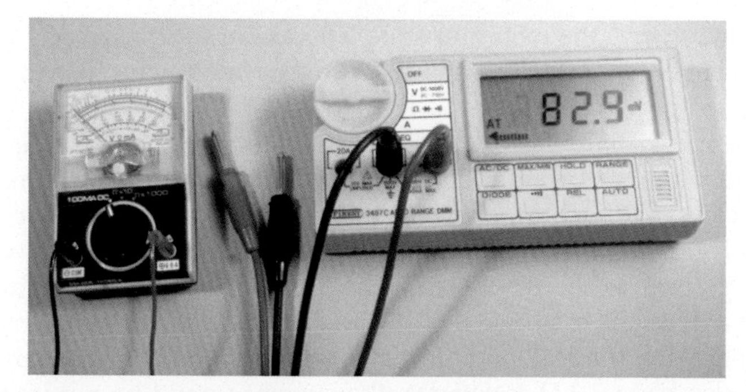

FIGURE 2.15 Moving coil meter (left) and digital voltmeter (DVM; right).

developed, which in turn moves the pointer. This means that a voltmeter *draws a bit of current out of the source* that it intends to measure the voltage of. The current drawn will reduce the voltage more or less, depending on the ratio of the internal resistance of the voltage source and the resistance of the meter coil. This is illustrated in Fig. 2.16. The circuit is shown at the top left. However, we are more interested in the electrical properties of the components. Any real-life voltage source has an internal resistance and so has the voltmeter. Therefore, we need to derive the so-called equivalent circuit. This is drawn at the top right. Here, R_{int} is the internal resistance of the voltage source and R_m, the resistance of the meter coil. Let us take a small battery such as the ones used in hearing aids and watches as an example. The voltage will be about 1.5 V and the internal resistance, about 1 Ω. If the voltmeter has a resistance of 1 kΩ, then the meter draws about 1.5 mA from the battery (1.5 V over 1001 Ω). In this case, the voltage drop across the battery's internal resistance is 1 mV. Thus, the voltage is underestimated by 1 mV. In all but the most demanding cases, this can be neglected. Nevertheless, a voltage is principally underestimated by the influence of the measuring device. The higher the meter's resistances, the better the approximation.

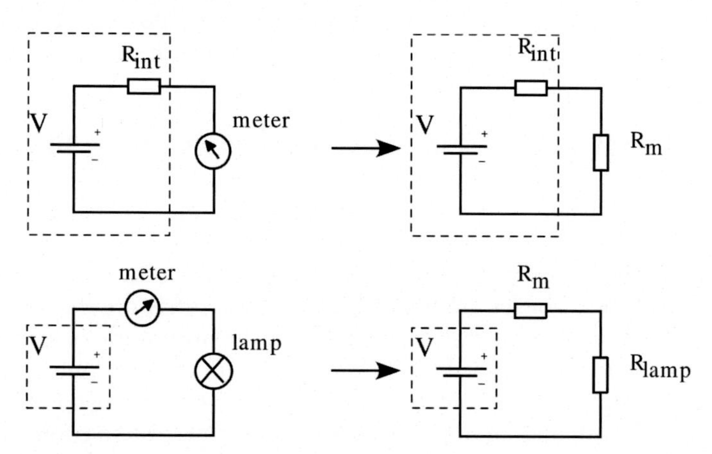

FIGURE 2.16 Errors in the measurement of voltage (top) and current (bottom). The real circuits are at left and the equivalent circuits at right. The dashed rectangle shows which properties belong to the battery.

For current measurement, the meter must be inserted in series with the other components of the circuit. We must break the circuit and insert the current meter to measure the lamp current in a flashlight for instance.

This is shown in Fig. 2.16, bottom row. Here, the demands are the other way round: The current that flows through the circuit must not be hindered by the current meter and so a very low meter resistance is needed. Suppose, for example, that the lamp in our example behaves as a resistance of 3 Ω. When connected to a 1.5-V battery, the lamp current would be expected to be 0.5 A. However, even without a meter in the circuit, the lamp current also flows through the battery's internal resistance. If we take this again at 1 Ω, the total resistance in the circuit amounts to 4 Ω. Therefore, the real lamp current would be 0.375 rather than 0.5 A. If our current meter would also have a 1 Ω resistance, the current would be reduced further to 0.3 A (1.5 V over $3 + 1+1 = 5\ \Omega$). Again, the measurement is an underestimation of the true value. Here, a lower meter resistance improves the approximation.

In the examples discussed above, the errors involved are not too serious: The battery voltage is estimated slightly too low and the current through the lamp is reduced a bit. For measurements in electronic and electrophysiological circuits, however, the errors would be far too high and might amount to almost 100%. Hence, electronic voltmeters having far higher input impedances must be used.

For any voltage measurement, the input impedance needed can be computed from the values of source impedance (i.e., the circuit under test) and meter impedance that together form a voltage divider circuit. The impedance of the meter should be about 50–100 times as high as the circuit's impedance to get a fair precision. In current measurement, the meter resistance must be lower than the other resistances in the circuit by approximately the same factor.

When monitoring the mains current drawn by a lamp or a similar appliance, losing a few hundred millivolt from the 230 V is no problem, but in low-voltage circuits, a similar voltage drop may be significant. Finally, electrophysiological signals are obviously too small to be measured in this way, being mostly less than 100 mV themselves. Therefore, we need electronic voltmeters and preamplifiers treated in Chapter 4. Next to the specialist equipment, however, a simple DVM that can measure voltage, current, and resistance (as most types do) comes in handy, e.g., for the checking of resistances, testing of cables, batteries, and so on.

COMPOSITION OF UNEQUAL COMPONENTS: FILTERS

The next circuit to be treated is composed of a resistor (R) and a capacitor (C). This important circuit behaves as a filter and is therefore called an RC filter. Other combinations of the basic components are the RL filter and the LC filter. The latter is very popular, although few people will realize that their houses are full of LC filters. LC filters are the universal, resonant filters that can single out one broadcast frequency from the multitude of signals that fills the ether. Tuning a radio or TV set is done with LC filters, but for electrical measurements and for the manipulation of electrical signals in general, RC filters are employed mostly.

Since a filter has an input and an output, the RC filter exists in two configurations, shown in Fig. 2.17.

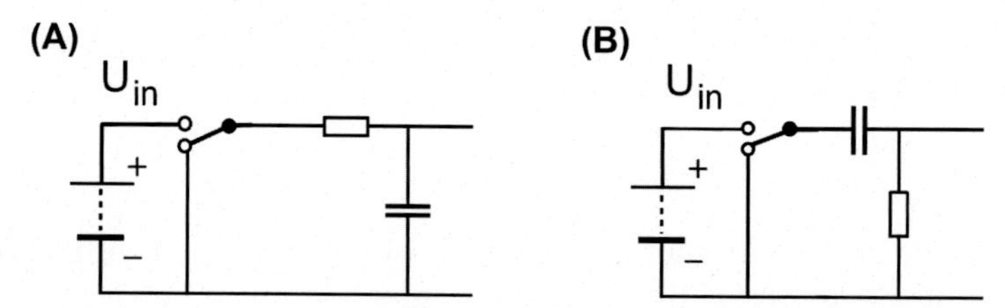

FIGURE 2.17 The two forms of RC filter: (A) low-pass and (B) high-pass.

The two forms are called low-pass and high-pass filters, respectively, names that will become apparent through the analysis of what happens when we feed signals into them.

Let us analyze the low-pass configuration first. We have seen earlier that a capacitor can be charged by feeding a current into it. This we will do by adding a voltage source (of, say, 1 V) and a switch (Fig. 2.17). The capacitor is shorted until the start of the experiment to make sure it is uncharged. As a simple example, we take a capacitor of 1 F and a resistor of 1 Ω. At time zero, i.e., the moment we have flicked the switch, the voltage appears at the input and so across the resistor (since the capacitor is uncharged, the output voltage is still zero). The charging current can be computed easily: $I = U/R$, that is, $1/1 = 1$ A. From the properties of electric current and the units chosen, this would charge the capacitor in 1 s, that is, if the current would remain constant. But it is easy to see that the current decreases continuously. After a certain time, the capacitor is charged to 0.5 V. At that time, the voltage across the resistor is reduced to 0.5 V, which in turn reduces the charging current by the same factor. In other words, charging gets progressively slower and the input voltage is approached asymptotically, described by an exponential function. This is shown in Fig. 2.18.

The mathematical description of this process is fairly simple and follows from the above discussion. We saw that the charging rate, that is the change of the output voltage

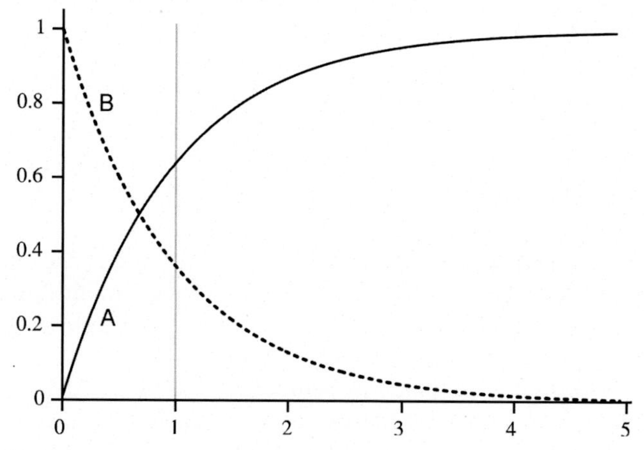

FIGURE 2.18 Step responses of RC filters (for explanation, see text).

(dV/dt), is proportional to the voltage across the resistor and this is equal to the difference of output and input voltage ($U_{out}-U_{in}$):

$$dV/dt = K(U_{out} - U_{in})$$

where K is a constant that determines the timescale of the charging process. The solution of this simple differential equation is the exponential function given below:

$$U_{out} = U_{in}(1 - e^{-Kt})$$

It will be clear that the constant K must depend on both R and C. With a higher resistance, charging would be slower because less charge is transferred per time unit, and the charging would also be slower whereas with a higher capacitance, this time because more charge is needed to develop a certain voltage. The quantity determining the timescale is the product RC, usually called the time constant (symbol τ, the Greek lower case tau). The process is analogous to the charging of a water butt through a (thin) tube. Thus

$$K = 1/RC \quad \text{or} \quad K = 1/\tau, \quad \text{hence} \quad U_{out}/U_{in} = 1 - e^{-t/\tau}$$

By dividing through the input voltage, a normalized, or dimensionless, quantity is obtained that runs from zero to one in an asymptotic way. This is shown in Fig. 2.18, curve A.

Mathematically, this is the same process as radioactive decay: When half the original amount is left, decay is also half as fast. Therefore, the characteristic used to describe the speed of the decay is the half-life or the time span in which half of the substance falls apart. We could also compute the half-voltage time for capacitor charging, but in practice, the base of the exponential function (e) is used (grey vertical line in Fig. 2.18). Thus, the time constant (τ) represents the time in which the output voltage is changed to (1−1/e) times the voltage step at the input.

The exponential curve shown is called the step response of the RC filter. Other filters will, in general, show other step responses so that a number of filter circuits can be characterized by their step responses. If the switch is flicked again, the capacitor will be discharged. Since the current flows through the same resistor, the discharge curve is similar to the charging curve, except that it is inverted, going from +1 to 0 V (Fig. 2.18, curve B).

What about the response to sinusoidal signals? Most often, a filter is fed with a composite signal that may consist of a sine wave of a certain frequency or a combination of sinusoids with different frequencies. It is not difficult to understand what happens when a sine signal is fed into the low-pass filter described above. We have seen with the step response that the output voltage lags behind the input voltage because it takes time to charge the capacitor. The same happens when we feed a sine signal into a low-pass filter: the output voltage lags behind. In other words, the phase of the sine wave is changed. In addition, the sluggishness of the response causes the amplitude to be lower than the amplitude of the input signal.

This amplitude decrease or attenuation depends on the frequency of the signal, and a frequency-dependent attenuation is just the definition of a filter. At very low frequencies, there is plenty of time to charge the capacitance so that it has hardly any effects on the output signal: The output signal is virtually identical to the input signal. At progressively higher frequencies, however, the effects of the capacitor become more prominent, causing both a phase delay and a lower output amplitude. At very high frequencies hardly any output signal is left; these frequencies are said to be filtered out. The degree of filtering, plotted as a function of the frequency (more precisely the logarithm of frequency), is shown in Fig. 2.19. This is called the frequency response or frequency characteristic. A is called the amplitude characteristic, and φ (Greek lower case phi), the phase characteristic.

The figure shows that the transition from passing the signal unaltered to filtering is a gradual one, sloping down with increasing frequency. The phase delay runs from 0 to 90 degrees ($\pi/2$ radians) also in a very smooth way with increasing frequency. Despite this wide area in which the filter affects the input signal, we can choose an arbitrary but practical point as the boundary, called corner frequency, also called roll-off or cut-off frequency. The frequency that is best taken as a standard, shown as a dotted line, is the point at which the signal is attenuated by 3 dB (stated otherwise, the gain is -3 dB). The phase is delayed 45 degrees ($\pi/4$).

Since all RC low-pass filters show the same *form* of filtering behavior, the cut-off frequency is a complete and sufficient description to characterize it. One is free to choose R and/or C, as long as the product RC remains the same (however, not entirely free, since the impedance of such a filter depends on the values chosen).

The amplitude is expressed in the well-known logarithmic measure called the decibel. The decibel is a relative measure and originated in telephone technology, where power rather than voltage is the interesting quantity. Hence, the unit of sound power, bel (B, after Alexander G. Bell, inventor of the telephone), was defined as a factor of 10 (one order of

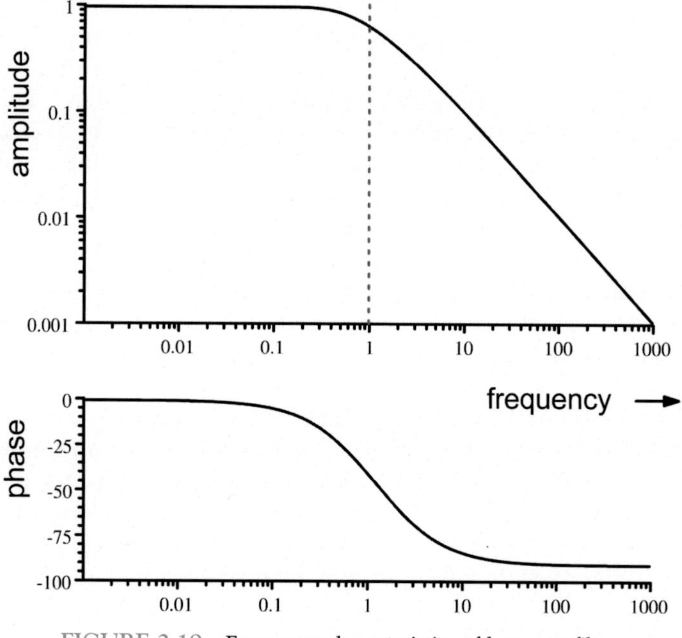

FIGURE 2.19 Frequency characteristics of low-pass filter.

magnitude) relative to some convenient standard. Because most measurements span a few bels and because about 0.1 B is a just perceptible change, the decibel (dB) became the standard unit. The decibel is often used to describe the gain (G) of amplifiers. In that case, the reference is simply the input voltage. For filters and attenuators, where the output voltage can only be equal to or less than the input voltage, the gain values in dBs are zero or negative, respectively.

For the description of filter behavior, we are discussing voltage rather than power ratios. Therefore, the equation for the conversion of the "gain" of a circuit into the decibel form is:

$$G = 20\log(U_{out}/U_{in})\mathrm{dB}$$

Here, U_{out} and U_{in} are the voltages to be converted, and the factor of 20 arises because (1) a bel is 10 dB and (2) power is the square of voltage, contributing a square under the log-sign or the equivalent, a factor of two in front of it.

The gain at very low frequencies is virtually 0 dB and becomes more and more negative with increasing frequency. The gain is −3 dB at the cut-off frequency, therefore often stated as the −3 dB point. Check that here the output voltage is $1/\sqrt{2}$ times the input voltage, or about 70.7%. The ratio of output to input powers is 1/2 or 50%. Thus, as a simple rule of thumb, the roll-off frequency is where both power transmission and phase shift are "one half."

Having described the low-pass filter by its time behavior and its frequency characteristic, it is easy to extend the discussion to the high-pass version: The characteristics of the high-pass filter are the complement of the functions described above. Thus, the step responses of the high-pass filter can be taken from Fig. 2.18 again, but in this case curve B depicts the charging behavior (starting at the moment the input voltage is switched on), whereas curve A is the discharging curve (from switching off). Thus, contrary to the response of the low-pass filter, the voltage step at the input is transmitted unattenuated after which the voltage returns to zero. The voltage step is, as it were, "forgotten." This is also known as adaptation and is seen very frequently in responses of sense organs and other neurophysiological structures. Note that a negative peak develops after the input step back to zero. This is also seen in neural signals, for instance, in the response of photoreceptors, where the spontaneous activity is lowered or even suppressed immediately after switching an excitatory light off.

The frequency characteristics of the high-pass circuit are also complementary: The highest frequencies are passed unattenuated, whereas amplitude reduction and phase shift occur at increasingly lower frequencies (Fig. 2.20). The cut-off frequency, here taken at unity again, is again the −3 dB point.

The phase shift occurs also in the opposite direction, running from 0 (at infinite frequency) to 90 degrees ($\pi/2$) lead rather than lag with 45 degrees at the cut-off frequency.

Hence, the output leads 90 degrees at very low frequencies, that is, the peak of the output signal occurs earlier than the peak of the input signal. At a first glance, having an output signal before the input may seem a violation of causality. This is not so, however, because we are describing a steady-state situation, where the input signal already existed for a

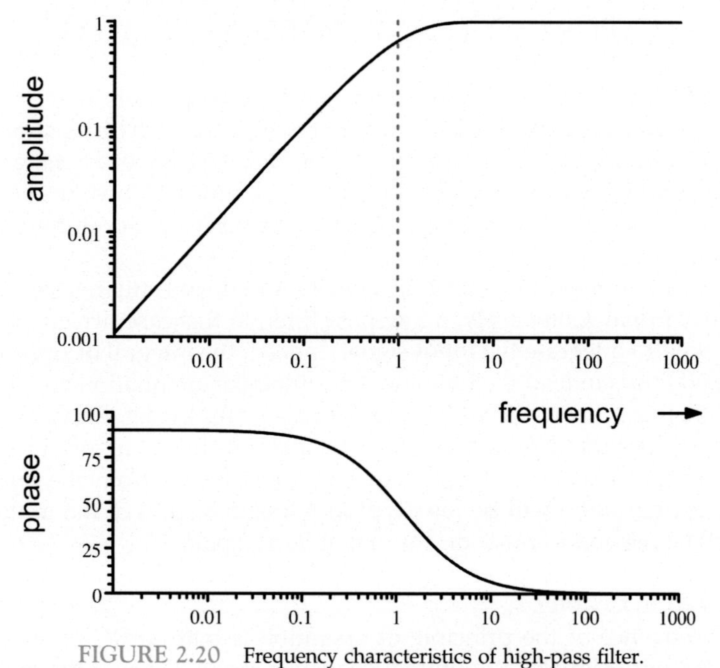

FIGURE 2.20 Frequency characteristics of high-pass filter.

number of cycles. We will see that causality is preserved, and that the phase lead builds up within one or two cycles. This is illustrated inFig. 2.21 for both the low-pass and high-pass filters.

Note that at 0 Hz, i.e., at a pure DC, the response of a high-pass filter is essentially zero, which one expects because of the insulation by the dielectric layer of the capacitance. At DC, the notion of phase ceases to be useful, although mathematically it is still defined.

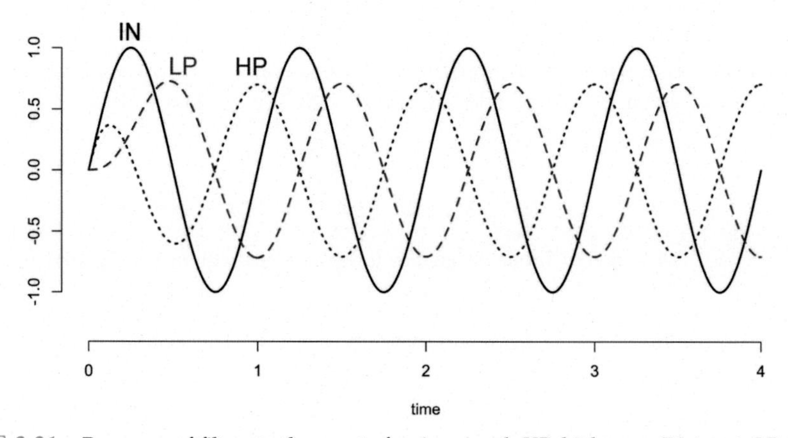

FIGURE 2.21 Response of filters to the onset of a sine signal. *HP*, high-pass; *IN*, input; *LP*, low-pass.

INTEGRATION AND DIFFERENTIATION

Apart from filtering, the action of low-pass and high-pass filters may be thought of as mathematical integration and differentiation, respectively. This will be illustrated with the responses of filters to sine signals just described. Consider the frequency characteristics of the high-pass filter again. At very low frequencies (say a factor of 10 below the cut-off frequency), the response to $U_{in} = \sin(t)$ is $U_{out} = \cos(t)$, which is the derivative or differential quotient of the input function.

This behavior is not limited to sinusoidal signals: A high-pass filter is said to differentiate any low-frequency signal. Conversely, a low-pass filter, in this case driven at frequencies far above cut-off, is said to integrate the input signal. Indeed, the integral of sin(t) with respect to time is $-\cos(t)$ and this is indeed the function describing the attenuation and 90°phase lag we saw earlier. Again, this property is not limited to sine signals. In fact, RC filters can be used to convert time signals into their time derivatives or integrals but in a very inconvenient way: To get the desired property, the signal frequency must be far from the cut-off frequency and so the output amplitude will be low, typically less than 10% of the input signal. Thus, an amplifier will be needed to push up the signal level again.

With the use of amplifiers, however, better differentiators and integrators can be built. These are described in Chapter 3.

A second consequence of the principle of operation is that an RC differentiator or integrator must be fed with signals in a limited frequency range. Outside the mentioned range, a filter either passes the signal unaltered or distorts the signal into a strange hybrid of the input signal and its time derivative or integral. This can be derived from the step responses shown in Fig. 2.18. The input step signal can be extended to a square wave signal, which is a series of steps from negative to positive and back. The response of different RC filters to such a square signal is shown in Fig. 2.22 (low-pass) and Fig. 2.23 (high-pass).

Note that these graphs may also be read to show the response of one low-pass filter to square signals of increasing frequency and of one high-pass filter to square signals of decreasing frequency. In that case, the subsequent graphs have different timescales.

It can be seen from Fig. 2.22 that a square wave is flattened more and more into a signal resembling the true time-integral, which is a triangle (check a square wave is simply a constant in the interval between two steps, and that the time-integral of a constant is a linear function). A linearly ascending or descending voltage is called a ramp, whereas a linearly rising and falling signal is called a triangle (although it is not, it has no base).

FIGURE 2.22 Distortion of a square signal by low-pass filters with increasing time constant.

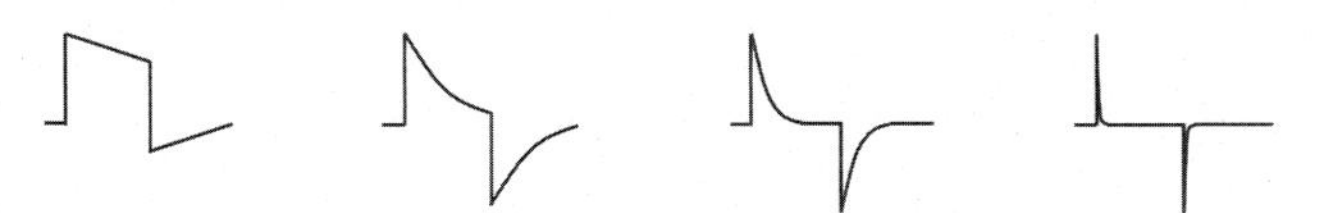

FIGURE 2.23 Distortion of a square signal by high-pass filters with decreasing time constant.

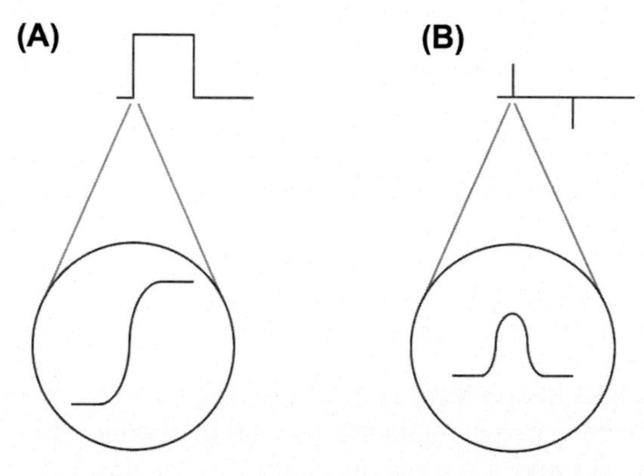

FIGURE 2.24 Rising edge (A) and derivative (B).

The situation for a high-pass filter is similar, although it might not be obvious at first why the peaks of the "differentiated" square signal resemble a time-derivative of that square. To clarify this issue, we must examine the rising edge of the input signal on a much faster timescale. Since the rising edge of any signal is not infinitely fast, we will find the sigmoidal shape shown in Fig. 2.24A. The concomitant output signal is shown in Fig. 2.24B and can be seen, on the same stretched timescale, to be approximately equal to the time derivative of the input signal.

LC FILTERS

We will digress briefly on LC filters for completeness, without going deeply into the theory. The two possible forms of combining a capacitance with a self-inductance are illustrated in Fig. 2.25.

An LC circuit comprises of two components that show reactance in opposite ways: A capacitor passes only changes, as we have seen, whereas the magnetic properties of

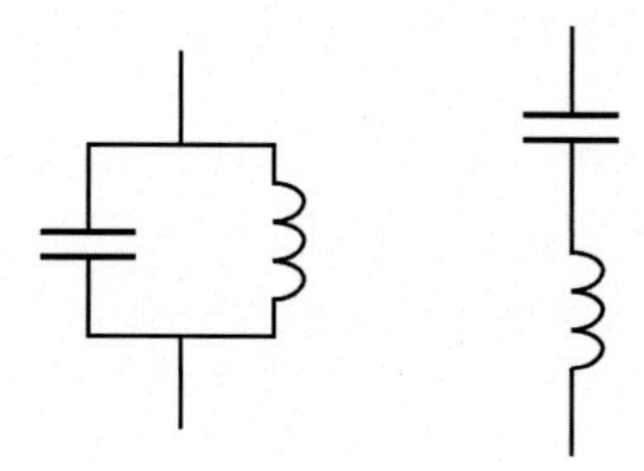

FIGURE 2.25 Parallel (left) and series (right) forms of LC circuits.

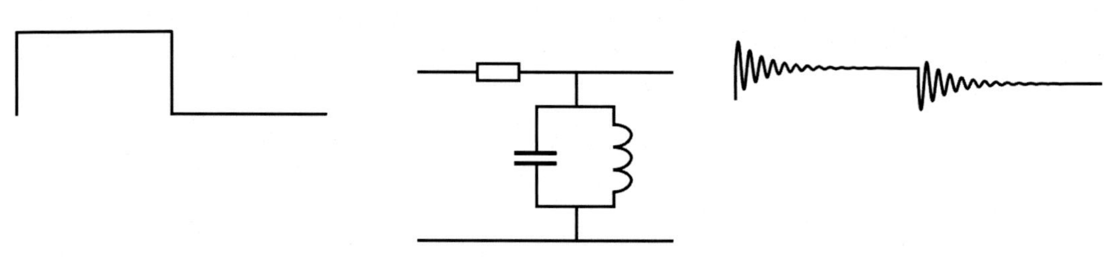

FIGURE 2.26 Ringing by an LC circuit.

inductances oppose changes in an electric current by the induced magnetic field. The combination shows resonant behavior, just as its mechanical equivalent: the combination of a mass (mechanical self-inductance) and a spring (mechanical capacitor).

The resonant frequency of an LC filter follows from:

$$f_{res} = \frac{1}{2\pi\sqrt{LC}}$$

where f_{res} is the resonant frequency in Hz; L, the self-inductance in H; and C, the capacitance in F. An LC filter is said to be tuned to this frequency. This means several things depending on the way it is used. The resonance shows up prominently when fed with a step, impulse, or pulse signal and is known as ringing. An example is shown in Fig. 2.26.

Sinusoidal signals are treated as follows. At the resonant frequency, the impedance of the parallel circuit is maximal, i.e., higher frequencies are shunted by the capacitor, lower frequencies by the self-inductance. An LC circuit in combination with resistors can be used to single out a narrow frequency band around its resonant frequency and this is why it is employed as tuning circuit in radio receivers and the like. The series configuration has minimum impedance at the resonant frequency, so this configuration can be used to eliminate a narrow frequency band from any input signal. Therefore, the series LC circuit is called a suction circuit.

In practice, the LC circuit must be completed with a resistor either implicit (a property of the coil) or explicit (connected in series as in Fig. 2.26 or in parallel) to introduce damping (the equivalent of the dashpot in the aforementioned car wheel suspension).

In electrophysiology, it may be used to get rid of interference signals with known and constant frequency, such as the 50 (or 60) Hz mains frequency, or the carrier frequency of a nearby broadcast station.

This ends the chapter on electric processes. The manipulation of electrical signals is enhanced enormously by the use of what is called electronics. This is the topic of the next chapters of this book.

CHAPTER

3

Electronic Devices

OUTLINE

ACTIVE ELEMENTS

So far, we dealt exclusively with passive components, which do not add energy to the signal (for the moment, consider any wanted quantity or message as a signal; later we will define the notion of a signal in a more precise way). What is "amplification"? Historically, mankind has found several ways to amplify its powers literally and figuratively and these may be compared to find both the possibilities and the limitations. The first step was to harness animal power and to exploit it for human purposes. In this way, a man (say 70 kg) controls the power of a bull (say 1000 kg) to plow the land. This example from antiquity illustrates the principle of amplification: The farmer uses his muscle power as a signal that controls a much stronger force, the bull's muscles. It also shows the expenses: The bull needs to be taken care of, housed, and fed on a regular basis. More recently, the same

Introduction to Electrophysiological Methods and Instrumentation, Second Edition
https://doi.org/10.1016/B978-0-12-814210-3.00003-7

principle was employed in engines, where the energy from an exothermic substance (fuel) is converted into mechanical energy that could drive a revolving shaft used for milling, pumping, traffic, and the like. Note that in both cases the energy is delivered in a bulky form that by itself does not perform the wanted operations.

Active elements in electronics perform according to the same principle. The weak signal from a microphone or an electrode must be amplified to be able to feed a loudspeaker, to show it on an oscilloscope, or just to get it through a long cable. The necessary energy comes from a power supply. In a few cases, this may be an alternating current tapped directly from the mains outlet (lamp dimmers and variable-speed drills, for example), but most electronic instruments are powered by a direct current, either from a battery or derived from the mains AC through a so-called rectifier, which we will analyze later on.

Amplification must not be confused with transformation. In Chapter 2, we have seen that a transformer may output a higher voltage than was fed into it, but with a proportionally lower current so that power is conserved and not increased. In fact, there are losses of power into heat, making the power conversion always less than 100% effective.

Since the advent of vacuum tubes in the beginning of the 20th century, the principle of controlling a large bulk force with a smaller, "structured" one became feasible for electrical quantities. Later, mainly after the Second World War, semiconductors such as the transistor took over; so a large part of this chapter is devoted to them. Today, vacuum tubes, including picture tubes, are well-nigh extinct, so we will deal only with semiconductors and start with the most elementary forms—diodes and transistors. The different types of semiconductors, each with their specific virtues for electrophysiology, will be dealt with later on.

SEMICONDUCTORS

In modern electronics, the active elements are usually transistors and similar "semiconductor" components. The basic structure of a semiconductor is a silicon crystal or rather a sandwich of crystals of silicon to which slight amounts of other elements are added. Unfortunately, the processes in such crystals cannot be understood without a thorough knowledge of quantum mechanics. Hence, we will treat the mode of operation of transistors in a rather superficial way.

Because electric currents in semiconductors flow in crystals, rather than in vacuum (as in vacuum tubes), these devices are also called "solid-state" devices. The materials suited for the wanted phenomena are neither conductors (such as metals) nor insulators (such as glass, ceramics) but a category in between: semiconductors.

In an ideal insulator, all electrons are bound to their respective atomic nuclei. Metals, to the contrary, have free electrons that may move from atom to atom, thereby transporting electric charge. In other words, they may sustain an electric current. Semiconductors hold an intermediate position: At absolute zero temperature, they behave as insulators, whereas at room temperature, they have a few free electrons. The materials with these properties are four valued: mostly germanium and silicon. These elements have four electrons in their outermost shell. Germanium was used first, but to date, silicon is used most often and will be used in subsequent examples. In very pure silicon, however, the current that can flow is far too low to be of practical value. This type of silicon is called "intrinsic." The interesting properties only arise by "doping": the addition of very small concentrations of other elements, called "impurities." The addition of a five-valued element, such as arsenic or antimony, to a silicon crystal causes more free electrons, since only four do fit in the silicon's crystal lattice. The fifth electron is bound so weakly that it may be considered as free. Therefore, pentavalent elements are called "electron donors." The excess electrons can move freely through the crystal and turn it into an almost-conductor. Silicon prepared in this way is called n-type silicon or "n-silicon" for short. In a similar way, trivalent elements, such as indium or gallium, cause a shortage of electrons and are called "electron acceptors." By the movement of the existing electrons, the deficit of an electron is also mobile and called a "free hole." It acts like a positive particle. Current flowing in such a semiconductor is called a "hole current." Silicon treated in this way is known as "p-silicon." The ability to use charge carriers of both signs enhances the freedom of design of electronic circuits. Many useful devices arise from the combination of p-type and n-type silicon. Occasionally, the intrinsic form ("i-silicon") is used.

DIODES AND TRANSISTORS

The simplest electronic element is the diode. The word stems from ancient Greek and means "two ways." In summary, a p-type semiconductor contains many free holes but hardly any free electrons, whereas an n-type semiconductor contains many free electrons but hardly any free holes. If these two opposite types of semiconductors are joined together to form a p–n junction, a device results that conducts an electric current only in one direction. This is shown schematically in Fig. 3.1.

If such a junction is connected so that the p-side is positive with respect to the n-side, the following happens. In the n-side, electrons are rushing toward the p–n junction, whereas in the p-side, holes are rushing equally in the direction of the junction. At the junction, the two types of charges, electrons and holes, combine. If the junction is connected in the reverse direction, hardly any current or no current at all will flow because in the p-zone there are insufficient free electrons, whereas in the n-zone there are insufficient free holes. Only

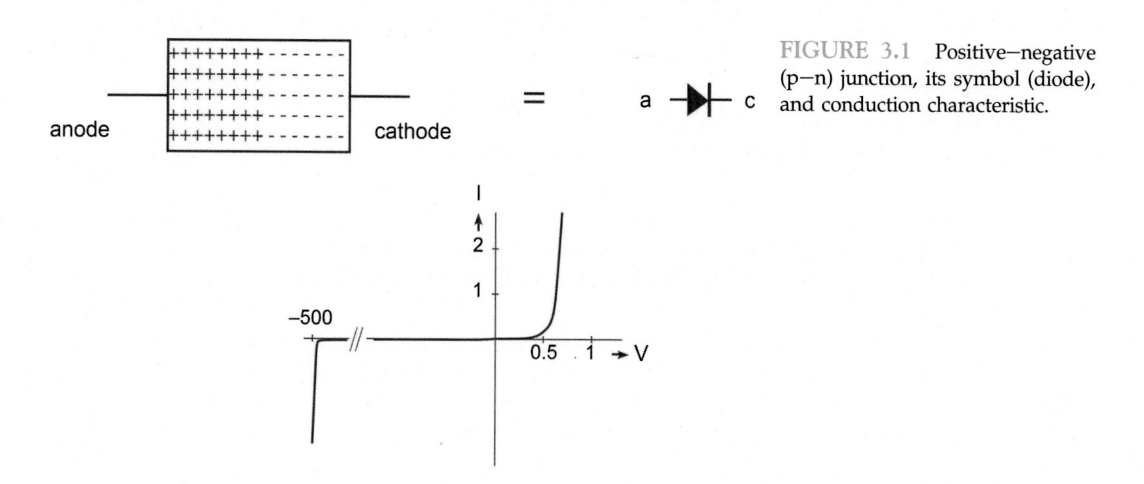

FIGURE 3.1 Positive–negative (p–n) junction, its symbol (diode), and conduction characteristic.

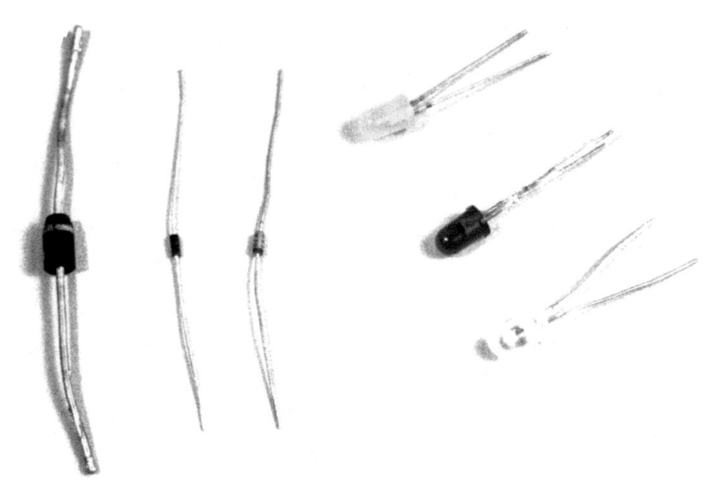

FIGURE 3.2 Practical diodes. The leftmost one is a power diode, suited to rectify currents up to about 1 A. The following two are signal diodes, useful up to the milliamp range and the three at the right are LEDs or light-emitting diodes.

at very high reverse voltages (hundreds or more), the junction will break down (Fig. 3.1, left side of I/V curve).

Diodes are very useful by themselves, e.g., to rectify the AC mains energy into the direct current that most electronic instruments need.

This is done in power supplies, dealt with later on. Practical diodes are illustrated in Fig. 3.2.

A disadvantage of semiconductor diodes is that conduction in the forward direction will take place only if the driving voltage exceeds a certain threshold value (see the I/V characteristic in Fig. 3.1). This threshold is close to 0.7 V for silicon (it is slightly dependent on the current strength). This means that very low voltages cannot be rectified.

Nevertheless, this device was truly revolutionary, since its inherently asymmetrical conductance performs the same task—rectification—as the vacuum tube diode, where it can be made far smaller, does not need a filament to heat the cathode, and works with lower voltages. If the polarity of a connected voltage is such that current flows, the diode is said to be connected in the forward direction, if the polarity is reversed, the diode is connected in the reverse direction. To build a better rectifier, silicon diodes can be joined into an array of four, which performs "full wave rectification": see Fig. 3.3.

This so-called "diode bridge" circuit is the universal solution for rectification in power supplies.

Note that most diodes are passive elements passing current without amplifying. The Gunn diode is an exception, since this type can be used as the active element of an oscillator circuit. The applications fall in the high-frequency and microwave ranges and so fall outside the scope of this book.

The most important invention in semiconductor devices, however, is the amplifying element, usually named "transistor" (a contraction of transduction and resistor). A transistor has three zones of p- and n-silicon.

This implies there are two possible configurations: pnp and npn. The two types of devices are known as "pnp transistor" and "npn transistor." They are said to be "complementary."

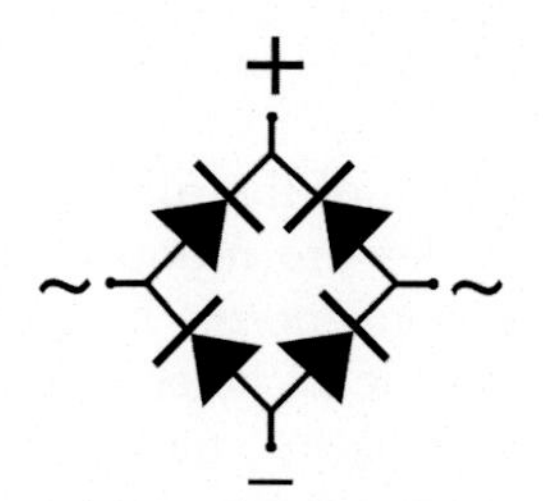

FIGURE 3.3 Diode bridge.

The outer zones are fabricated by doping the middle zone, a thin silicon layer called the "base." This functions as the control electrode, controlling the current through the other two layers. At one end, charge carriers are "injected" into the crystal and exit through the other end.

For this reason, the two outer zones are called "emitter" and "collector," respectively.

Schematics are shown in Fig. 3.4.

Again, our explanation of the mode of operation of transistors has to be superficial and qualitative, since a true explanation of the underlying principles would need a thorough knowledge of solid-state physics and hence of quantum mechanics. Fortunately, such knowledge is not necessary to use semiconductors in a proper way, as long as one

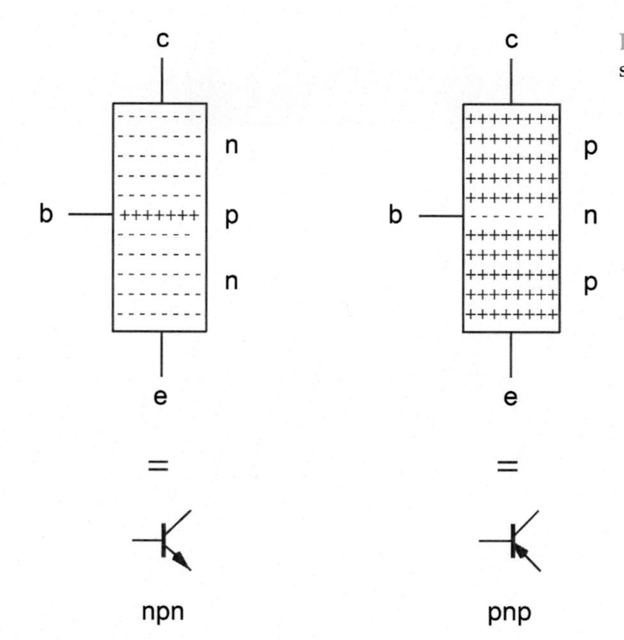

FIGURE 3.4 Types of transistors and their symbols. b, base; e, emitter; c, collector.

has a good notion of voltages and currents in the device and the relationships between them.

A popular and simplified explanation of the function of an npn transistor is shown in Fig. 3.5. The emitter and collector sides are n-doped, thus containing mostly electrons. The thin layer between the two outer layers, the base, contains p-doped material, holes. In the absence of a voltage on the base lead, holes and electrons in the interface between p and n materials cause two zones devoid of charge carriers because holes and electrons neutralize each other (Fig. 3.5A). Things become different when a positive voltage difference of more than the junction potential of about 0.7 V between base and emitter is applied. This potential attracts negative charges from the heavily doped emitter region to such an extent that the number of negative charges in the base segment becomes dominant. Their presence causes a current to flow between collector and emitter (Fig. 3.5B). Only a relatively small current from base to emitter flows to maintain the base potential. The final effect is that the middle lead controls the current through the transistor. As can be seen from Fig. 3.5, most of the effects governing the transistor function occur in the thin p-doped base layer. For this reason and simplicity, the outer layers are presented only by their name in the diagrams Fig. 3.5C and D and in some of the following figures. The explanation of the function of a pnp transistor follows similar lines; only positive and negative, holes and electrons, are to be inversed.

Practical forms of transistors are shown in Fig. 3.6.

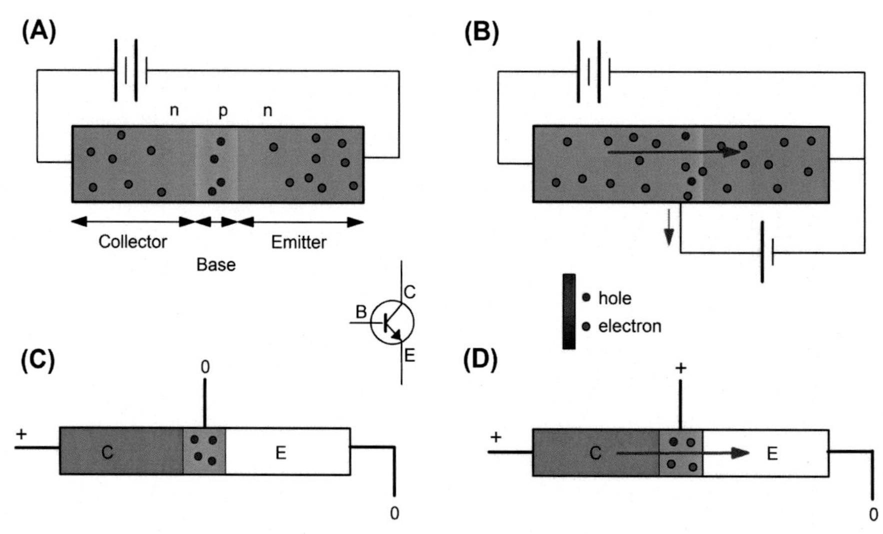

FIGURE 3.5 The npn transistor. No collector to emitter (C-E) current flows in the absence of a base voltage (A and C). The base region is invaded by electrons upon application of a positive base-emitter voltage, causing a C-E current to flow. (C and D) are simplified versions of (A and B).

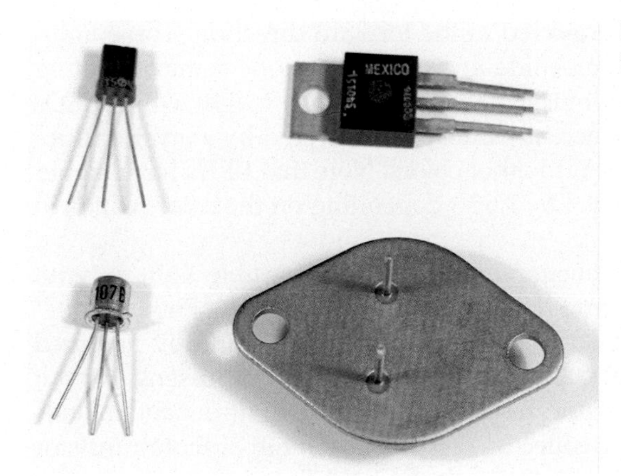

FIGURE 3.6 Practical transistors. The size reflects the maximum current that can be handled.

OTHER SEMICONDUCTOR TYPES

The many properties of semiconductor devices, governed by solid-state physics, have given rise to a number of other, very useful, types of semiconductor elements, which we will treat briefly here.

To start with, a "Zener diode" is a kind of diode in which applying a voltage in the reverse direction leads to an abrupt breakdown at a very sharply defined voltage. Not only is this breakdown reversible, the precise value depends on the composition of the semiconductor layers and may be chosen between about 2.5 and 25 V. Therefore, Zener diodes are used as voltage stabilizers by forcing a current in the reverse direction and using the maintained, well-defined voltage across the terminals as reference value for power supplies, voltmeters, AD converters, etc. Note that in this case, the cathode is the positive terminal. By the way, Zener diodes designed for 5.6 V Zener voltage proved to have the best properties as to stability, temperature (in)dependence, and the like and are often used as a basis for the derivation of other voltages. Fig. 3.7, left, shows the symbol for a Zener diode.

The Schottky diode is a special type of diode having a junction between silicon and a metal. Certain metals, such as platinum and molybdenum, form interesting junctions with n-silicon, having a lower threshold voltage in the forward direction. The advantages are faster switching and lower losses at rectifying low voltages. Fig. 3.7, right, shows the symbol for a Schottky diode.

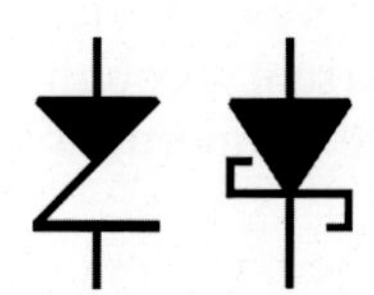

FIGURE 3.7 Symbols for zener diode (left) and Schottky diode (right).

A special type of diode emits light when connected in the forward direction. These light-emitting diodes (LEDs), made out of gallium arsenide and/or other exotic semiconducting materials, are used frequently to replace colored signal lamps. Large, bright white LEDs are suited for illumination purposes. They are actually blue LEDs covered by a layer of fluorescent paint that converts part of the blue light into other colors. Note that LEDs have higher threshold voltages than rectifying diodes: about 1.6–3.6 V, depending on the color, which in turn depends on the materials used.

Many semiconductors show photoelectric phenomena, either with visible light or with infrared radiation, so diodes may be used either to detect light or to emit light. "Photodiodes" and "phototransistors" resemble normal diodes and transistors but are specifically designed for the reception of light. In principle, normal transistors are intrinsically light-sensitive but are packaged to avoid light reaching the transistor. In photodiodes, the light-dependent current flows in the reverse direction, whereas the collector–emitter current in a phototransistor flows as if it were caused by a base current proportional to the light flux. Like normal transistors and diodes, these components need a power supply to utilize the light sensitivity.

Photovoltaic cells and PIN photodiodes, however, generate a current, which is approximately linearly dependent on the light flux when the cell is short-circuited. PIN stands for positive, intrinsic, negative; this diode has a layer of intrinsic (not doped) silicon between the two sides of the junction, where charge carriers kicked by photons may "leak through." The photo current may be measured with a sensitive current meter or amplifier, where the meter itself acts as the ground connection. When left open, the light-induced voltage of these components may reach about 0.5 V in bright light. Solar panels are simply stacks of photovoltaic cells that generate 12 V or more, useful as energy source to charge batteries or power other apparatus. The photovoltage generated by a PIN photodiode, if left open, is approximately proportional to the log of the light intensity. Because of this property, they may be used to read density units or photographic "stops" directly if connected to an amplifier with a high input impedance.

FIELD-EFFECT TRANSISTORS

The semiconductor type that is no doubt the most important one in electrophysiology is the field-effect transistor or FET. The basic form consists of a silicon strip or channel between two conducting terminals. In the middle is a p–n junction which, contrary to the normal transistor, is connected in the reverse direction, that is, if the third terminal forms an anode, it is made positive with respect to the channel so that no "base current" will flow. The important process is that the electric field in this reverse-biased junction causes the resistance of the channel to change.

In the most familiar form, the n-channel depletion FET, the channel is made of n-silicon, and the electric field in the junction is able to reduce the current in the channel, hence the term depletion. When a voltage is applied to the middle terminal, the current through the channel is modulated. It is therefore appropriately called the gate. The connections to the channel are called source and drain; see Fig. 3.8.

FETs have a number of advantages over junction transistors: first of all, the high input impedance, the property most important to electrophysiology. The only junction of this

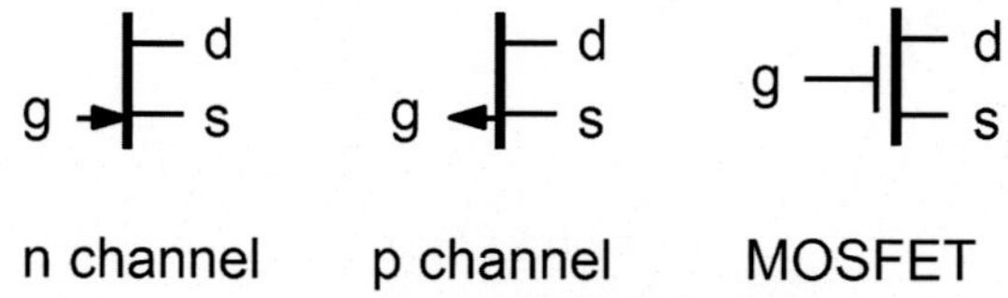

n channel p channel MOSFET

FIGURE 3.8 Field-effect transistors. d, drain; g, gate; MOSFET, metal-oxide semiconductor field-effect transistor; s, source.

transistor is reverse biased so that virtually no current flows: The gate voltage only controls the current through the channel. Now, this must be corrected before you think we have reached Utopia. Any junction, operated in reverse bias, conducts a very small but nevertheless important leakage current and this current has to be provided by the input circuit. The great advantage of the FET over the normal so-called junction transistor is the amount of current needed at the input (base and gate, respectively). As we have seen, transistors need about 1 μA of current to function properly, whereas the leakage current of an FET is orders of magnitude lower: about 1–10 pA (10^{-12} to 10^{-11} A)! These current values may be translated into the input resistance or input impedance, by applying Ohm's law. If, in a normal transistor, 1 μA of base current flows at 0.1 V of base voltage, the equivalent input resistance would be $0.1/10^{-6}\ \Omega$, or 100 k Ω. As input stage of an electrophysiological amplifier, these transistors would be useful only if the electrodes used had a far lower resistance. This is the case for the large metal plates used in electrocardiography and similar techniques, but it would be utterly useless for monitoring the electrical life of a single cell, where currents in the order of nanoamps are all there is.

Here, the FET scores far better: 1 pA per 0.1 V, or 100 G Ω, of input resistance is no exception. Therefore, FETs are the main building blocks of electrophysiological preamplifiers and are used in a host of other situations where only small currents are available. And this is not even the maximum attainable: a variant called the MOSFET, short for metal-oxide semiconductor field-effect transistor, has a still higher input resistance, i.e., a still lower leakage current. This is made possible by doing away with the junction altogether and just "gluing" the gate electrode to the channel between source and gate or "body" (a.k.a. "substrate") via a thin metal-oxide layer. Metal oxides are very good insulators, and so the gate is effectively insulated from the channel. Consider the n-type MOSFET as shown in Fig. 3.9A. The source has been made positive with respect to the drain. The gate-body voltage is zero. Often the drain and body terminals are connected as in this example. The channel between the source and drain is p-doped. At zero gate voltage, the number of holes and negative sites is identical and no current flows. A positive gate to body voltage chases the holes creating a negatively charged layer (inversion layer) that carries a current between source and gate (Fig. 3.9B). In depletion-type MOSFETs, the channel (body) silicon is doped in such a way that the channel has an excess of electron sites with respect to holes (Fig. 3.9C). As a result, the channel conducts an S-D current in the absence of a G-B voltage. A negative G-B voltage attracts holes, restoring an equilibrium of holes and electrons in the channel region, thereby inhibiting the S-G current (Fig. 3.9D). The operation of the depletion type of MOSFET resembles a junction FET (JFET) in this respect.

The voltage on the gate controls the channel current again, making it a valuable component for high-impedance preamplifiers and electrometers used in many types of physical, astronomical (e.g., star light), chemical (pH meter), and biological (intracellular and patch-clamp) measurements. The MOSFET has a few hidden flaws, however, such as a higher

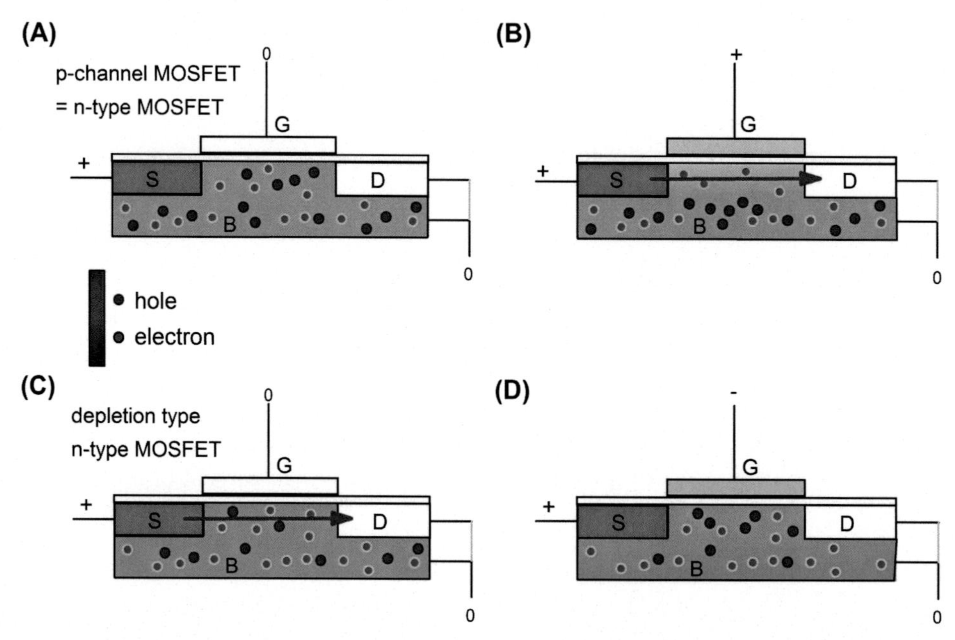

FIGURE 3.9 Enhancement-mode (A,B) and depletion-mode (C,D) n-type metal-oxide semiconductor field-effect transistors (MOSFETs) with a p-channel. Abreviations: S source, D drain, G gate; B body. The drain and the body are kept at 0 V. A positive gate potential creates a negatively charged region in the body of the enhancement-mode FET, thus alowing a current to pass between source and drain (B). The depletion type FET is conducting without applying a positive gate potential because electrons outnumber holes in the body. The channel region between source and drain becomes neutral by applying a negative gate voltage thus preventing a current to flow between source and drain.

noise level, in comparison with the abovementioned "normal" FET, somewhat confusingly called the junction FET, or JFET.

In addition to the almost electrostatic control of the channel current, FETs may be made in different "flavors" to suit various needs. Like pnp and npn transistors, FETs come in complementary types, here called n-channel and p-channel, in which the polarity of voltages is reversed. Also, FETs may be made with a thin intrinsic region so that the channel conducts no current unless the gate electrode bears a certain voltage. These are called enhancement-type FETs. As to the properties, depletion and enhancement types perform equally well. Like normal transistors, the FET is symmetrical in principle, but asymmetrical in practice, that is, source and drain should not be exchanged.

Warning: Because of the high gate resistances, FETs are vulnerable to electrostatic voltages and can be destroyed by handling or merely by touching with a finger or a plastic ballpoint. Especially MOSFETs must be handled without touching the gate connection, or with grounded tools only. Note that this holds also for devices that use FETs as input transistors, such as many electrophysiological preamplifiers.

ION-SENSITIVE FIELD-EFFECT TRANSISTORS

A special type of transistor is the ion-sensitive electrode. It is based on the MOSFET, dealt with above. If the metallic gate surface is left out, the insulating oxide layer is exposed to the external world, viz. an electrochemical solution. The oxide can be SiO_2 and a few others, such as aluminum oxide (Al_2O_3) and tantalum oxide (Ta_2O_5).

If the rest of the transistor "chip" is covered with insulating plastic, the device is water tight, except for the tiny "naked" gate surface and so can be dipped in a water solution. The ionization of the metal oxide depends on pH. For silicon dioxide, an equilibrium between SiOH, $SiOH^{2+}$, and SiO^{-} at the surface of the channel between source and drain is established. Changing the ionization of the metal oxide influences the distribution of holes and electrons in the semiconductor channel and hence the source to drain current. Fig. 3.10 shows the principle of the ion-sensitive FET (ISFET).[1–3]

Like with conventional hydrogen electrodes, the reference electrode for pH measurements with an ISFET must be stable obviously, and not ion-sensitive, so silver/silver chloride (Ag/AgCl) is the best choice (see the glass electrode, Chapter 5).

Such electrode pairs are feasible and are in use for semimicro measurements.

However, Ag/AgCl electrodes are not easily miniaturized, and the necessary KCl store (e.g., in gel form), also tiny, would be exhausted rapidly. This problem can be solved by incorporating a second naked gate FET, called a REFET (reference FET), on the same chip, this one being explicitly non–ion selective. In this way, the entire pH sensor can be miniaturized. To render the exposed gate of the REFET, (almost) ion-insensitive has been a challenge. Although REFETs are still experimental, the most successful way is to cover the gate area with a conducting polymer membrane. A commercially available form is shown in Fig. 3.11.

Next to sensing pH, modified ISFETs can be harnessed to measure other (electro) chemical quantities, such as DNA, proteins, and enzymes. This can be done, e.g., by attaching biomolecular receptors to the insulated gate surface, usually immobilized in a polyacrylamide membrane.

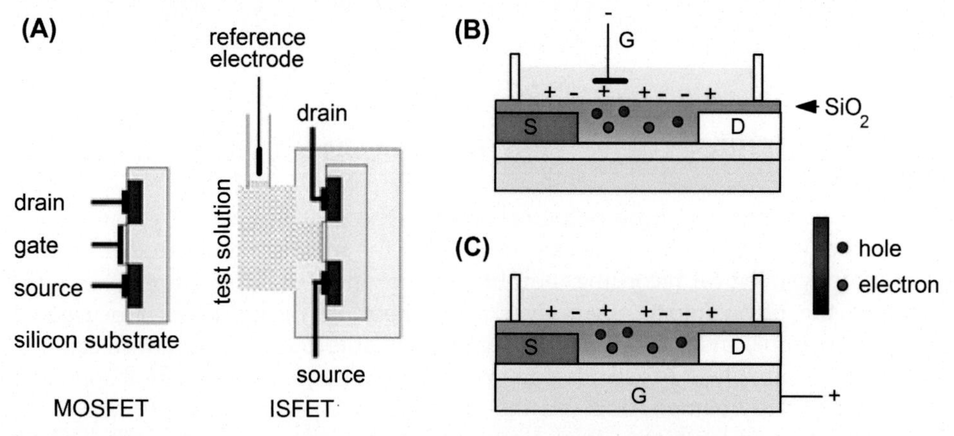

FIGURE 3.10 (A) Metal-oxide semiconductor field-effect transistor (MOSFET) and ion-sensitive field-effect transistor (ISFET) juxtaposed. This shows the principle of an ISFET, showing the "naked" gate oxide layer exposed to the liquid environment, fitted with an AgCl reference electrode (note: drawn not to scale, see text). Two ISFET configurations with the gate electrode in the bath (B) or situated under the channel (C). In both cases the gate voltage is chosen as to create an offset current between source and drain that can be modulated by the pH.

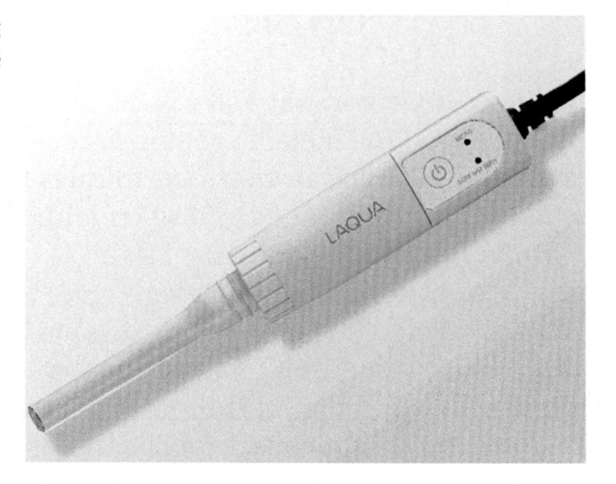

FIGURE 3.11 A commercially available pH probe using an ion-sensitive field-effect transistor. It has the reference electrode built in.

Naked gate MOSFETs are also the basis for the "electronic nose," a device sensing volatile substances ("odors"), and hence called OSFETs (odor-sensitive). Cells can be grown on (arrays of) naked gate MOSFETs to monitor changes in size, electrical activity, and the release of ions or other molecules. Sensitivity and durability of the devices used in cell culture has been improved by separating the MOSFET, whose transconductance is optimal if the channel between source and drain is thin, from the sensing area that is preferably large (about the size of a cell). This configuration or floating gate MOSFET is shown in Fig. 3.12.

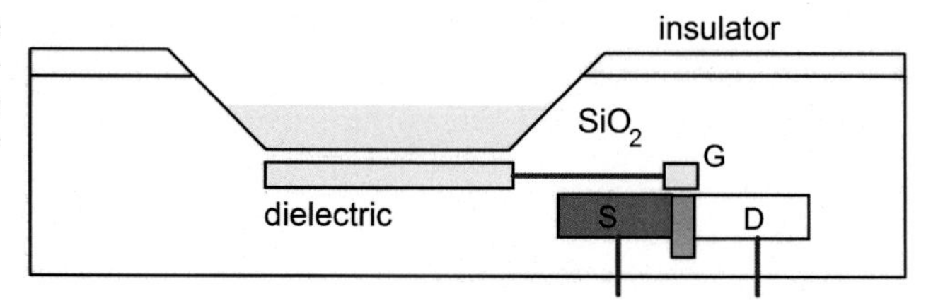

FIGURE 3.12 A floating gate metal-oxide semiconductor field-effect transistor (MOSFET). The gate is not directly exposed to the medium and may be large. Blue lines represent metal leads.

AMPLIFIERS, GAIN, DECIBELS, AND SATURATION

The fulcrum of every electrophysiological setup is the preamplifier, or preamp for short, because it is the first apparatus connected directly to the measuring electrodes. In the early days of electrophysiological recording, amplifiers were made with vacuum tubes and were often built by the physiologists themselves. Nowadays, semiconductors are used almost exclusively in preamps, and many types of preamp, versatile and almost perfect, can be delivered off the shelf from a host of reliable companies. This does, however, not absolve the user from the obligation to maintain and use the apparatus properly, especially since a lack of knowledge about one's tools may lead to the publishing of erroneous results, which is a waste of time, money, and intellectual energy. Therefore, despite the streamlined technology, the many electronic devices and computer algorithms available for filtering or postprocessing of the signals and for the presentation of the data, all students of electrophysiology must gain proper insight in the working principles of their tools and more specifically of the most vital stages, viz. the electrodes and the preamplifier. In planning experiments, and before purchasing instruments, one has to know the possibilities to choose from and the consequences for the validity of the measurements. What are these important properties of preamplifiers?

Gain

This is the most conspicuous property, not necessarily the most important one. Usually, one means the voltage gain factor, defined as the ratio of output voltage to input voltage. But how much gain is necessary for an experiment depends on the ratio of the voltages to be measured over the voltage that can be "seen," or measured, by, e.g., voltmeters or by the most universal indicator, the oscilloscope. Usually, one takes 10–100 mV as measurable, although many oscilloscopes and electronic voltmeters may be useful down to 1 mV (of course, they have amplifiers built in, which makes the notion of "necessary" gain a bit vague). The consequence is that for intracellular recording of membrane potentials, being tens of millivolts, no extra gain is needed. Thus, microelectrode preamps often have a gain switchable between 1 and 10. Why we cannot do without preamp will be clear later on.

The opposite case arises if the electrical life of one or more cells is measured from the outside of the cell or even from outside the specimen or body, such as electrocardiograms, electroencephalograms, and the like. Here, a high gain factor of 10^4 to even 10^6 is necessary, since the signal voltages in question may be no more than a few microvolts. For these large values, it is custom to express the gain factors in decibels. The gain factor, expressed in decibels, is simply called the gain. By the log transform, a large range of values may be expressed as fairly small numbers. In addition, when connecting two amplifiers in series, the gains just add, whereas gain factors must be multiplied.

The voltage gain (G) is computed using the following equation, taken from Chapter 2:

$$G = 20\log(U_{out}/U_{in})\,\mathrm{dB}$$

Thus, an electrophysiological preamp with a gain factor of 10,000 has a gain of 80 dB. The same relation holds for current gains or, used frequently in acoustics, power gain.

The power gain follows from:

$$GP = 10\log(P_o/P_i),$$

where GP is the power gain, and P_o and P_i, the output and input power, respectively. In audio circuits, one is still interested in powers. For instance, the power emitted by a microphone, is about 10^{-9} W (1 nW). In order to be audible, the loudspeakers emit, say, 10 W. Thus, a record player's amplifier has a power gain of $10\log(10/10^{-9})$ or 100 dB.

Note that in computing power gain, one has to take the input and output impedances (resistances) into account. Take a microelectrode preamplifier with a *voltage* gain factor of unity (i.e., 0 dB). Since the input impedance will be in the order of 1 TΩ (teraohm, or 10^{12} Ω), and the output impedance might be 10 Ω, the power gain factor is 10^{11}, or 110 dB. No wonder we cannot do without a preamp!

Bandwidth

The bandwidth tells what frequencies are amplified. The well-known audio amplifier has to amplify sound frequencies, i.e., about 20 Hz to 20 kHz. Both lower and higher frequencies are not perceived by the human ear and so need not be amplified or passed at all. Bats and crickets, by the way, would need audio amps with higher frequencies.

In other words, amplifiers have to filter the incoming signal and pass only the wanted frequencies. Fundamentally, all amplifiers are low-pass filters, in that an infinite frequency is not physically possible. The capacitance of any wire to its environment (the instrument case, the shield of the cable, etc.) acts as a parallel capacitor that attenuates the signal at high frequencies. At the highest frequencies, the wires themselves behave as self-inductances, enhancing the low-pass effect. This should come as no surprise, since "in no time, nothing can change." Design and purpose determine which frequencies will be admitted, but there is always an upper limit.

The upper limit of the bandwidth of an amplifier, indeed of any circuit, is limited by parasitic capacitance, also called stray capacitance. Parasitic capacitances are everywhere: Any conductor has capacitance with its environment, even the shortest wire between two components. Especially coaxial cables, where the shield is close to the conductor all the way long, have a capacitance of about 100 pF/m. Special foam insulation cables may have 30 pF/m but are more vulnerable. Keeping wires as short as possible and as far as possible from other conductors are measures used to reduce parasitic capacitances, both in the design of instruments as in wiring up an electrophysiological setup. In addition, parasitic capacitances arise internally in resistors, transistors, switches, and so on, so that full knowledge of the used components may help to keep stray capacitances in control and thus to prevent the bandwidth from getting lowered unintentionally. Especially a preamp input circuit with a connected microelectrode is prone to unintended loss of bandwidth.

Often, however, the bandwidth of amplifiers is limited intentionally to allow the wanted signal only to pass unattenuated, whereas disturbing signal components such as hum and noise must be eliminated as far as possible. The simplest way of filtering, and the one most frequently used, is to incorporate the familiar RC filters, treated in Chapter 2. Thus, an amplifier fitted with a high-pass and a low-pass filter can be considered as a "band-pass filter with gain." Like with these filters, the bandwidth (BW) of an amplifier is defined as the frequency band between the −3 dB points. It must be kept in mind that, despite this convention, signal frequencies outside this passband are not fully suppressed, and one must always be aware of the frequency characteristics of the mentioned filters. In many designs, higher-order filters may be used that filter more sharply outside the wanted passband. Adjustable electronic higher-order filters are sold as separate instruments. The principles are discussed later in this text. Even with the most sophisticated filter design, however, passing signals inside the passband completely and at the same time rejecting signals outside it completely is a physical impossibility.

It is also feasible to build band-reject filters in an amplifier, e.g., to reduce hum, but the same limitations hold here: It is impossible to reject the 50 Hz hum fully while keeping signal frequencies of, say, 49.99 and 05.01 Hz.

So, a high-frequency limit is taken for granted but is there also a lower limit? In a strict sense, a frequency of 0 Hz does not exist either: This would mean a current or voltage that has been there for an infinite time and will remain forever. For practical purposes, however, a direct current, or DC voltage, may be considered to have zero frequency: within the period of interest it will not change (polarity). The voltages of batteries, power supplies, and the like fulfill this criterion, and there are no physical objections at all to measure them. A voltmeter is in principle capable to measure the 1.5 V battery voltage as long as it exists. Early electrophysiological amplifiers had a lower limit to the passband. In this case, the filtering was caused by capacitors used as coupling elements between the amplifier stages. A coupling capacitor between the collector of one stage to the base of the next stage simplifies the design because usually the collector bears a higher voltage (e.g., about 6–30 V) than the base (about 0.7 V). This is still used in all cases where the DC value of a signal is not necessary or not even wanted. Such amplifiers are called AC amplifiers, as opposite to DC amplifiers, the bandwidths of which reach down to zero.

Most electrophysiological amps have rather broad passbands, whose position depends on the function. For example,

for nerve membrane potentials:	about 0–3000 Hz,
for electrocardiograms:	about 0.1–30 Hz,
for electroencephalograms:	about 1–50 Hz,
for nerve or muscle spikes:	about 300–3000 Hz,
for plant action potentials:	about 0–1 Hz,
for the analysis of spike shape:	about 0.1–100 kHz,
for single-channel recording:	about 0–50 kHz.

Amplifiers with very narrow passbands are used mainly in other branches of electronics, such as radio communication, where one wants to tune in on a single broadcast frequency. This is seldom necessary in electrophysiology and is outside the scope of this book.

It is important to note that the passband of an amplifier must be chosen on the basis of the frequency contents of the signal, not merely the repetition frequency of a signal. To be sure, in order to allow a 1 Hz sine wave, one needs a passband centered on 1 Hz, but to pass a spike train, one needs to pass the fast changing process, the action potential, irrespective of the repetition frequency. Thus, an amplifier to record nerve action potentials needs a passband centered at about 1 kHz, whether a neuron fires 100 spikes per second or 1 spike per hour!

A pulse signal, such as a nerve action potential, is said to have a "frequency content" of, say, about 300 Hz to a few kilohertz, irrespective of the repetition rate.

A second consequence of the use of filters in an amplifier is that outside the passband, signals are (linearly) distorted, in the sense that frequencies lower than the passband are differentiated, and frequencies above the passband are integrated. This is treated in our analysis toolkit (Chapter 7).

Input and Output Impedances

These are crucial quantities that determine the usefulness of an amplifier in a particular situation. With regular electronics, however, the output impedance of any device is between

1 and 50 Ω and so will cause no problems. The only demand is that the output impedance of a preamp is about 30–50 times lower than the resultant input impedance of all devices that are connected to it. Since oscilloscopes, recorders, computer inputs, and other electronic circuits have relatively high input impedances (100 kΩ–1 MΩ), this demand is easy to fulfill. Passive voltmeters (moving-coil meters), some recording devices, and audio amps may have lower input impedances so that one has to keep track of impedances when wiring any setup. And never connect a loudspeaker (4–8 Ω) directly to any device that is not an audio amplifier designed to drive them.

People accustomed to wiring up their own audio chain will contend that input and output impedances must be matched. Indeed, in power-transferring circuits, impedance matching is often best: an 8 Ω amplifier must be used to drive one 8 Ω loudspeaker. This is true, but with voltage measurement circuits, things are different. This can be shown as follows.

If an amplifier with an output impedance of 8 Ω is not loaded at all (the load impedance is infinity), the voltage at its output is maximal, but since no current flows, no power is developed. If the same amp is short-circuited (load impedance is zero), the maximum current flows, and the maximum power is developed, but unfortunately, the output voltage is zero and the power is dissipated within the amplifier in the form of heat. The optimum energy transfer is at impedance matching, where exactly half of the power is transferred to the load impedance (in this case, the loudspeaker). This may seem a poor result but is nevertheless the best attainable. At the same time, the output voltage is exactly half the maximum voltage mentioned earlier, and this is the main reason to do it differently in circuits that have the purpose to measure voltages. If the membrane potential of a certain neuron is −60 mV, we want to measure this value fully, not half of it.

In addition, impedance matching would be far too critical in this situation: If the cell impedance would change, the input impedance of our preamp would have to be changed accordingly. Therefore, one adopts the mentioned high-impedance rule: if each instrument has an input impedance that is about 30–50 times as high as its predecessor output resistance. The signal is attenuated a few percent or less, and this is usually considered acceptable. The ratio has to be raised accordingly for precision measurements.

A notable exception to this rule is when one wants to measure current rather than voltage, such as in the patch-clamp recording. In this case, the ideal current amplifier (usually a current-to-voltage converter) has zero impedance so that the flow of current is not hampered. Here one has to obey the rule that the input impedance must be 30 to 50 times as *low* as the impedance of the preparation.

Maximum Signal Strength, Distortion

The maximum signal strength (in our case voltage) an amplifier can generate depends on a number of constructive details, the power supply voltages being the most important limitation. Transistor circuits operate with low voltages, although a few special (and expensive) types may handle a few hundred volts. Typically, modern amplifiers for measuring purposes run on 5–15 V and that limits the output voltage to these values, even often less, as will be explained below.

Remembering the properties of conventional transistors, about 0.5–0.7 V, is lost inherent in the way transistors function so that in circuits with cascaded transistor stages, some 2–5 V may be lost. Although FETs do not have such a threshold and so may be operated from zero up to the power supply voltage, most so-called FET amplifiers also use conventional transistors. As a guideline, it is safe to assume that an amplifier powered with 15 V may deliver about 10 V of signal voltage. Since most professional electronics operate on dual power supplies, i.e., + and −15 V, the voltage swing at the output of an amp will range from about −10 to +10 V. Some amplifiers, especially battery-operated ones, may be limited even to a few volts (plus and/or minus). In any case, the amplifier's specifications sheet must be consulted to learn the maximally sustained output voltages.

What about the input voltage? This quantity seems to be derived simply from the output voltage and the gain. In principle, an amp that sustains ±10 V and amplifies 1000 times (60 dB) will be saturated if the input voltage exceeds ±10 mV. Things get more complicated, however, if filter circuits are involved. In this case, the input stages of an amplifier may be saturated by a DC voltage (electrode polarization!) which goes unnoticed because later on in the circuit it is filtered out. Note that this means that the user may be kept unaware of the malfunctioning because the output voltage is neatly zero. Therefore, electrophysiologists have to have sufficient knowledge of the internal structure of their amps and of the recording conditions to prevent this situation from arising in an experiment. This may for instance happen when electrode polarization voltages rise slowly, as they often do, under aging, changing ion concentrations and the like.

What else may happen when an amplifier is operated with too strong signal voltages? In most cases, one is warned because the shape of the signal changes: A sinusoid of increasing amplitude "bangs its head" against the maximum output voltage and is said to be distorted. This so-called clipping is shown in Fig. 3.13: The lower trace fits between the two saturation limits and so is undistorted. The middle and upper trace show increasing grades of distortion.

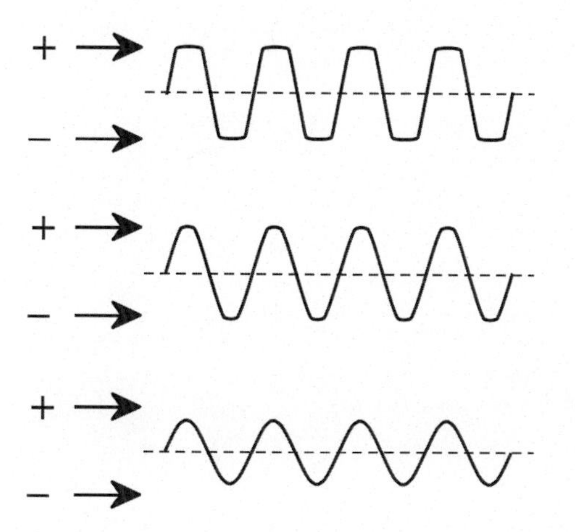

FIGURE 3.13 Clipping, a form of nonlinear distortion.

It must be kept in mind that the transition to saturation is not always as sharp as is shown here so that the distortion "sneaks in" at slowly increasing amplitudes, leaving the user unaware again. Contrary to the effects of filtering, this type of distortion is nonlinear.

If the signal is a small AC, such as a spike train from a neuron, saturation at one side, caused by the abovementioned increase in polarization voltage, may result in an apparent decrease in spike amplitude. Mistrust any such signals unless you are absolutely sure that no amplifier malfunctioning is the real cause.

In some cases, amplifier saturation is used to convert a sinusoidal into a square wave signal, but apart from sound effects in pop music, distortion should be recognized and avoided.

NOISE, HUM INTERFERENCE, AND GROUNDING

By the virtues of electronic circuitry, weak signals arising in biological preparations may be amplified to drive oscilloscopes, audio amps, and a host of instruments for recording and analysis. Obviously, there will be a limit to the gain or gain factor that may be employed before one reaches a fundamental or practical limit, where some kind of "interference" or unwanted signal dominates over the wanted one. In a broad sense, any unwanted signal component is occasionally called "noise," but that term is best reserved for the random fluctuations of electrical quantities that make the familiar hissing sound if fed to a loudspeaker. Noise, hum, and other forms of interference can be coped with successfully but stem from different sources and must be treated accordingly. Noise is most fundamental because it is omnipresent in all physical, chemical, and biological systems at temperatures higher than absolute zero and so cannot be eliminated entirely. However, noise can be—and must be, in most cases—minimized by carefully designing a recording setup (i.e., kept close to the theoretical minimum) for the given situation. Amplifier noise is illustrated in Fig. 3.14.

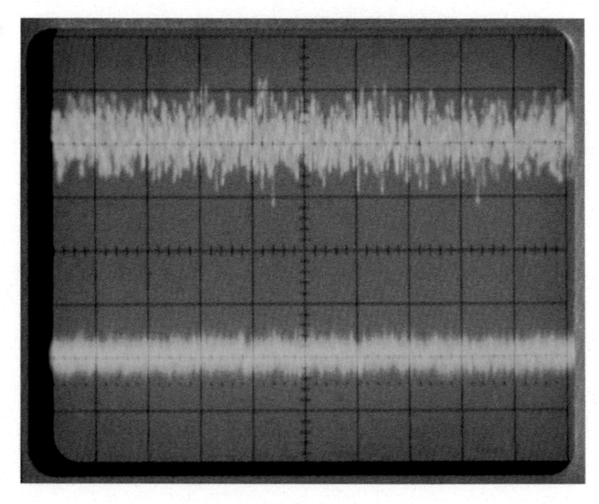

FIGURE 3.14 Two examples of amplifier noise as it may show on an oscilloscope screen.

Note that not the absolute noise amplitude but rather the noise level relative to the signal strength is what determines the success of an electrophysiological recording. This quantity is called signal-to-noise ratio, or S/N ratio, and is usually expressed in decibel. By a few not too complicated calculations, one may assess the possible S/N ratio of a certain situation before even attempting an experiment. Usually, the signal strength depends on the way of recording and can be estimated in advance using the following rules of thumb, in decreasing order as follows:

Intracellular potential and spike recording:	c. 10–100 mV;
Electrocardiogram (ECG):	c. 100 μV–1 mV;
Extracellular spike recording:	c. 10–100 μV;
Electroencephalogram (EEG):	c. 1–10 μV;
Extracellular current measurements (with the so-called vibrating probe):	c. 100 nV–1 μV.

The patch-clamp recording techniques measure current rather than voltage:

On-cell or excised patch recording:	c. 1–10 pA.
Whole-cell patch-clamp recording:	c. 100 pA–10 nA.

Since the exact signal voltages depend much on the detailed properties of the preparation and the recording geometry, these are only indicative values that give nevertheless a good idea about the order of magnitude involved and will be used in the subsequent discussion.

What about noise amplitudes? The most basic form of noise is thermal noise or Johnson noise and appears in any electrical circuit or component that has resistance. It arises by the random movement ("Brownian motion") of charge carriers in any conducting body. It can be shown from statistical mechanics that the power is dependent on temperature and bandwidth:

$$Pn = 4kTW$$

where, Pn is the noise power, k is Boltzmann's constant (c. 1.38×10^{-23} J/K), T is the absolute temperature (room temperature being about 293K), and W is the used frequency bandwidth (i.e., $f_{max}-f_{min}$). As an example, the noise power more than a 10-kHz bandwidth at room temperature amounts to 1.62×10^{-16} W.

With the help of the relations $P = U^2/R$ and $P = I^2 R$, we can compute the voltage across or the current through a resistor. This yields equations for noise voltage and current:

$$U_{eff} = 2\sqrt{(kTRW)} \quad (3.1)$$

$$I_{eff} = \sqrt{(4\,kTgW)} \quad (3.2)$$

where, U_{eff} and I_{eff} mean the effective value of the voltage and current, respectively, which will be explained below, whereas R is the resistance (in Ω) and g, the conductance (1/R; in S).

It is important to keep these equations in mind, especially the fact that noise amplitude is dependent on resistance and on bandwidth: Both can be controlled to a certain extent in an electrophysiological experiment. The noise voltage equation is used most because most electrical measurements are voltage measurements. However, in configurations for current measurement, such as in patch-clamp recording, the noise current must be computed from the conductances involved.

Next, let us explain what is meant by effective voltage. We explained the notion of effective value of an alternating current in Chapter 1 and found it to be $1/\sqrt{2}$ times the peak value. Now, most types of noise have a Gaussian amplitude distribution, which means that values around zero occur most and values farther from zero (both positive and negative) are increasingly rare. Thus, a noise signal has no formal "peak amplitude" like any deterministic signal, but has an effective value, that would again be the value of a DC that gives the equivalent amount of energy (heat). For Gaussian noise, this is the well-known root mean square (RMS) value, or standard deviation, which is the value from zero to one of the points of inflection of the Gauss curve. One may wonder in this context what a noise voltage looks like on an oscilloscope. Low-frequency noise may look like irregular bouncing of the trace, but high-frequency (or white) noise simply broadens the trace into a fuzzy horizontal band. In practice, the width or amplitude of this visible noise band is about 4–6 times the effective value, depending somewhat on the brightness of the oscilloscope trace intensity relative to the ambient light level.

What about the bandwidth of thermal noise? A noise signal has a continuous spectrum, i.e., it can be considered as a mixture of an infinite number of frequencies, each with infinitesimally small amplitude. Therefore, the greater the admitted bandwidth, the greater the noise amplitude. It can be seen from Equation 3.1 that the noise voltage is dependent on the square root of the bandwidth. Since power is proportional to the square of the voltage, the power spectrum is flat: Equal frequency bands hold the same noise power, whether from 0 to 1 kHz or from 1.000 to 1.001 MHz. Therefore, thermal noise is also known as white noise, a term derived from the analogy with white light that contains all visible wavelengths at (approximately) the same intensity. Although the parallel is somewhat sloppy (in what we perceive as white light, all wavelengths in fact do not carry the same power density), the notion of a "white" noise spectrum is nevertheless well defined.

Equation 3.1 for voltage noise leads to a fundamental rule for electrophysiological and indeed all other measurements: Do not use a wider bandwidth than necessary to preserve signal shape. A 10 times higher bandwidth results in a $\sqrt{10}$, or about three times, higher noise level. Note that the upper limit to the bandwidth is usually the most important one in this respect. Reducing the bandwidth by a high-pass filter from, say, 1–1000 Hz to 10–1000 Hz leaves still 990 Hz of noise-admitting bandwidth.

Note also that, since frequencies just outside the formal bandwidth do contribute a bit of noise, the noise bandwidth of an amplifier is slightly larger than the −3 dB limit suggests, usually (i.e., if first-order filters are applied) $\pi/2$ (approximately 1.57) times as high. Since noise calculations are usually intended to monitor orders of magnitude, the effect of this correction is of minor practical importance.

Armed with the knowledge just acquired, we may compute the thermal noise of a glass microelectrode before even having tried one. Let us assume that the required bandwidth is 5 kHz. Since the resistance is the other important quantity, the tip resistance of the pipette is the main cause of thermal noise. Micropipettes intended for intracellular recording usually have a resistance of about 100 MΩ. Entering these values in the Johnson equation yields $V_{eff} \approx 100\ \mu V$. This means that signals to be measured must be substantially larger than this value. Fortunately, intracellular signals are usually in the millivolt region. Extracellular (semi/micro) electrodes have larger tips and hence lower resistances than intracellular ones. Unfortunately, there are more types of noise that play a part in electrical measurements.

The second type to deal with is shot noise. It arises from the random movement of charge carriers through a barrier, such as a p–n junction or a cell membrane. Therefore, shot noise occurs not only in all kinds of devices such as amplifiers but also in living matter. Shot noise arises because even in a steady (DC) current, the number of charge carriers passing the border at any moment is a statistical quantity. Thus it can be compared to the noise hail stones make on a metal roof: Each hail stone contributes a minute tap, imperceptible in the whole hissing sound that is nevertheless made up of myriads of these tiny sound impulses. In the same way, the electrons or holes crossing the p–n junction of a semiconductor or the ions crossing the membrane of a cell form a current of which the random fluctuations are a form of shot noise. The magnitude of shot noise can be derived by statistical arguments, in which the elementary charge, the total number of charges (hence the total current), and the time or its inverse, the bandwidth, play a part. The current noise amplitude follows from:

$$I_{eff} = \sqrt{(2eIW)}$$

where e is the elementary (electron) charge (about 1.6×10^{-19} C), I is the current, and W is the bandwidth again. Note that the fluctuations become more prominent at lower current strengths. Thus, a membrane current of 1 μA using a bandwidth of 10 kHz has a shot noise component of 57×10^{-12} A, or 57 pA, which is only a minute fraction of the total current. At 1 pA, the order of magnitude of ion channel currents, the noise amounts to about 6×10^{-14} A, or 6%, and at 1 fA (10^{-15} A) the current would exist mostly as shot noise: 1.8×10^{-15} A or 180%!

Like Johnson noise, shot noise has a "white" spectrum and a Gaussian amplitude distribution. It must be mentioned that a certain type of electrophysiological signal resembles shot noise. This is the case when one records with relatively large electrodes from a whole nerve bundle. With this gross-activity recording, each spike is a minor click, barely detectable if at all, whereas the amplitude of this noiselike signal depends nicely on the average spike activity of the fibers in that bundle. The spectrum, however, is not entirely white, since a spike is not an (infinitely short) impulse. Hence, the spectrum of a gross-activity signal will be the same as a single-unit or population spike recorded under comparable circumstances.

Note that, if an electrophysiological signal has the characteristics of noise, extra care must be taken to discriminate it from instrumental noise.

A third type of noise is the excess noise, also called pink noise or 1/f noise. This is often the largest noise component in a signal, but unfortunately, the hardest to grasp. The name is derived again from the power spectrum, which in this case is inversely proportional to frequency. The term pink is a loose parallel with light again and alludes to light red light, where long wavelengths (and hence low frequencies) are more abundant than shorter ones. Again 1/f noise originates in semiconductors and presumably also in the cell membranes we study. However, it is not as fundamental as the other two noise types. Therefore, the amount of 1/f noise in transistors and hence in amplifiers may be reduced by certain aspects of the design and thus may be influenced by the designer (but not by the user) of scientific instruments. Because of the mentioned power spectrum, excess noise is most conspicuous at low frequencies. At higher frequencies there is always a point beyond which the white noise types are dominant. The transition point at +3dB above the white noise level is called the 1/f corner (Fig. 3.15) and is a test characteristic of commercially available preamplifiers.

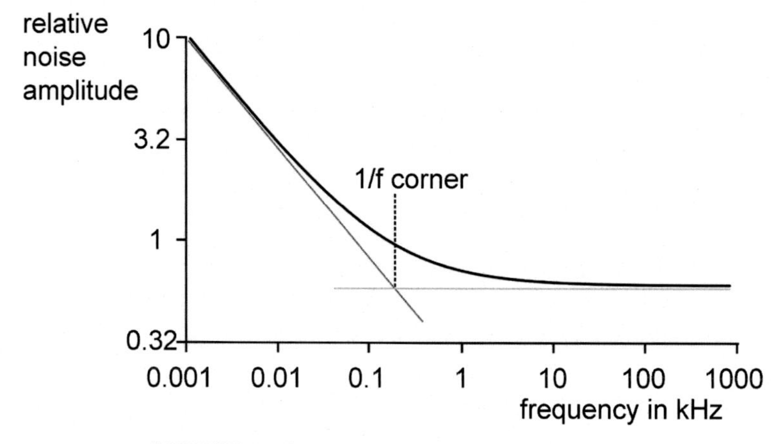

FIGURE 3.15 One-over-f (1/f) noise and corner.

The other forms of interference with an electrophysiological signal are less fundamental, yet often hard to combat. This holds especially for hum, the often persistent periodical disturbance (see Fig. 3.16) that stems from the mains lines, abundant in any house or laboratory. The frequency will be 50 or 60 Hz depending on the continent: 50 Hz in (among others) Europe and 60 Hz in (among others) the Americas.

Careful design of sensitive electronic apparatus and a careful and "clean" way to build a setup may help to minimize the detrimental influence of hum on electrophysiological signals. The main therapy against hum trouble consists of grounding and shielding to combat two forms of hum.

The first one is electrostatic, since it is caused by the electric field emanating from the mains lines. These have to be kept as far as possible from an electrophysiological preparation, since they carry relatively high AC voltages. The electric field emanating from mains

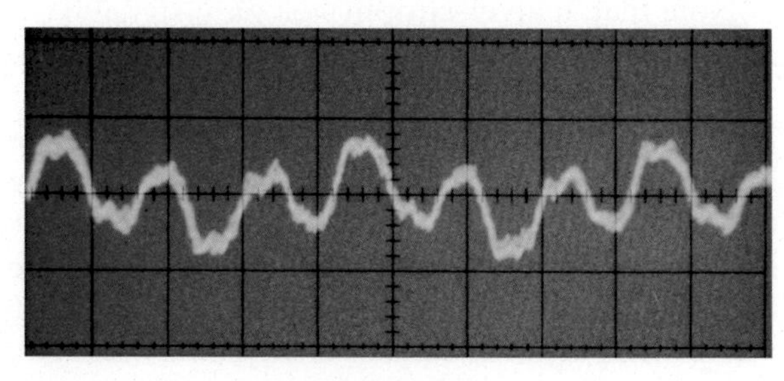

FIGURE 3.16 Hum as it may show on an oscilloscope screen. Since the shape is not a pure sinusoid, we conclude that it is at least partially caused by magnetic fields from the mains.

lines is sinusoidal and so causes a hum signal that has the same shape. A simple and effective cure is shielding, that is, covering instruments, cables, or the whole setup by a grounded, conducting layer of metal. Most scientific instruments are built in a metal case, and signals are transported via shielded or "coaxial" cables (see Chapter 2) between the instruments. The preparation itself and any other "open" parts of the setup may be enclosed in a so-called Faraday cage. Complete rooms for EEG recording in hospitals may be shielded this way.

The other type of hum is caused by the *current* flowing through the mains lines, through instruments, cables, etc., causing a magnetic field that may reach the setup. Contrary to the "static hum" described above, a "magnetic hum" signal is often peaked, caused by the bumpy current flowing through gas discharge tubes, such as fluorescent lights, power supplies, and motors in refrigerators, centrifuges, etc., and from apparatus switching on and off, such as thermostats. Here, the only cure consists of keeping distance, together with a proper grounding technique. A proper way of grounding is absolutely vital for the recording of weak signals encountered in electrophysiological research and consists of several measures, explained below.

The abovementioned grounding of all instrument cases, cables, and shields suffices to keep static voltages from building up.

If one grounds too eagerly, however, ground loops may emerge, and these are the cause of much trouble. This can be explained by recalling the principles of electromagnetism. By mutual induction (the same principle that makes transformers work) an AC current through a mains or ground cable induces a voltage in any nearby wire. Whether this AC voltage will be accompanied by a current in this wire depends on the impedance of the loop: An open circuit carries no current, and the voltage appears simply at the ends of the wire. If this is, say, 1 mV, it will be relatively harmless in most electrophysiological setups. If, on the other hand, the loop is closed, making the loop impedance, say, 10 mΩ (milli-ohm), a current of 100 mA would be the result. This current may induce yet another hum voltage in other nearby cables and so on. Unfortunately, this situation arises if a chain of measuring instruments is wired up in the normal way, i.e., by plugging all instruments in a grounded AC wall outlet and interconnecting the instruments by means of coaxial cables. This is illustrated in Fig. 3.17A. The ground loops, indicated by circular arrows, may span more than a square meter and so can pick up numerous stray fields from mains cables and transformers, especially from the instruments themselves. Several remedies are in use. The first solution, shown in Fig. 3.17B, consists of interrupting the shields of the coaxial cables connecting the individual instruments so that the safety ground connections of the instruments provide the sole ground connection. A complication is that the safety ground supplied with the mains outlets is often too "dirty" to be used with delicate measuring instruments. This is because these wires carry ground leakage currents from any apparatus connected to them, such as refrigerators, centrifuges, heating baths, and other laboratory instruments. These large and often peaky currents may cause many mV to develop across the ground circuit notwithstanding its very low internal resistance. A better instrument ground is often mandatory for electrophysiological measurements and should be provided in every laboratory. In this case, the best ground "mecca" is at the input circuit of the preamplifier, i.e., at the specimen ground connection (Fig. 3.17C, left part).

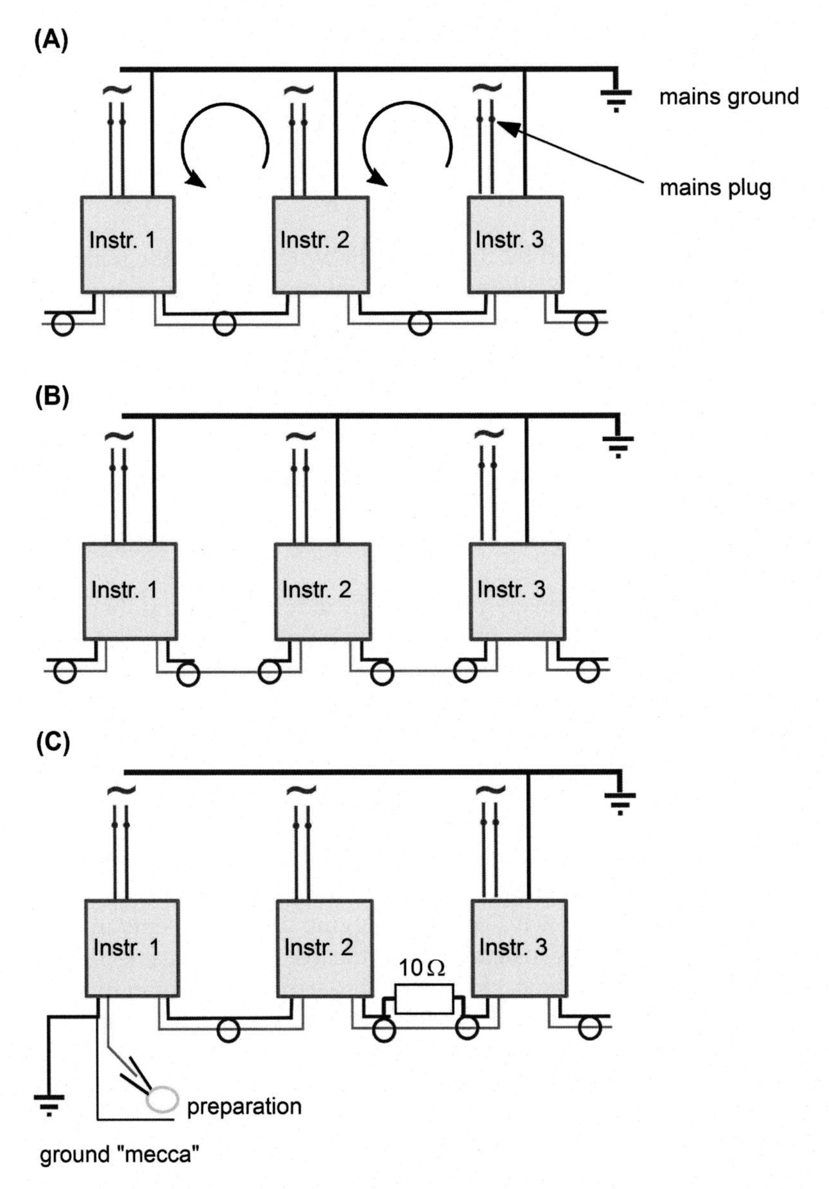

FIGURE 3.17 Grounding a chain of measuring instruments. See text for details. *blue lines*, grounding wires; *Instr*, instrument; *red lines*, mains power lines.

This introduces another problem, however, since the solution sought by many electrophysiologists, which consists in removing the mains grounding wires from the instruments, is on bad terms with security. This is especially hazardous because in building a setup, most people connect the power cords first, then wire up ground and signal connections. The problem is aggravated by the fact that in computers and other instruments employing MOSFETs, omitting the ground connection may blow all chips at once and so must be avoided completely.

The best solution to this grounding dilemma is either to use only the security ground (in case it is sufficiently clean; Fig. 3.17B) or to separate the ground circuits of digital (computers, counters, etc.) and analog (preamps etc.) apparatus. The two ground circuits may be connected by a resistance of, say, 10 Ω. In that case no significant loop current can flow, whereas from a static viewpoint all components are connected to "ground" potential. This is shown in Fig. 3.17C on the right.

Here, introducing a 10 Ω resistor into the ground loop reduces the loop current by a factor of up to 1000 (because the loop resistance jumps from about 10 mΩ to 10 Ω). Some measuring instruments, such as preamps, even have the ground circuits of their input and output separated by such a 10 Ω resistor. This is particularly useful, since an electrophysiological preparation may have "false" ground connections through thermostats, saline flow circuits, and the like. In many practical situations, one has to fiddle around with grounding wires to find the best configuration, i.e., the way that minimizes hum and interference. This may resemble more an art than a science.

The ultimate solution to the multiple ground syndrome is the use of isolation amplifiers. Although not standard in every lab yet, this solution must be considered seriously, since it is not unduly expensive and may prevent a lot of trouble. Electronic musical instruments are already coupled through optically isolated (MIDI) interfaces.

Occasionally, one may find a form of hum that is 100 (120) Hz rather than 50 (60) Hz. Usually, this stems from insufficiently filtered power supplies and so has an obvious remedy: increase the ripple filter capacitor or buy a better power supply.

The last form of interference that may vex electrophysiologists is radio-frequency interference (RFI) mostly from the strongest broadcast stations, occasionally from a nearby radio amateur, or even from the campus alarm network.

In principle, the remedies against these disturbances are the same as described for static hum, i.e., shielding and Faraday cages. In reinforced concrete buildings, where the steel grids provide some shielding, it often helps to keep electrophysiological setups far from the windows, i.e., closest to the center of the building, or even in the basement.

A final form of interference stems from the many wireless instruments that surround us: DECT and cell phones, remote controls, WiFi and Bluetooth connections, and so on.

DECT phones and remote controls use relatively weak radio waves (in the range of hundreds of megahertz to a few gigahertz microwaves), but cell phones, capable of communication over about 10–20 km, are often a nuisance, even in audio circuits used by reporters and radio stations.

Cell phones operate in the gigahertz range, using several frequencies, with each frequency up to eight time slots. Therefore, the digitally coded voice data are sent in "packages" that sound as sequences of short, ugly pulses when picked up by a sensitive instrument (see Fig. 3.18).

Note that, apparently, the gigahertz signal is rectified to a certain extent to create the low-frequency interference signal.

The remedy is again the same: keep cell phones far from sensitive instruments.

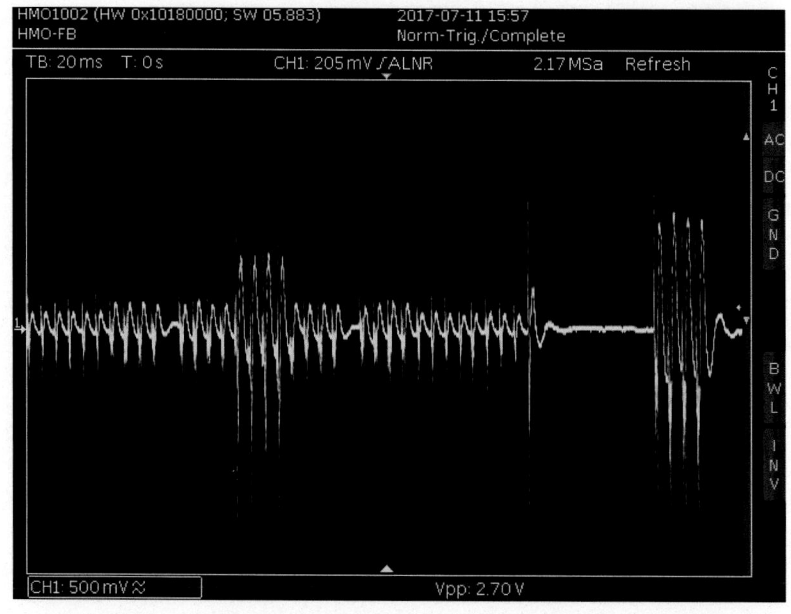

FIGURE 3.18 Signal from a cell phone penetrating a measuring setup.

DIFFERENTIAL AMPLIFIERS, BLOCK DIAGRAMS

The basic form of amplifier we discussed so far has only one input. Thus, the output voltage, U_{out}, can be described as the input voltage with respect to ground, U_{in}, multiplied by a gain factor, M:

$$U_{out} = M \times U_{in}$$

However, it is often necessary to measure the voltage between two points neither of which is connected to ground. In electrophysiological recording, one has for instance one electrode in or near a nerve fiber (the measuring electrode), and a second one nearby (the reference electrode). This is called differential recording and needs a differential amplifier.

Such a preamp has two input terminals, usually marked "+" and "−" or "A" and "B," respectively, and the circuit is such that the output is dependent on the difference between the two input signals, again multiplied by a gain factor, M:

$$U_{out} = M\,(U_A - U_B)$$

This implicates that a common-mode voltage, V_C, which means *present on both inputs*, would not appear at the output because it would be canceled by the subtraction:

$$U_{out} = M\,((U_A + V_C) - (U_B + V_C))$$

In electrophysiological recording, the relevant signals are often as follows:
V_A—or signal, voltage at the recording electrode;
V_B—or reference, voltage at the reference electrode; and
V_C—common-mode voltage (hum, RFI, etc.).

Ideally then, a differential amplifier would just amplify the differential signal without being influenced by the interference. In a practical situation, the signal might be a 1 mV ECG (left arm minus right arm), whereas radio waves or mains cables cause an interfering common-mode signal of, say, 200 mV. Without a differential preamp, then, the ECG would be totally obscured by the interference.

In practice, however, differential amplifiers are not perfect, i.e., mathematically exact, so that traces of a common-mode signal can be found at the output if it is relatively large, as in the previous example. The degree to which a preamplifier suppresses a common-mode signal is called the common-mode rejection (CMR) and is usually expressed in dB. The CMR is defined simply as the gain of the amp for differential signals (the "normal" gain) minus the gain for a common-mode signal. As a practical test for preamps, one can (and must!) determine the CMR by connecting a signal source such as a sine wave generator to both A and B inputs at the same time. If, for example, the gain of an amp is 60 dB (i.e., 1000×), and a common-mode signal is attenuated by 20 dB (i.e., the common-mode gain is −20 dB), the CMR is $60-(-20)=80$ dB. Well-designed electrophysiological preamps should have a CMR of at least 100–120 dB.

To discriminate between the two types, an amplifier having only one input is called a single-ended amplifier. In fact, a differential amplifier can be considered as two amps in one: if a single signal voltage is applied to input A, while grounding input B, the amp behaves as a normal, single-ended amplifier. If input B is connected to the signal source, and A, grounded, we have again a single-ended amp that reverses the signal polarity:

$$U_{out} = -MU_{in}$$

This configuration is called an inverting amplifier. This is often useful by itself and is the reason that the A and B inputs are often labeled + and − inputs, respectively. In words, the inputs are called inverting input (B or −) and noninverting input (A or +). Note that the + input does not need to be fed with a positive voltage and that the − input has not necessarily to be kept negative: The symbols stand only for the polarity of the output with respect to the corresponding input.

Block Diagrams

We need a convenient shorthand to distinguish the different types of amplifiers to discuss differential amplifiers any further. Up to now, we have seen the schematic symbols used to represent components like resistors, capacitors, and semiconductors that together may make up an amplifier or other electronic instrument. These symbols are useful to elucidate the functioning of simple circuits, but it will be obvious that most instruments are so complicated that it is neither necessary nor feasible to draw all components separately. Thus, new symbols were defined to represent complete instruments rather than components. These include

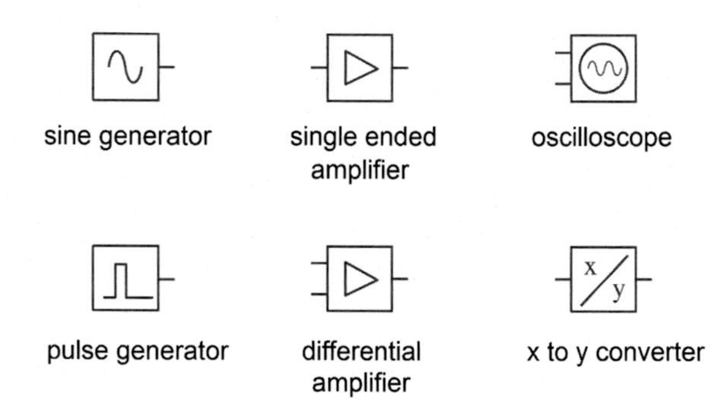

FIGURE 3.19 Symbols for block diagrams.

both single-ended and differential amplifiers, voltmeters, oscilloscopes, signal generators, filters, audio amps, and so on. Schematics composed of these meta-symbols are known as block diagrams and are indispensable in describing one's setup. The most frequently used and best standardized block symbols are depicted in Fig. 3.19. As you see, most symbols consist of a rectangle with input and output connections as well as a schematic indication of the function or contents of the box. New ones can be invented easily to describe new or special functions. In the most general form, a block symbol consists of a box containing the name or type number of an electronic apparatus. The boxes must be connected with straight lines as short and neatly arranged as possible. By adding (arbitrary) symbols to reflect measuring and ground electrodes experimental subject, tissue preparation or cells, and the like (Fig. 3.20), the fundamentals of any electrophysiological recording setup can be made clear.

Every experimenter should make a habit of drawing a complete block diagram of the setup used before doing experiments.

Differential amplifiers are often used in electrophysiology in situations where interference voltages are high—or signal voltages are low—such as is the case in most laboratories doing extracellular nerve recording, EEG, or patch-clamp experiments. The advantages of differential recording are explained in Fig. 3.21. Here, the cell or nerve to record from is indicated as the "object," whereas the rest of the animal, tissue, and/or saline is indicated as "animal."

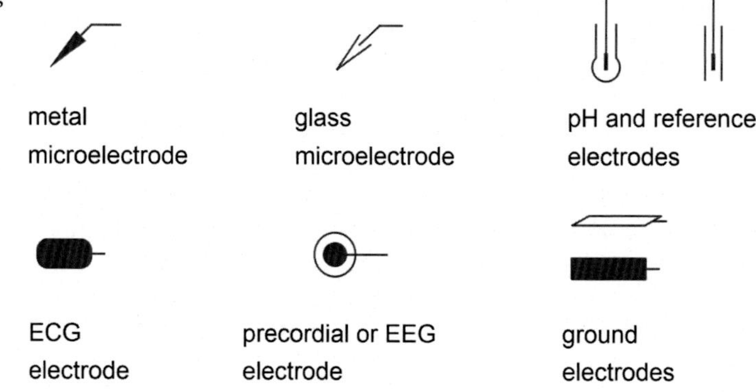

FIGURE 3.20 Suggested symbols for electrodes.

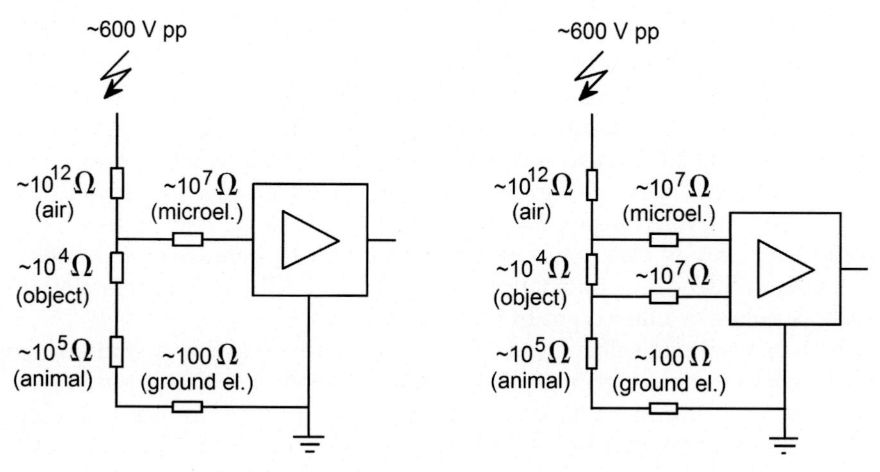

FIGURE 3.21 Single-ended versus differential recording.

To analyze the situation properly would need to include the geometry of the electric fields involved, but the simplification used in Fig. 3.21, although not entirely correct, may help to grasp the principle. In reality, the "animal" has a lower resistance than the "object" but is spatially larger, picking up more interference.

Why not use the second electrode as ground? The reason is that grounding electrodes need to have a low resistance (or impedance, since most forms of interference are alternating currents), say 1 kΩ or less, whereas measuring electrodes have to be small and hence have a much higher impedance: mega-ohms or more. If one would use a 10−100 MΩ microelectrode as a ground electrode, even the relatively weak currents stemming from nearby mains cables or the local broadcast station would give rise to high interference voltages, masking the signal to be measured.

OPERATIONAL AMPLIFIERS, FEEDBACK

The so-called operational amplifier, or op-amp for short, was the basic building block of the analog computer, the large signal processing machine that was used in laboratories before the digital computer made its soaring rise. Despite the flexibility of digital circuits, operational amplifiers are still used frequently and have become so cheap and so small that they replace transistors in most applications. The name still refers to the mathematical "operations" that can be performed with these devices, such as subtraction, summation, and integration explained below. The first operational amplifiers, built in the 1940s with vacuum tubes, were inverting amplifiers and, although nowadays most op-amps are differential by design, many functions are performed with the noninverting input grounded, thus reducing it to the inverting amp described earlier. The clue to understanding the functioning of op-amp circuits is to understand the principles of feedback, since this is what makes them work in a so well-defined way. Feedback comes in two flavors, positive feedback and negative feedback, and the classical example of the latter form is how cyclists and car drivers can keep their vehicles on a straight path despite all kinds of diverting forces: wind, irregular pavement, sloping road surfaces, and so on. A gust of wind from the left, for example, would push the vehicle to the right and the driver counteracts this force with an equally strong force in the opposite (or "negative") direction. As a result, the driver continues the wanted straight path. In other words, by compensating all diversions continuously, the wanted situation can be maintained indefinitely. Positive feedback on the other hand is like steering with crossed hands: any disturbance will be amplified instead of corrected and will likely end in disaster. Feedback is used in the same way in electronics: negative feedback to stabilize any wanted situation against unwanted influences or fluctuations and positive feedback to generate electrical "disasters" such as pulses or other types of oscillations.

Let us return to negative feedback and see how this is applied with operational amplifiers. A simple one-transistor amplifier might have a gain of about 50–300, depending on several properties of the used transistor, on temperature, aging, and so on. Gain may be pushed easily to over one million times (120 dB) by cascading transistors, but the uncertainties grow proportionally. Moreover, if such a component would have to be replaced, the properties of the whole device might get altered in an intolerable way. So, most transistors are convenient but unreliable components, having wide ranges of gain, temperature coefficients, and so on. As we will show, negative feedback may be used to stabilize gain so that an amplifier may be built that has distinctive properties independent of temperature, aging, and the properties of unreliable components.

An op-amp is an amplifier with a very high gain and a very high input impedance. A feedback circuit is provided by a few added components, often resistors. A simple circuit suitable to analyze the mode of operation is depicted in Fig. 3.22. The op-amp is used as an inverting amplifier (+ input grounded). The feedback is effected through R_2 (10 kΩ), whereas input is delivered via R_1 (1 kΩ). Let us analyze this circuit for the case in which the amplifier would have a gain factor of 1000 and let the input impedance be so high that virtually no current flows through the input terminals. Let us assume further that there are no DC offsets, which means that an input signal of zero yields an output signal of zero. If we connect an input signal of +1 V, a current will tend to flow through R_1 that makes the inverting input positive. Obviously, the gain of the amplifier will tend to drive the output to very high negative values, but by the action of R_2, the voltage at the inverting input is reduced again. This is the principle of negative feedback and will yield an equilibrium state at approximately the

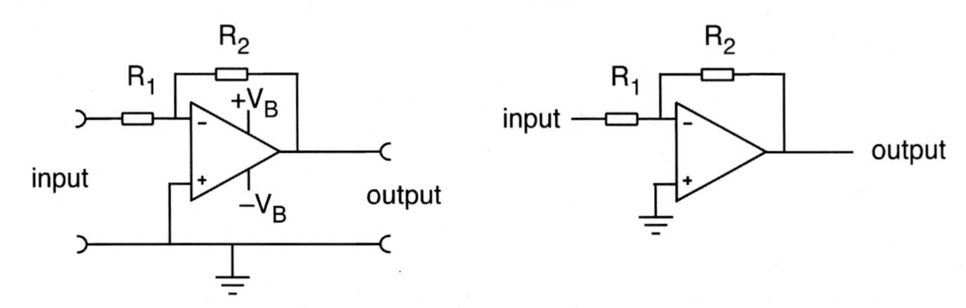

FIGURE 3.22 Basic op-amp circuit (left version). R_1, input resistor; R_2, feedback resistor. $+V_B$ and $-V_B$ are the positive and negative power supply lines, respectively. The connectors, power supplies, and ground wires are often omitted in op-amp schematics (right version).

following values. The output signal will be about −9.9 V, the voltage at the inverting input will thus be about +9.9 mV. The end result is an amplifier with a gain of almost 10 times (and inverse polarity).

Now what will happen if we increase the gain of the op-amp? If the gain factor is increased to one million, the output voltage will be closer to 10 V, while the voltage at the inverting input will be a mere 10 μV. Thus, the circuit is now a nearly perfect times-10 amplifier. This leads to the simple rule of thumb to understand op-amp functioning: If the gain approaches infinity, and the input impedance too (no current through the inputs), the output voltage of the op-amp will be such that the inverting input is held at (very, very nearly) the same potential as the noninverting input. If the noninverting input is grounded, such as in our example, the inverting input is kept *actively* at ground potential. Therefore, it is called the virtual ground. In this case, the output voltage follows simply from the current involved: an input voltage of 1 V drives a current of 1 mA through R_1. Because of the infinite input impedance, the same current must flow through R_2, yielding an output voltage of $-R_2/R_1$ volt.

Our amplifier has a modest gain factor, but it is virtually independent on the gain of the used op-amp, i.e., the gain without feedback or open-loop gain. The closed-loop gain, although only 10, is much more reliable and depends mainly on the values of the added components. In real op-amps, the open-loop gain may drift and be temperature dependent between, say, 100,000 and 250,000, but this has only negligible consequences for the properties of the entire circuit. Needless to say that the used resistors must be precise enough and have a good stability, but this is much easier to accomplish: Metal film resistors have usually a tolerance of 1% or better and may be selected to match within 0.1%. They have a very low temperature coefficient: about a 1000 times better than most semiconductors. Practical op-amps are cheap (a few dollars to about 100 dollar), have open-loop gains of at least 100 dB (100,000×), and input impedances of at least 1 GΩ. The bandwidths reach into the megahertz range. So, by designing op-amp feedback circuits, we can harness these near-perfect devices to perform all kinds of tasks, or operations, necessary to collect or process electrophysiological signals.

A few examples are given below. Fig. 3.23A shows an op-amp as adder. Since the inverting input behaves as the virtual ground, currents into this point simply add in a linear way: the output current is the sum of the input voltages. By choosing different values of R_1 and R_1', the inputs can be given different weights. Subtraction is done with the circuit of Fig. 3.23B. This is a way to build a differential amplifier. Note, however, that the input resistors R_1 determine the input impedance of the whole circuit, so it is not suited as differential input stage for electrophysiological amplifiers. Finally, the principles of addition and subtraction may be extended to any number of input signals by adding extra input branches. The use of op-amps is not limited to amplification or summation, as is shown in the following circuit (Fig. 3.24).

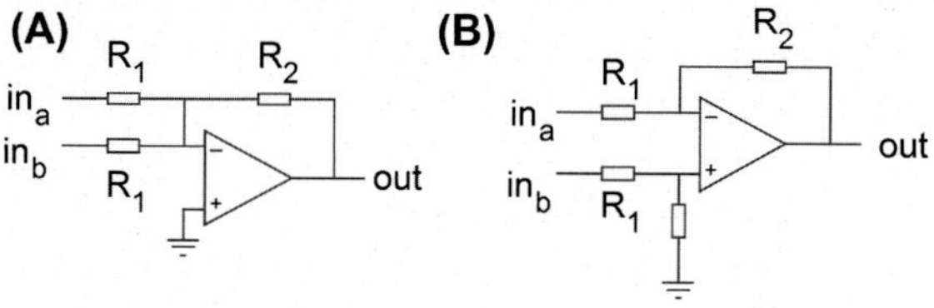

FIGURE 3.23 (A) Op-amp adder. (B) Op-amp subtractor.

A close look at the circuit of Fig. 3.24 shows that by adding a capacitor, we have built an electronic integrator: the output is the integral of the input voltage, with RC as the familiar time constant:

$$U_{out} = -1/RC \int U_{in} dt$$

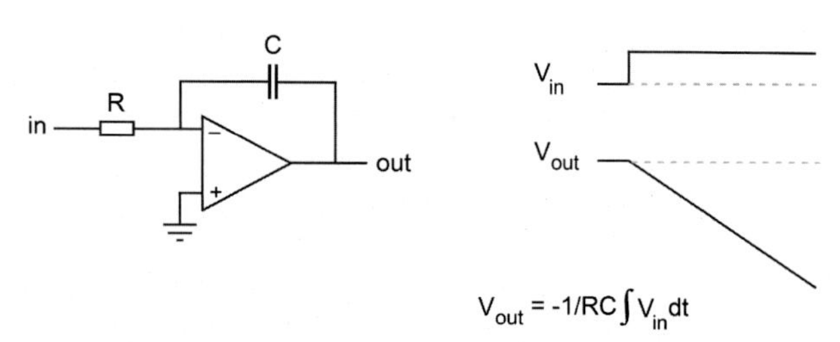

FIGURE 3.24 Op-amp integrator.

We will analyze the integrator by letting R = 1 MΩ, and C = 1 μF, and feeding the input again with 1 V DC This yields an input current of 1 μA that charges the capacitor with 1 V/s. Thus, the output voltage runs away with (minus) 1 V/s. Clearly, measures have to be taken not to get stuck at the maximum output voltage. This means that practical integrators will need extra components to prevent build-up. The simplest one is a reset button across the capacitor to discharge it at will, but in most applications, the signal voltages are limited electronically, for instance by reversing the input signal. This is the basic principle of function generators, which will be dealt with later on.

By exchanging the resistor and the capacitor of the integrator, the circuit is turned into a differentiator that renders the derivative of the input current. We will leave the analysis of this circuit to the reader. As with the case of the integrator, practical differentiators need additional components to keep some properties within bounds. As a hint, bear in mind that the input impedance of a differentiator is a pure capacitance and that the input current will increase with increasing frequency. Integrators and differentiators can be considered as RC filters, made ideal by the action of the op-amp.

Whereas passive RC filters have a roll-off frequency, above or below which the 6 dB/oct slope flattens off, the op-amp circuits described above maintain their +6 dB (differentiator) or −6 dB (integrator) slope over their whole functional bandwidths. Likewise, the output of a differentiator leads the input 90 degrees in phase, whereas the output of an integrator lags 90 degrees. Finally, the output voltage may be higher than the input voltage.

We will meet these circuits again, since they play an important part in constructing electrode test and compensation circuits in intracellular and patch-clamp amplifiers.

All former circuits with the noninverting input grounded inverted the input voltage and so were an inverting amplifier, an inverting adder, and so forth. The circuit of Fig. 3.25A is surely the most simple op-amp circuit, since it has no external components other than a piece of wire: the output is connected to the inverting input. Using the same rule of thumb, the output voltage turns out to be the same as the input voltage. Therefore, it is called a follower or voltage follower. What would be the advantage of an amplifier that does not amplify as in Fig. 3.25A— remember the high input impedance of op-amps)? Even better still, the intrinsic

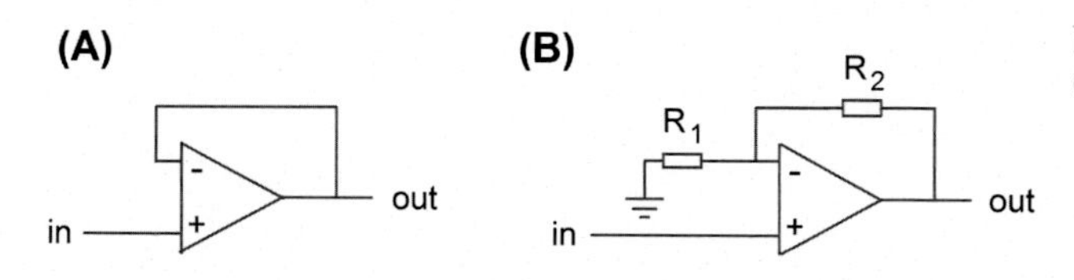

FIGURE 3.25 Simple voltage follower (A) and follower with gain (B).

input impedance of the device, which depends on the type of transistor used, is multiplied by the open-loop gain. Because of this useful feature, the basis of any intracellular microelectrode amplifier is a follower circuit. Fig. 3.25B shows how to add gain to such a noninverting amplifier. Installing a number of resistors, together with a rotary switch, turns this circuit into a versatile amplifier with a selectable gain factor of, e.g., 1, 2, 5, 10, 20, 50, and 100. Practical forms of the op-amp are shown in Fig. 3.26.

The last op-amp circuit we will discuss is the current-to-voltage converter (CVC): a simple but important tool in electrophysiology. Fig. 3.27 shows that the op-amp is fitted with only a feedback resistor, and that the input is directly connected to the virtual ground. This implies that the circuit connected to it is effectively grounded: the input impedance is near-zero. But except providing a ground connection, this circuit measures the input current and so can be used to ground an electrophysiological preparation and to measure the stimulus current flowing through the preparation. Therefore, a CVC is also part of the famous voltage-clamp amplifier. If a good, high-impedance op-amp is used, feedback resistors as high as 10^{10} or even 10^{11} Ω can be used. Under certain precautions, currents as low as 0.1–1 pA can be measured. Therefore, a CVC with a feedback resistor in the abovementioned range is the heart of any patch-clamp amplifier, which is discussed in Chapter 4.

This end our review of op-amp circuits. More complex applications, such as microelectrode preamplifiers, are discussed in the next chapter. The realm of applications of op-amps is virtually unlimited. Using the same electronic building block, the feedback circuit,

FIGURE 3.26 Practical integrated circuit (IC) op-amps.

FIGURE 3.27 Current-to-voltage converter.

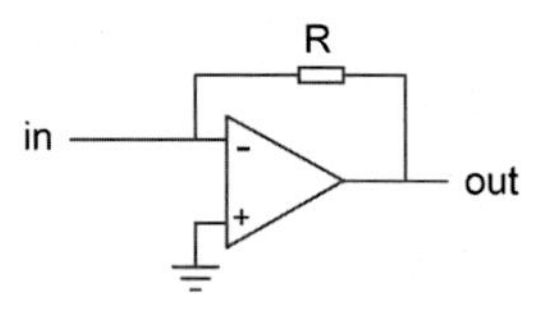

usually consisting of a mere two to three parts, determines which of a host of functions will be performed.

But what would happen if the feedback circuit is left out, so if an op-amp is used with open-loop? In that case, the almost infinite gain drives the output to the maximum voltage (about the power supply voltage), either the positive or the negative one, depending on minute differences between the input voltages.

Although such a configuration is likely to be unstable, and might cause a lot of problems for the designer, it is nevertheless useful, and known as a comparator. The output voltage signals a difference between two voltages with a high precision. If the open-loop gain of such a comparator op-amp is 100 dB, the difference between the voltages at the inverting and the noninverting input needs only to exceed 100 μV to yield a 10 V output voltage. Thus, if the voltage at the inverting input is at least 100 μV higher than that at the noninverting input, the output is fully negative, if it is 100 μV lower, the output is fully positive. In fact, special op-amps that sustain stable, open-loop performance are sold explicitly as "comparators."

ELECTRONIC FILTERS

Operational amplifiers are the ideal building blocks for electronic filters.

In many recording situations, one needs more filtering than the low-pass and high-pass RC filters dealt with in Chapter 2 can provide. These basic, so-called first-order filters have a filtering slope of 6 dB (high-pass) or −6 dB (low-pass) per octave outside their passbands. They can be cascaded to get higher orders. The simplest design would be to combine two low-pass sections into a second-order low-pass (Fig. 3.28). This combination yields a −12 dB/oct roll-off slope, accompanied by a 180-degree phase lag asymptote.

This passive solution to a higher-order filter suffers from two serious flaws. In the first place, the two sections are coupled directly, and so their actions are mutually influenced. Different R and C values have been chosen for the two sections to minimize this influence in this example, but in general, one would need an amplifier between the stages (and preferably in front of the first section, as well as after the last one). However, op-amps have the possibility of using filters as feedback paths. This provides for the necessary separation and makes the filter characteristics far more flexible. A second-order low-pass with a single op-amp is shown in Fig. 3.29. Frequency and damping (the latter determining the peak in the frequency response curve) can be varied by adjusting the components marked f and d, respectively.

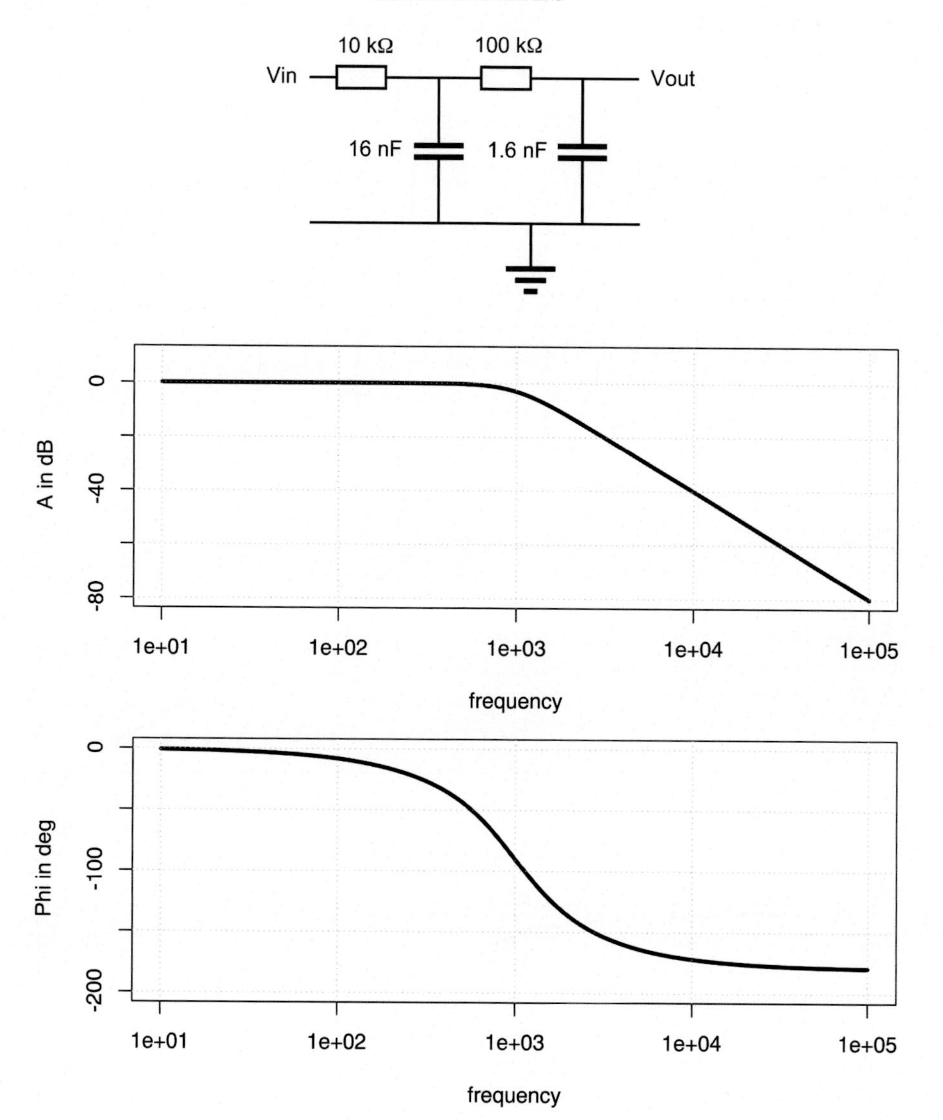

FIGURE 3.28 Two low-pass sections cascaded to obtain a second-order low-pass filter. Top: circuit diagram, middle and bottom: frequency characteristics.

Higher-order filters (fourth, sixth order, etc.) are usually built out of second-order sections like the one shown here. For odd orders (third, fifth, etc.), one first-order section is added.

By altering the feedback circuits and the relative values of the different components, a number of filter characteristics may be obtained that differ widely in their frequency behavior. Some filters have a beautifully flat passband but a nonlinear phase shift. Others have ripples in the passband but cut far steeper beyond the cut-off frequency. In addition, high-pass, band-pass, and band-reject configurations can be built. The different versions

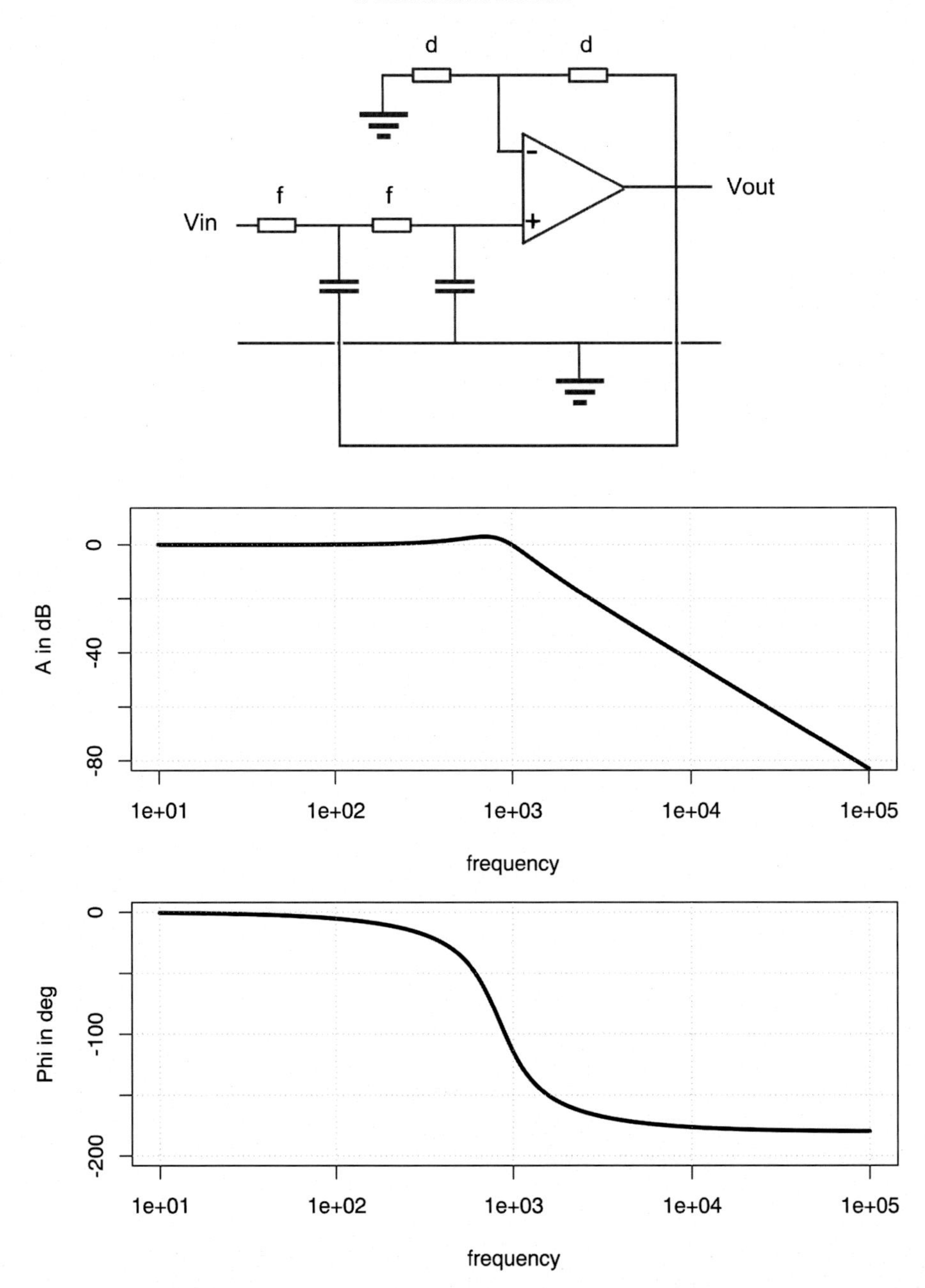

FIGURE 3.29 A second-order active low-pass filter with a single op-amp. The frequency is determined by the value of the capacitors and by the resistors, damping by the resistors marked d. Top: circuit diagram, middle and bottom: frequency characteristics.

TABLE 3.1 Higher Order Filters Listing Their Advantages and Drawbacks

Type	Passband Performance	Transition	Stopband Performance	Phase or Delay Performance
Butterworth	**Flattest**	Slow	Fair	Fair, nonlinear
Bessel	Decreasing	Slow	Fair	**Best (linear)**
Chebyshev I	Rippled	Fast	Monotonic	Nonlinear
Chebyshev II	Decreasing	Fast	Rippled	Nonlinear
Elliptic (Cauer)	Rippled	**Fastest**	Rippled	Nonlinear

are often named after their inventors, such as Butterworth and Chebyshev. Table 3.1 shows the properties of the most popular ones. The properties printed in bold type are often the reasons to use a particular design, provided the disadvantages can be tolerated.

References

1. Bergveld P. Thirty years of ISFETOLOGY what happened in the past 30 years and what may happen in the next 30 years. *Sensor Actuator B* 2003;**88**:1–20.
2. Lee CS, Kim SK, Kim M. Ion-sensitive field-effect transistor for biological sensing. *Sensors (Basel)* 2009;**9**(9):7111–31.
3. Bergveld P. ISFET, theory and practice. In: *IEEE Sensor conference* 2003.

CHAPTER

4

Electronic and Electrophysiological Instrumentation

OUTLINE

ELECTROPHYSIOLOGICAL PREAMPLIFIERS

Although specialized instruments such as microelectrode amplifiers are occasionally still made with separate transistors and other components, most electrophysiological instruments

Introduction to Electrophysiological Methods and Instrumentation, Second Edition
https://doi.org/10.1016/B978-0-12-814210-3.00004-9

are composed of op-amp circuits or can be represented by op-amps as functional units. Below are a few examples of circuits for the main forms of electrophysiological recording: extracellular, intracellular, and patch-clamp. Note that for flexibility, precision, and stability, practical amplifiers are usually more complex than those shown here.

AMPLIFIER FOR EXTRACELLULAR RECORDING

An amplifier suited for extracellular recording is shown schematically in Fig. 4.1. Here, we need a high gain (a gain factor of 1000 or more), a fairly high input impedance of at least 100 MΩ, the possibility to bar DC voltages ("AC coupling"), and filters to limit the bandwidth. Op-amps A1 and A2 form a differential pair with a high input impedance and a high gain, determined by:

$$G = 1 + \frac{2R_2}{R_1}$$

The signal is converted from differential to single-ended by A3. This 3-op-amp circuit is known as instrumentation amplifier. To provide further gain and/or filtering, amplifiers A4 and A5 are added. These amps are shown in the simple follower configuration for simplicity. They prevent mutual influencing of the filter sections. In practical designs, the

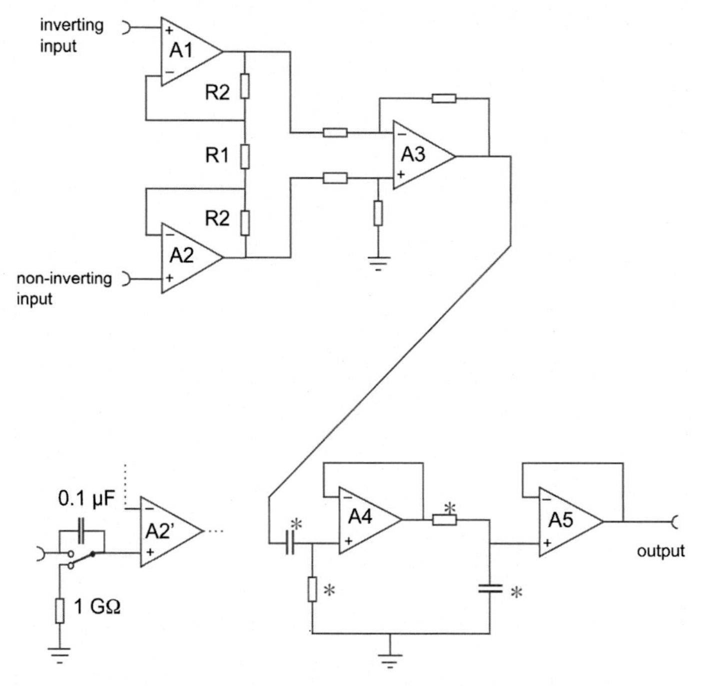

FIGURE 4.1 High-gain differential preamplifier. A1 and A2 preamplifier stages, A3 differential amplifier, A4 and A5 are for filtering and extra gain. A1 and A2 may be replaced by A2′ for AC-coupled amplification thus eliminating DC voltages at inputs and output.

resistors and capacitors indicated with asterisks are sets, selectable with rotary switches so as to provide flexible gain and bandwidth choices. Offset and calibration circuits are also omitted for simplicity.

The alternative input circuit at the lower left (op-amp A2′) is a way to implement AC coupling. It is usually provided at both inputs. The reader is invited to compute the cut-off frequency using the values shown. Unfortunately, such a circuit reduces the common-mode rejection at low frequencies, largely because the components used cannot be matched better than about 1% and because carbon may still be used for resistors in the GΩ range. An example of a low-noise, high-gain differential preamplifier is shown in Fig. 4.2.

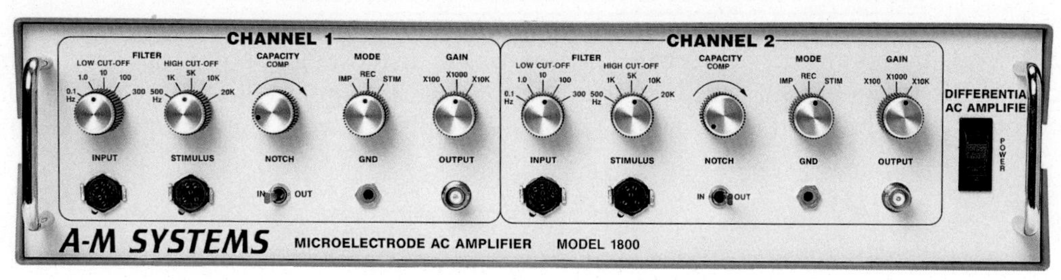

FIGURE 4.2 A practical extracellular amplifier, the A-M Systems Model 1800.

TESTING AMPLIFIERS

The human thumb is a simple test object for researchers to test a preamplifier. One only needs two sharp wire electrodes and a larger ground electrode to record (differentially) from the ball of the thumb. An example made from an audio plug is shown in Fig. 4.3. The large metal tube and flange function as ground electrodes to minimize mains frequency hum and other unwanted signals.

FIGURE 4.3 Amplifier test electrode, intended to be pressed against the ball of the thumb.

Pressed firmly against the skin, this electrode set picks up the activity of several muscle units if one moves one's thumb a bit (see Fig. 9.1). The amplitude of these extracellularly recorded spikes depends on the distance of the muscle fiber or motor unit from the electrode. Finger muscles ventrally in the lower arm can also be recorded from. These muscles lie a bit deeper, yielding a more gross activity like signal.

By playing with the filters of the amplifier, the best bandwidth for this type of myogram can be found.

AMPLIFIER FOR INTRACELLULAR RECORDING

A second, important way of recording is by means of an intracellular glass capillary microelectrode. An amplifier schematic is shown in Fig. 4.4A. Since intracellular voltages are relatively high, we do not need much gain: a factor of 10 suffices. A very high input impedance is

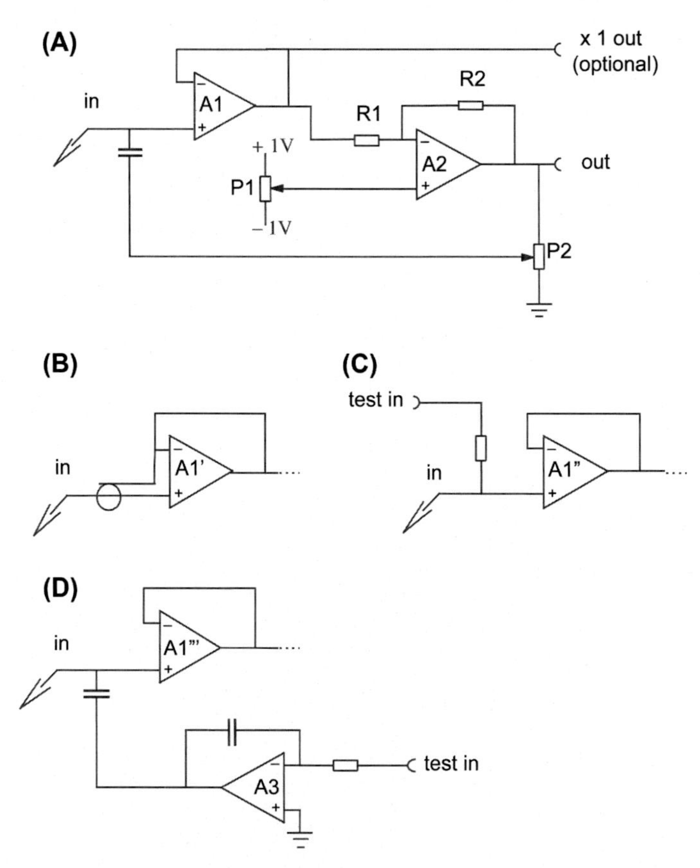

FIGURE 4.4 Microelectrode voltage amplifier. (A) Single-end microelectrode amplifier with offset compensation using potentiometer P1 and capacity compensation with P2. (B) The output of operational amplifier A1 can be fed to the cable shield thus eliminating capacitive losses in the cable. Test voltage pulses may be fed to the microelectrode via a high resistance (C) or a capacitor via an integrator-differentiator circuit (D).

mandatory because intracellular microelectrodes can have impedances of more than 100 MΩ. Therefore, the first op-amp (A1) is connected as a voltage follower. The second one (A2) provides some gain (R_2/R_1) together with a DC offset control (potentiometer P1). P2 is used to provide a controllable input capacitance compensation. The latter is often necessary because at the high impedance level used, the stray capacitances of pipette holder, cables, and/or input circuit will reduce the bandwidth too much. As an example, a 100 MΩ pipette together with 10 pF stray capacitance reduces the bandwidth to a mere 160 Hz, insufficient to record most nerve and muscle spikes. The capacitance compensation circuit is in fact a form of positive feedback, here for high frequencies, and must be used with care. Overcompensation leads to a very peaked response, ringing, or even to a strong continuous oscillation. A way to reduce the effect of cable capacitance is to drive the shield of the cable with the times-1 output of the follower amp. This technique, known as guarding and shown in Fig. 4.4B, diminishes the bandwidth reduction without the use of positive feedback and so without the risk of overcompensation. Here, the harmful effect of the cable capacitance is reduced to a very low value because the voltage across that capacitance is reduced. This is because the full input voltage charges the cable capacitance if the shield is grounded, whereas here only the op-amp error voltage, which is far lower, charges the cable capacitance. Guarding is often helpful when the use of relatively long cables is necessary but does not help to reduce the effects of other stray capacitances such as the capacitance of the pipette shaft and taper. Since glass micropipettes have a variable impedance that may vary between experiments and also during recording, the amplifier must be fitted with a circuit to measure the electrode impedance in situ. Fig. 4.4C and D shows two alternative ways to accomplish this. The first circuit (A1″) uses a high resistor of 1 GΩ or more to inject a small current into the electrode at the input. When operated with a square wave at the test input, the amplitude of the square signal at the electrode can be converted to read electrode impedance. A disadvantage of this circuit is that the input impedance of the amplifier is reduced to the value of the test resistor. A better way is shown in Fig. 4.4D (op-amp A1‴), where a small capacitance (about 1 pF or less) is used to feed a test signal in. Because the capacitor differentiates the test signal, it must be fed with a triangle rather than a square signal. So, an integrator circuit (IC) is built in, to convert the test square into a suitable triangle.

PATCH-CLAMP AMPLIFIER

The patch-clamp amplifier, illustrated in Fig. 4.5, is the last type of amplifier to discuss. Because membrane currents rather than voltages are measured, we have here a situation different from the voltage amp discussed above.

A patch-clamp amplifier measures the current flow from the electrode (pipette) to ground. Therefore, the core of a patch amp is a virtual ground or current-to-voltage circuit (see Chapter 3). Note that, although a current meter needs to have a low input impedance, the electronics used must have a very high input impedance nevertheless. This is because the currents to be measured are so small: in the range of 1 nA down to less than 1 pA. This also means that stray capacitances tend to spoil the bandwidth again. For this reason, a number of correction and compensation circuits are needed. Fig. 4.5 shows a somewhat simplified

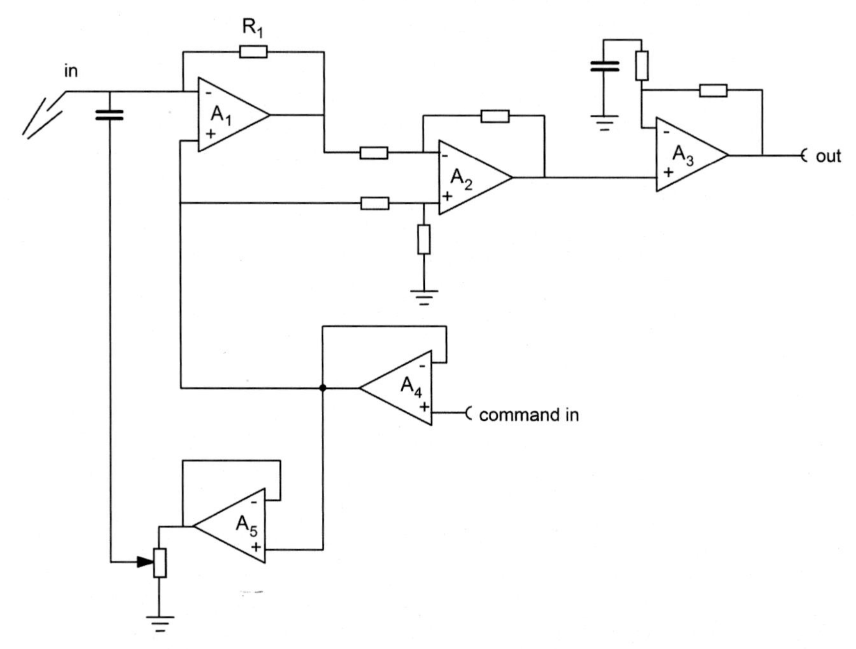

FIGURE 4.5 Simplified schematics of a patch-clamp amplifier. A1 through A5 are operational amplifiers.

circuit diagram of a patch amp suited to measure single-channel as well as whole-cell currents. Op-amp A1 is the current-to-voltage converter. It needs to have a very high input impedance but has excellent low-noise properties.

The feedback resistor R_1 is usually 1–10 GΩ, depending on the range of current strengths to be measured. The noninverting input is not grounded but is connected to an input circuit providing the command voltage. Thus, the pipette can be voltage-clamped, usually by step inputs to the command input op-amp (A4). The difference between the output voltage of A1 and the command voltage reflects the membrane current. This is accomplished by the difference amplifier A2. Op-amp A3 forms a circuit to correct the reduced bandwidth due to stray capacitances at the input. It has unity gain at low frequencies but enhances high frequencies. A5 improves the response to command voltage steps by charging the stray capacitances to the new voltage through a small, separate capacitor. In addition, practical patch-amps employ compensation circuits for the cell membrane capacitance and for the pipette resistance in the case of whole-cell recording. A practical patch-clamp amplifier is the Axoclamp 900 A, shown in Fig. 4.6. Many electrophysiological preamplifiers have a separate head stage which harbors the pipette directly (Fig. 4.7).

The advent of the field-effect transistor technology and ICs has greatly helped to reduce instrumental noise in patch-clamp amplifiers by grouping the elements of the first amplification stage (A1 in Fig. 2.25), including the high-resistance feedback resistor, which constitutes the principal source of noise, on a single chip. It has thus been possible to reduce noise power to below 10^{-30} A^2/Hz at 100 Hz and 10^{-29} A^2/Hz at 10 kHz, giving a usable bandwidth of approximately 10 kHz in optimal conditions. As explained in Chapter 3, at least part of the noise is due to random thermal movements of charge carriers in the material making up

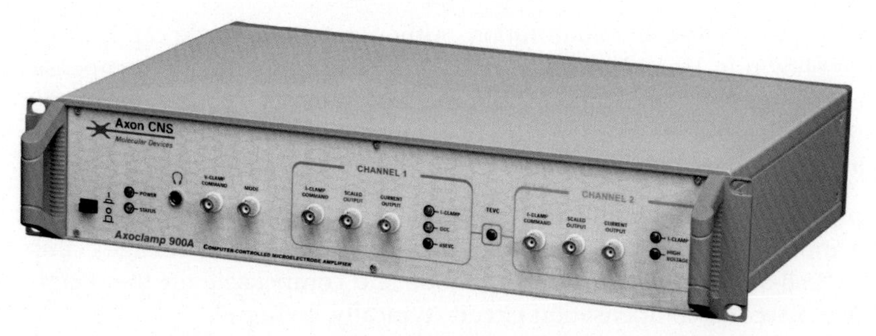

FIGURE 4.6 Axoclamp 900A patch-clamp amplifier.

the electronic components (Johnson noise). The feedback resistor of the first amplification stage can be replaced by a capacitor to avoid Johnson noise, converting A1 into an integrator (see Fig. 3.24). A disadvantage of this design is that the feedback capacitor needs to be discharged regularly (usually at a rate of 1 KHz) to prevent saturation of the signal. An additional improvement of the signal-to-noise (S/N) ratio is cooling of the head stage. The combination of these two techniques has led to a 10-fold improvement of S/N ratio and hence of bandwidth.

In practice an electrophysiology setup will not contain a separate amplifier dedicated to each of the experimental conditions that were discussed so far. In particular, a setup for patch-clamping will often have to do with the same patch amplifier to measure intracellular membrane potential. All commercially available voltage-clamp amplifiers have a so-called "current-clamp" option that allows for membrane potential measurements. In current-clamp, the current injected into the cell has to be exactly zero. This goal is achieved if the voltage difference across resistor R1 in Fig. 4.5 is zero, hence the output of A1 should be equal to the membrane potential (V_m). By connecting this output (A1) to the noninverting input of A4 (which in voltage-clamp receives the command voltage) this condition is met and the stages A2 and A3 will have zero output, indicating zero injected current.

The same problems with pipette capacitance mentioned for microelectrode amplifiers also apply here. To reliably record the membrane potential, the amplifier input capacitance and the pipette capacitance have to be compensated correctly before breaking the membrane

FIGURE 4.7 Head stage of a preamplifier.

and passing to the whole-cell configuration. Although only a single capacitance compensation circuit is shown in Fig. 4.5, usually at least three are provided to compensate for input capacitance, pipette capacitance, and membrane capacitance. Of course, the membrane capacitance is left uncompensated when in current-clamp mode. The sequence of events leading to the recording of the membrane potential is thus the following:

1. Fix the pipette in its holder and with the pipette still in the air compensate for the amplifier input or stray capacitance (typically 1–3 pF) while in voltage-clamp.
2. Go to the "cell-attached" voltage-clamp mode and compensate for the pipette capacitance using a second compensation circuit (typically 3–10 pF).
3. Then break the membrane and switch to current-clamp.

This works fine as long as the capacitances remain constant during an experiment, which is not always the case. If, for example, the level of the bath solution changes by a perfusion system, the pipette capacitance varies, which is especially dangerous if the solution level drops below the starting level. In that case, the pipette capacitance becomes overcompensated, the amplifier oscillates, and large currents are injected into the cell.

A second important inconvenience of voltage-clamp amplifiers if used in current-clamp mode is related to the high amplification factor of the head stage A1. As the amplifier is designed primarily to convert very small currents to large potentials, very small errors at the input, e.g., a very small bias current, inherent to all electronic devices, may lead to important voltage-offsets at the inverting input of A1 of as much as tens of millivolts when recording from small cells. Not all manufacturers of voltage-clamp amplifiers provide adequate nulling of bias currents. Hence, it is often difficult, although not impossible, to measure the membrane potential reliably using a voltage-clamp amplifier.

TWO-ELECTRODE VOLTAGE-CLAMP AMPLIFIER

Oocytes are large and easy to manipulate. For this reason they are often used to heterologously express proteins such as ion channels and transporters by injection of plasmids. Two glass electrodes are inserted to voltage-clamp an oocyte, one to record the transmembrane potential and the other to inject current into the oocyte in to obtain the desired potential. The electrodes used in two-electrode voltage-clamp (TEVC) are filled with 3M KCl for low resistance. First the electrode tips are immersed in the salt solution for a few minutes to allow the solution to creep up by capillarity, after which they are backfilled with a syringe. They should then have a resistance of 0.5 to a few MΩs. Like in most living cells, the oocytes' intracellular potential is negative with respect to the exterior (between −30 and −80 mV). The potential could very well differ from these values if the oocytes express foreign ion channels. Actually that is why the oocyte preparation is used most often. To do so, synthetic cDNA coding for particular ion channels or mRNA obtained from human or animal tissue is injected in mature oocytes (stages V–VI). A day to a few days later the proteins are expressed at the membrane and the oocytes are ready to record from. It is also possible to inject membrane vesicle preparations into the oocytes. A fraction of these membrane vesicles fuse with the oocyte membrane within 24 h.

The single-electrode voltage-clamp amplifier relies on the fact that the cell or membrane patch has a high resistance (typically 100 MΩ to 10 GΩ) with respect to the pipette resistance (1–10 MΩ) such that the voltage drop across the pipette is relatively small and can be compensated easily. If the cell resistance is low, as in the case of *Xenopus* oocytes, this condition is not met. In that case a two-electrode amplifier is more appropriate. In Fig. 4.8, one electrode is used to measure the membrane potential. The amplified difference (A2) between command potential and the membrane potential is fed into the cell by the second electrode. Amplifier A3 monitors the current injected.

An alternative configuration using a virtual ground circuit to measure voltage-clamp currents is very popular (Fig. 4.9) in current TEVC design. One electrode measures the membrane potential (V_m) and feeds into a voltage follower. V_m is compared with the command voltage (V_c). A current (I_m) flows through the second electrode to maintain $V_m = V_c$. The same current (I_m) flows through the virtual ground electrode to maintain the bath at 0 V. I_m generates a voltage drop over the resistor at the output of the virtual ground amplifier (V_{out}) that is proportional to I_m.

Amperometry

The use of ICs containing arrays of electrodes is an advance that makes electrophysiological methods easier to use and augments data throughput. In the following paragraphs the amperometric method for the measurement of impedance changes in cell culture or the detection of interactions between biomolecules on a CMOS chip will be discussed. An application for the detection of molecules is schematically shown in Fig. 4.10.[1]

The configurations shown in Fig. 4.10 represent electrodes in a biological solution. Molecules such as enzymes are grafted, or entire cells are grown on a substrate electrode or "working electrode" that is usually the ground electrode. Ligands released in the medium may in the end interact with the grafted molecules or cells, thereby changing the electric properties of the biolayer. These properties could be a change of impedance or the generation of substances that can be reduced or oxidized at the working electrode. To measure, for example, glucose concentrations, the enzyme glucose oxidase may be grafted on a chip. This enzyme catalyzes the reaction of glucose into gluconolactone and hydrogen peroxide. A current is generated if

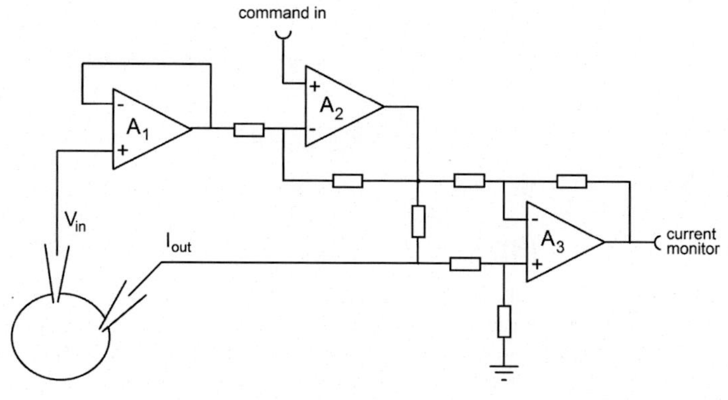

FIGURE 4.8 Two-electrode voltage-clamp setup.

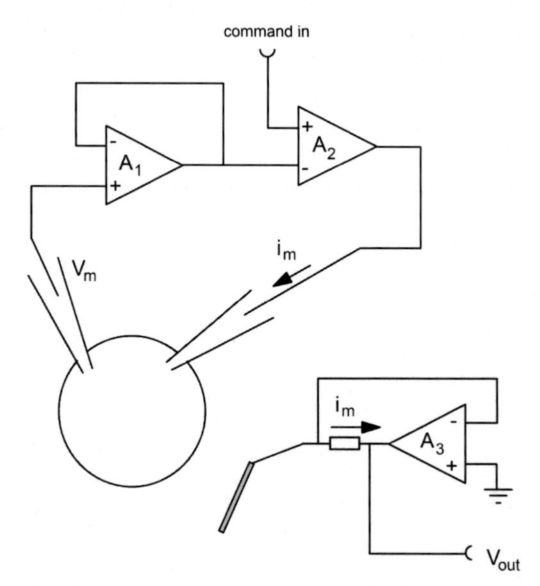

FIGURE 4.9 Simplified schematics of two-electrode voltage-clamp using a virtual ground.

the latter substance is reduced at the working electrode. This current reflects therewith the glucose concentration in the bath. A second electrode is used to impose a voltage difference across the biolayer such that changes in current can be measured as in Fig. 4.10A. In this configuration the actual voltage drop over the biolayer is not exactly known and is certainly less than V_{ref} because the impedance of the medium also causes a voltage drop, which, moreover, may change with time. The equivalent circuit is shown in Fig. 4.10B. V_{ref} is the reference electrode potential, Z_{fluid} and Z_{bio} are the fluid and biolayer impedances, and V_{st} is the actual, but unknown, stimulation voltage gradient over the biolayer. The two-electrode

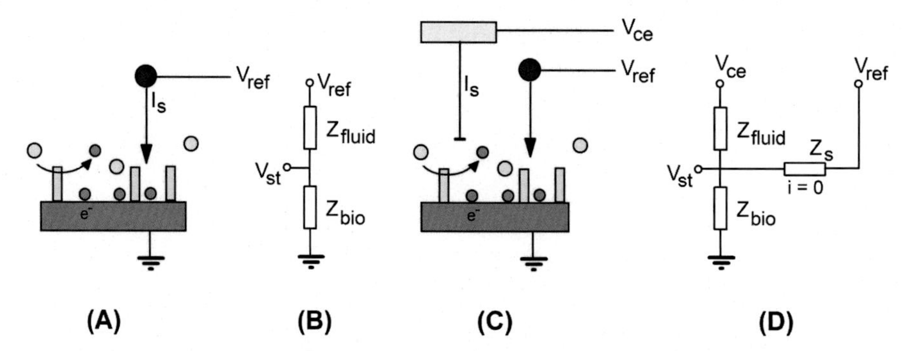

FIGURE 4.10 Amperometry across a bilayer. Enzymes grafted on the ground electrode convert molecules into an oxidized species that is subsequently reduced at the surface of the working electrode, thus generating a current. A and B show the two-electrode configuration; C and D the three-electrode configuration. The latter compensates for the voltage drop across the fluid. B and D represent the equivalent electric circuits of A and C respectively. *Adapted from Li H, Liu X, Li L, Mu X, Genov R, Mason AJ. CMOS electrochemical instrumentation for biosensor microsystems: a review.* Sensors (Basel) *2016;17(1) (Creative Commons).*

configuration is hardly an option in applications where V_{st} needs to be close to a potential where a redox reaction occurs. The problem of the voltage drop over the fluid layer can be solved by using a third electrode, the counter electrode, that is set to a potential, V_{ce}, such that the current flowing through the reference electrode is zero (Fig. 4.10C and D). Because the current through the path represented in Fig. 4.10D by the resistor Z_s equals zero, the voltage drop over Z_s is zero and hence $V_{st} = V_{ref}$.

The circuit that maintains V_{ref} constant irrespective of changes in Z_{fluid} is the simple voltage follower that we have already encountered in Fig. 3.24. The counter electrode is usually large with respect to the reference electrode and finds itself at a certain distance from the working electrode where the reactions take place. The reference electrode to the contrary is brought close to the working electrode.

The voltage follower clamps the trans-biolayer potential at $V_{st} = V_{ref}$. The working electrode is often made of inert metals such as gold, platinum, or carbon. The reference electrode is preferably an Ag/AgCl electrode. The circuit in Fig. 4.11 needs to be complemented to measure the changes in the current needed to maintain this potential, since that is what we are interested in. A circuit that can do this is the so-called current-mirror circuit followed by the current-to-voltage converter (see also Fig. 3.26).

Two p-channel MOSFETs are driven in parallel by the output of the voltage follower. These should be identical such that they draw the same current. The two twin MOSFETs are usually proposed in monolithic form.

It is clear from inspection of the figure that only positive reference potentials can be applied and that therefore only reactions leading to reduction of molecules can be used. This is fine for the measurement of oxygen, for example, which is reduced by setting the reference electrode to 650 mV.

Oxygen measuring probes usually follow the design of L. Clark (Clark electrode). In this design a platinum electrode and an Ag/AgCl reference electrode are placed in a glass housing whose tip is separated from the medium by an oxygen-permeable membrane made of Teflon or a similar material. It is based on the reduction reaction:

$$O_2 + 4e^- + 2H_2O \rightarrow 4OH^-$$

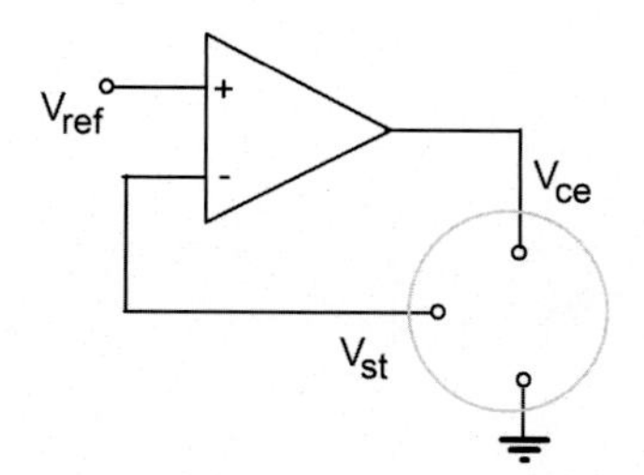

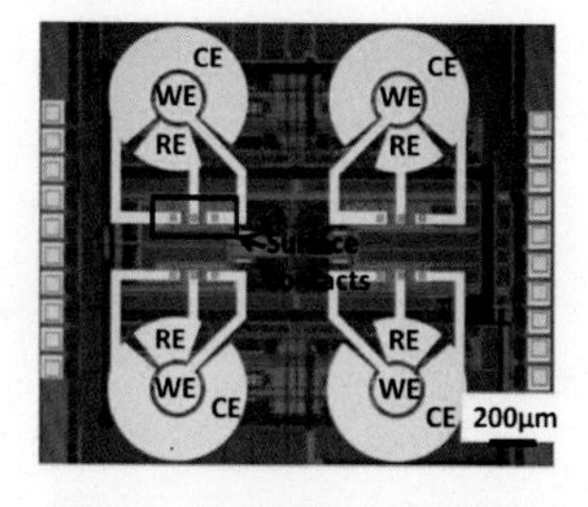

FIGURE 4.11 A voltage follower clamps the test potential to $V_{ref} = V_{st}$ (left). A configuration using four circuits on a CMOS chip is shown on the right. *From Adapted from Li HH, Liu XX, Li LL, Mu XX, Genov RR, Mason AJ. CMOS electrochemical instrumentation for biosensor microsystems: a review.* Sensors (Basel) *2016;17(1) Creative Commons.*

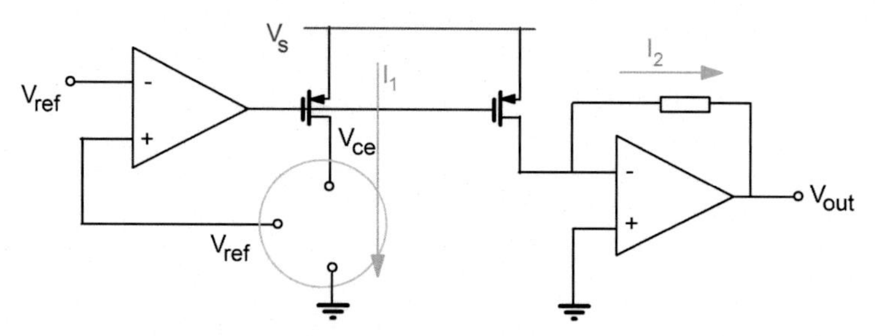

FIGURE 4.12 V_s is the supply voltage, V_{ref} is the reference voltage. I_2 is the mirror current and is identical to I_1. Note that the inputs to the follower have been inverted with respect to the previous figure.

Miniaturized Clark-type O_2 sensors of the same design as in Fig. 4.12, but fitted with an O_2-permeable membrane, have been developed for automated recording of oxygen consumption in cell culture.[2]

To measure oxidation reactions instead of reduction reactions, the design of Fig. 4.12 needs to be inverted. In Fig. 4.13, the working electrode is connected to the positive supply voltage and the p-channel MOSFETs are replaced by their n-channel counterparts.

The layouts of the circuits in Figs. 4.12 and 4.13 are very basic, demonstrating the principles. The requirement of two different circuits, one for positive and one for negative reference voltages, is not very practical. More involved circuits exist that, among other improvements, combine the circuits shown in Figs. 4.12 and 4.13 in one design.[3,4]

MEASUREMENT OF MEMBRANE CAPACITANCE IN VOLTAGE-CLAMP

After having recorded responses from a series of cells, the need is sometimes felt to normalize the responses with respect to the size of the cell, especially if the size varies widely from one cell to another. As the membrane surface area of a cell is proportional to the

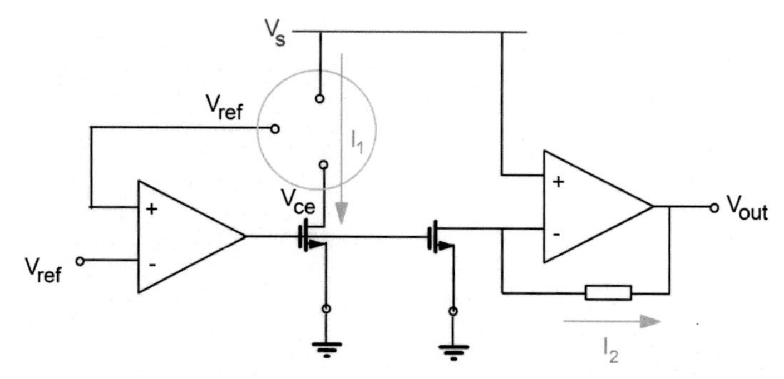

FIGURE 4.13 The circuit shown in Fig. 4.12 is modified to measure oxidation reactions using n-channel MOSFETs.

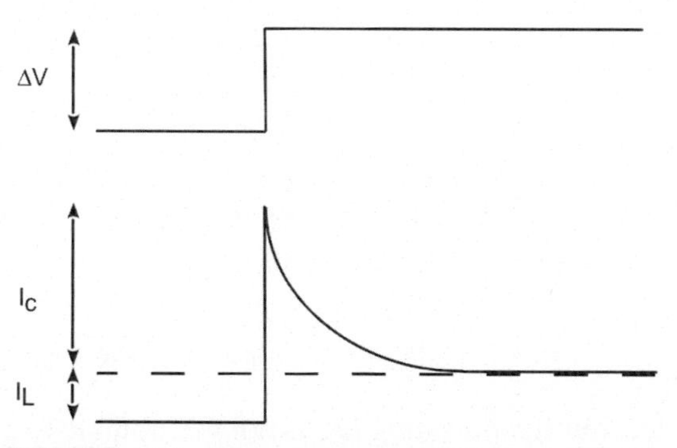

FIGURE 4.14 Measurement of membrane capacitance.

membrane capacitance, it suffices in general to normalize the data with respect to the membrane capacitance. The latter can be easily measured by applying a voltage step and recording the current in response of this step. This response contains two components: a steady-state component due to the ohmic membrane conductance (I_L in Fig. 4.14) and a capacitive component (I_C).

The time-integral of I_C then gives the capacitive charge Q.

$$Q = \int I_C dt$$

By dividing by the voltage step, the membrane capacitance C_m is obtained:

$$C_m = \frac{Q}{\Delta V}$$

POWER SUPPLIES

After the foregoing relatively complicated circuitry, the structure of power supplies is simple and fairly straightforward. In the early days of electronics, huge piles of galvanic cells (i.e., batteries) were the only sources of direct current. Since the invention of diodes, the alternating mains voltage can be converted into direct current, yielding a strong and inexhaustible power source. Today, more elegant forms of battery are still used to power portable instruments, and in addition, some electrophysiological instruments use batteries because keeping them disconnected from the mains may help to reduce interference.

Most apparatus, however, need mains power to furnish the needs of op-amps and other building blocks: usually +15 and −15 V. Digital circuits such as counters, frequency meters, and computers need an additional single 3.3 V or +5V power supply. The mains voltage of 230 or 110 V can be transformed up or down easily to furnish higher or lower voltages as required.

An AC is converted into a DC by a so-called rectifier, which in its simplest form consists of one or more diodes and a storage capacitor (Fig. 4.15, top). The signal passed by a diode is

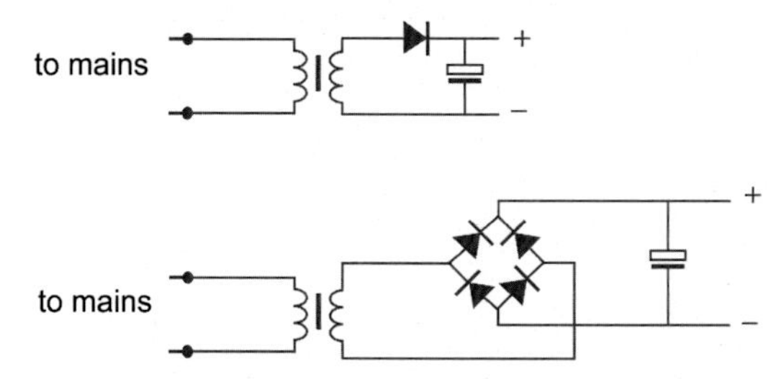

FIGURE 4.15 Half-wave (top panel) and full-wave rectifiers (bottom panel).

rectified in the literal sense, but the pulsating DC is not yet suitable to power our instruments. Therefore, a large storage capacitor is added as a water butt that stabilizes the water level of a source that delivers in gusts.

The voltage on the capacitor is formally a DC, but has a ripple caused by the charging of the capacitor each cycle of the mains voltage, hence at 50 or 60 Hz. Therefore, this basic half-wave rectifier is used seldom. A better form is the full-wave rectifier, employing a diode bridge (see Fig. 4.15, bottom). The result is shown in Fig. 4.16. By sending both the positive and the (inverted) negative peaks to the capacitor, the ripple is substantially lower, since the capacitor is now charged twice each mains cycle, hence at 100 (120) Hz.

To reduce the ripple further and to make the output voltage independent on mains voltage fluctuations, most instruments are fitted with a stabilized power supply, in which electronic circuitry (in fact, a kind of powerful op-amp) is added to flatten the voltage further. If a 100 or 120 Hz hum signal is found in an electrophysiological setup, it most probably stems from an ill-dimensioned power supply.

Apart from built-in power supplies, separate devices are sold to deliver direct current, usually with two or three adjustable voltages (see Fig. 4.17). In the laboratory, they are used not only to deliver power to operate special or homemade circuits but also to administer DC stimuli, such as (the DC component of) the command voltage to a voltage-clamp amplifier.

Switching Power Supplies

With electronics getting more and more flexible and affordable, most power supplies today have the bulky, heavy, and expensive transformer replaced by switching circuits.

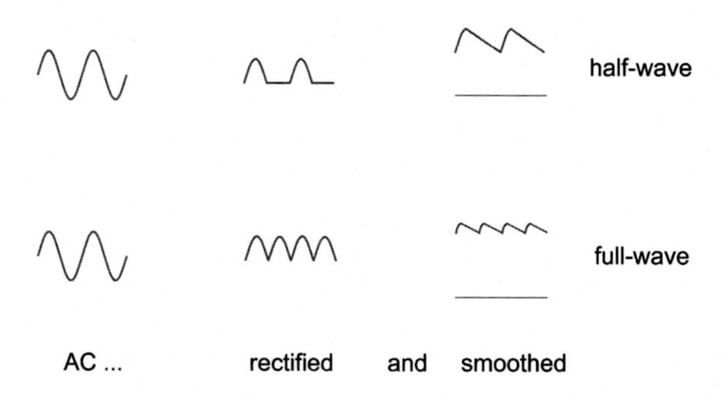

FIGURE 4.16 Ripple of rectifier circuits.

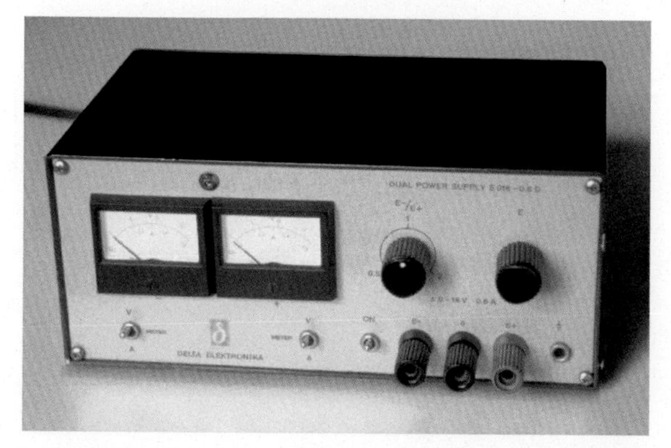

FIGURE 4.17 A dual adjustable laboratory power supply.

The mains voltage can be rectified first, stored in a capacitor, and subsequently switched on and off in short pulses, filling another capacitor to maintain a constant DC voltage as power to the rest of the equipment.

In this way, it is possible to convert the AC mains power into a low DC voltage in a very light and compact form, such as 5 V in the USB chargers.

But there is a downside to switching power supplies: Since the necessary current is drawn from the mains in short pulses, interference by electrical transients is a serious threat. Therefore, most switching power circuits are fitted with small coils and/or capacitors to dampen the peaks.

Even then, users of sensitive apparatus must be aware of the threat and either keep power supplies far from preamplifiers or revert to the old-fashioned transformer type of power supplies. The same holds for lamp dimmers, since these work with the same switching principle. And the increasingly popular LED illumination lamps use switching inherently because a white LED needs about 3 V and 20 mA or more to operate (a number of such LEDs can be connected in series to have such a chain operate at a higher voltage, but some form of current switching remains necessary). The simplified principle is shown in Fig. 4.18.

The mains voltage is rectified first and the resulting high-voltage DC stored in capacitor C_1. The current is then fed to the other capacitor, C_2, by means of closing the switch in brief pulses. Note that the switch symbol shown stands for a high-voltage transistor circuit rather than a mechanical switch. Note further that the low-voltage circuit is connected directly to the mains so that it must be isolated and not touched. This is fine for LED lamps and other totally insulated devices, but in other applications where one needs to touch the low-voltage safely (lab power supplies, USB devices, etc.), a galvanic separation circuit must be included. This is usually a radio-frequency link using a transformer. However, it can be very small (see Fig. 4.19; the transformer is the yellow thing in the middle).

SIGNAL GENERATORS

In addition to direct currents and voltages, electrophysiologists need several sources to deliver test and measurement signals, such as sine and square waves, pulses, and ramp or

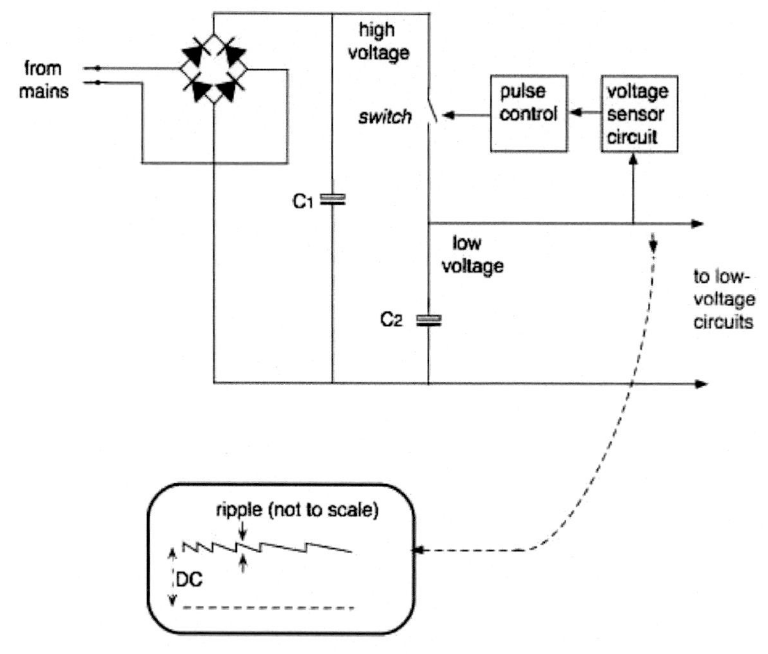

FIGURE 4.18 Principle of the switching power supply. Top: Simplified circuit. Bottom: Output waveform. The ripple is exaggerated.

sawtooth-shaped voltages. The instruments used are therefore called generators, so sine generator, pulse generator, and so on.

In principle, sinusoidal signals ("pure tones" when converted into sound) can be generated by an active element (e.g., a transistor) using an LC circuit as the frequency-determining

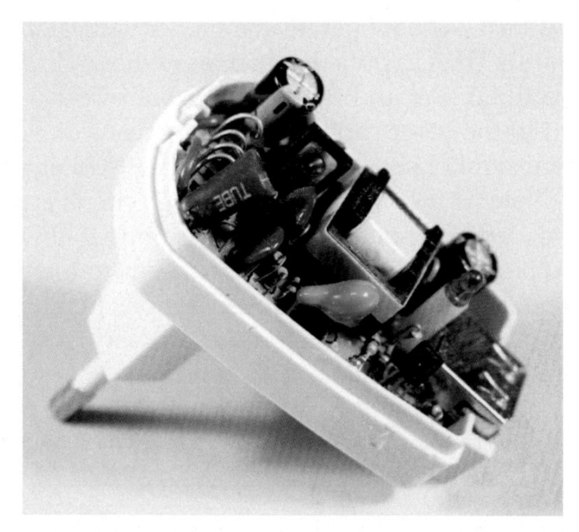

FIGURE 4.19 A small mains to USB power supply, showing the RF transformer to separate the mains circuit from the low-voltage output.

elements. This is an electrical "analogue" of mechanical devices such as swings, pendulums, and spring-and-mass systems, in general.

However, for low frequencies, very large self-inductions (coils) and capacitors would be needed. Therefore, most sine generators, usually called function generators, are built around another electronic principle, comprising of a multivibrator (square wave generator) and an integrator. The triangular waveform this circuit produces is "bent" into an approximately sinusoidal shape by a network of diodes and resistors. Audio-range and still lower frequencies can be made in this way using fairly small components. These instruments have, in addition to the sine wave, the synchronous triangular and square functions available. Hence the term function generator. A disadvantage of this "analog" principle of signal generation is the low stability. A chosen frequency of, say, 1.00 kHz may drift more than a few percent within minutes. These instruments are still in use today but are more and more replaced by digital signal sources.

A digital waveform generator consists of a dedicated computer (you would have guessed), together with the appropriate conversion, output and control circuits (DACs, amplifiers), and a nonvolatile memory containing a number of signal shapes. These include the mentioned sine, triangle, and square but may generate many other types of signal.

A so-called "arbitrary function generator" has the ability added to let the user edit and/or enter almost any waveform and "play" it with adjustable frequencies, amplitudes, delays, pauses, etc. An example is shown in Fig. 4.20.

This instrument has two channels, so it can emit two different waveforms simultaneously. In addition, the user can compute almost any desired signal form and feed it into the generator through a USB memory stick. In this way one can store preprogrammed voltage-clamp steps in the signal generator, repeating the sequence started by hand or automatically.

Controlling a digital signal generator is both more versatile and more precise. Parameters such as signal shape, frequency, and amplitude can be controlled through menus and keyboards, through rotating knobs, or remotely, via a connected computer.

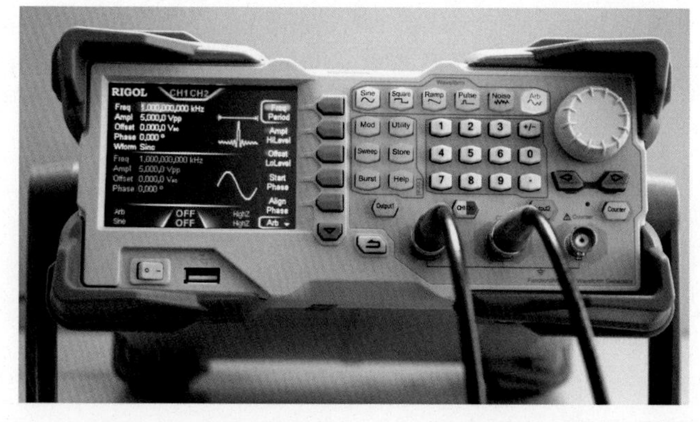

FIGURE 4.20 The Rigol 1022Z arbitrary-function generator. Channel 1 outputs a custom function, channel 2 a sine. Channel 1 is selected for editing (*highlighted*).

Since all waveforms are in fact series of numerical values, all principles and caveats of digitizing and digital signals hold, such as time and amplitude resolutions, aliasing, etc. (see below: Fig. 4.37).

For pulse generators, the same holds. Analogue types use capacitors, resistors, and op-amps to generate pulses of adjustable polarity, amplitude, and duration. But again, digital versions such as the mentioned arbitrary function generator are more stable and more versatile.

Properties of Signals

The most important characteristics of sine and square signals (period, amplitude, peak-to-peak amplitude, and RMS amplitude) are illustrated in Fig. 4.21, together with the characteristics of pulses. Here, T indicates the period, A the amplitude, and R the effective value (see below). The peak-to-peak value is 2A. In a pulse series, or pulse train, the ratio of on-time to total time is called the duty cycle and is usually expressed in percent. In the example, the duty cycle is about 25%. A square signal can be considered a bipolar pulse of equal positive and negative durations.

A further characteristic of sine signals is the effective value or root mean square value (RMS). The mains voltage of 230 V is such an RMS value, since the peak voltage in that case is about 325 V. The reason is that the power (work) delivered by an alternating current is proportional to the square of the mean value. Hence, a 325 V peak sine wave generates an amount of heat (work) that corresponds with a direct current of 230 V, the root of the mean square. For a sine, the peak value is $\sqrt{2}$ times the RMS value. In the case of pulse signals, the RMS value depends on the duty cycle. If a pulse train has a very low duty cycle, the RMS value is far lower than the peak value. The ratio of peak to RMS value is called the crest factor.

This is an important specification of instruments that have the readout calibrated in RMS value.

The peak voltage of a series of short pulses, even if its RMS value is low, may saturate a receiving device. Most signal generators have an output impedance of 50 Ω (to comply with the 50 Ω telecommunication impedance standard). Some older apparatus may still use the older standard of 600 Ω. Here one needs to be careful because if many instruments are connected (audio amps, computers, pen recorders, long cables, etc.), the output voltage might decrease slightly. At the other end, some function generators may drive an 8-Ω loudspeaker directly.

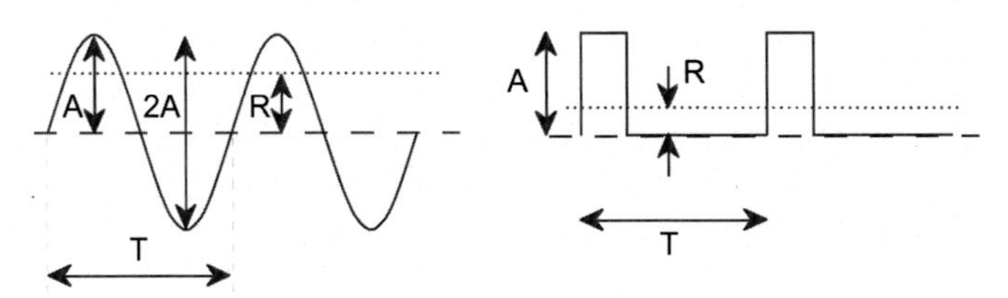

FIGURE 4.21 Properties of sine (left) and pulse signals (right).

ELECTRONIC VOLTMETERS

The passive voltmeter, based on the "microammeter" or moving-coil meter described in Chapter 2 has a rather low input impedance and low sensitivity. It draws a current from the source, thereby attenuating the signal to be measured. Therefore, (analog) electronic voltmeters and AVO meters (ampere, volt, ohm) consist essentially of a moving-coil meter with a preamplifier. In this case, the input impedance for voltage measurements can be sufficiently high and for current measurements sufficiently low to minimize the meter's influence on the measurement.

A modern, electronic voltmeter has usually a digital readout, turning it into a digital voltmeter, or DVM, which is convenient and reduces the chance of misreading values. The principal components of a DVM are shown in Fig. 4.22.

The input "conditioners" include input attenuators and shunt resistors, rectifier and current source, necessary to measure DC and AC voltage, resistance, and current. For more explanation see digital techniques described below. For the moment, it is important to note that DVMs may be made more precise than those employing a moving-coil meter, which has a precision hardly better than 2.5%. Digital meters can be designed to have a precision of 0.05% or even better and may show up to five or six significant digits. Needless to say that the most precise types are the most expensive and that the most common types of meter are definitely not better than 0.5%−1% error tolerance. In addition, precision instruments must be recalibrated regularly, at least once per year, to warrant their high precision. Note that the practical accuracy is often lower than the nominal one. In addition, the stated precision usually holds only for the DC ranges. Thus, the ohms and AC ranges are often less precise than the stated overall precision.

Measuring AC signals has another caveat: Voltmeters that measure RMS value must pass the peak amplitude without getting saturated or else it will be underestimated. Thus, true RMS voltmeters intended to measure pulse trains must have a high crest factor (see "properties of pulses" described above). However, the most common voltmeters do not measure true RMS value but attenuate the measured AC voltage by a factor of $\sqrt{2}$, so that the indicated value is calibrated for sine signals. Note: only for sine signals.

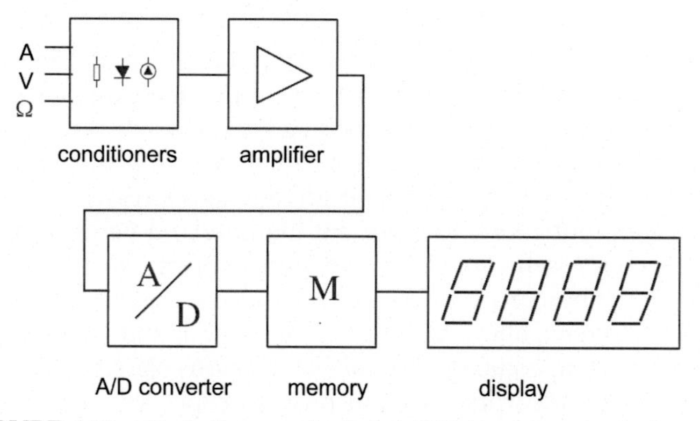

FIGURE 4.22 Block diagram of a digital AVO (ampere, volt, ohm) meter.

Some DVMs can store a reference value and display differences of the actual voltage and the stored reference. In addition, many DVMs can send the measured values to a computer through a USB interface. Usually, a computer program for Windows computers is available to record and/or analyze the sent values. However, some DVMs still sport the venerable old RS232 "serial port." These older interfaces can be interrogated with the aid of text terminal programs, like back in the 1980s. Adapters exist to convert the RS232 output to the current USB standard.

Note that, although the input impedance of electronic voltmeters is relatively high, 1 MΩ or more, it must still be kept in mind that a measuring instrument can influence the quantity to be measured. Next to input resistance, unwanted capacitance of the input, cables, etc. can also deteriorate (high-frequency) signals to be measured.

ELECTROMETERS

A special type of voltmeter called electrometer is essentially an AVO meter with an extremely high input impedance. This is accomplished by an MOSFET input stage. By the use of built-in shunt resistors and current sources, an electrometer can be used to measure voltages from about 100 μV to 10 V, currents from about 10^{-14} to 10^{-5} A, and resistances of 10^5 up to 10^{14} Ω. Electrometers are indispensable for the testing of amplifiers, electrodes, etc. In fact, an electrometer can be used as a high-impedance preamplifier for electrophysiological recording. In fact, a very high input impedance is always beneficial.

The reason that MOSFETs are not used everywhere is that these transistors produce more noise than junction FETs. For certain measurements, however, MOSFETs are the only way. Apart from the use as a separate test instrument, electrometers are used in conjunction with pH electrodes and ion-selective electrodes, in general. This is necessary because these electrodes use a membrane of glass or a liquid ion exchanger as the ion-sensitive part, which have impedances of 10^9 up to 10^{12} Ω. At such high electrode impedances, the use of guarding is mandatory. This holds especially for the intracellular versions.

OSCILLOSCOPES

Most (electrophysiological) signals are fast varying voltages or currents and so cannot be read from a voltmeter. Some DVMs can record or output the measured values, but this is mostly useful for slowly varying voltages. Some viewing and/or recording device is necessary for faster signals. The oldest way to view a signal in time is the oscilloscope, an instrument that "draws" the signal with an electron beam onto a fluorescent tube screen. It has two ways to deflect the beam: horizontal to represent the time and vertical to show the voltage.

An oscilloscope is most useful for repetitive signal forms, since brief transient voltage changes are too fast for the human eye to follow. Storage oscilloscopes have a way to "freeze" the written trace on the screen, allowing longer observation, and two and more channel oscilloscopes make it possible to compare several synchronous signals. These "analog" oscilloscopes are still in use, but not surprisingly most oscilloscopes are digital instruments today.

A standard digital oscilloscope, such as the one shown in Fig. 4.23, has an liquid crystal display (LCD) display rather than a picture tube, usually in color, and of course memory channels that can be read out and transferred to a computer for further processing and evaluation. The example model displays its two channels in yellow and cyan so that overlap between the traces is no problem (Fig. 4.24A). A stable image on the screen of a periodic signal such as a sine wave is obtained by triggering: after having written a signal trace, the sampler waits until a starting signal, or trigger, is generated. The trigger signal may be derived from the vertical signal (e.g., a certain voltage level) or from a separate external trigger signal. For instance, when we stimulate the eye of an animal with light flashes, we want to record an electrical signal from the start of the flash. When showing sinusoidal signals, the positive zero-crossing is often converted into a trigger signal.

Digital oscilloscopes have many more possibilities for acquisition and processing of signals, such as signal averaging and pretrigger recording. Signal averaging in such instruments is not performed batchwise (sampling the signal n times, taking the mean of the corresponding samples, and stop). Rather, the signal is averaged continually by taking a fraction of the "old" signal and adding the complementary fraction of the new signal; for example, 90% of the old signal plus 10% of the new one.

The terminology is important: An oscilloscope may take about 2000 samples (point values) of an electrical signal in 1 s (the sampling frequency, or rate, is then 2000 Hz or 2000 samples/s). This is sometimes called a sample too but is better called a "trace," or "sweep," to avoid confusion. It is stored in memory and can be put on a storage medium such as a USB memory stick.

Signal averaging means that the corresponding samples (point values) of more than one trace are averaged, usually in the exponential way just described.

This is in fact a way of digital low-pass filtering dealt with later on (see IIR filters, Chapter 7) but does not smooth the signal in time, since all sample moments are treated separately. However, a clean (not noisy) synchronous signal (trigger) must be available to horizontally align the periodic signal to be averaged. The result is shown in Fig. 4.25.

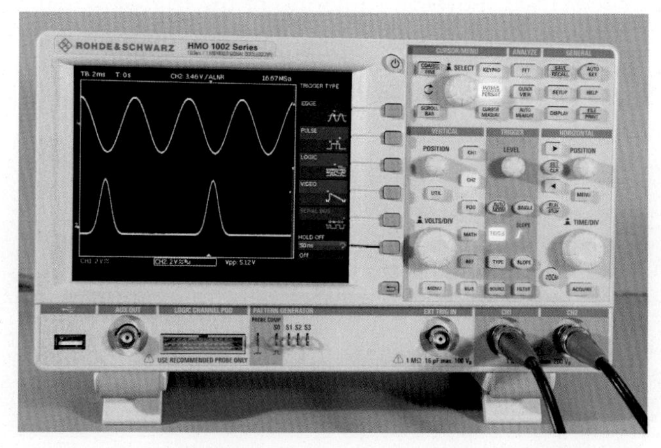

FIGURE 4.23 Rohde & Schwartz HMO1002 2-channel digital oscilloscope with a color liquid crystal display.

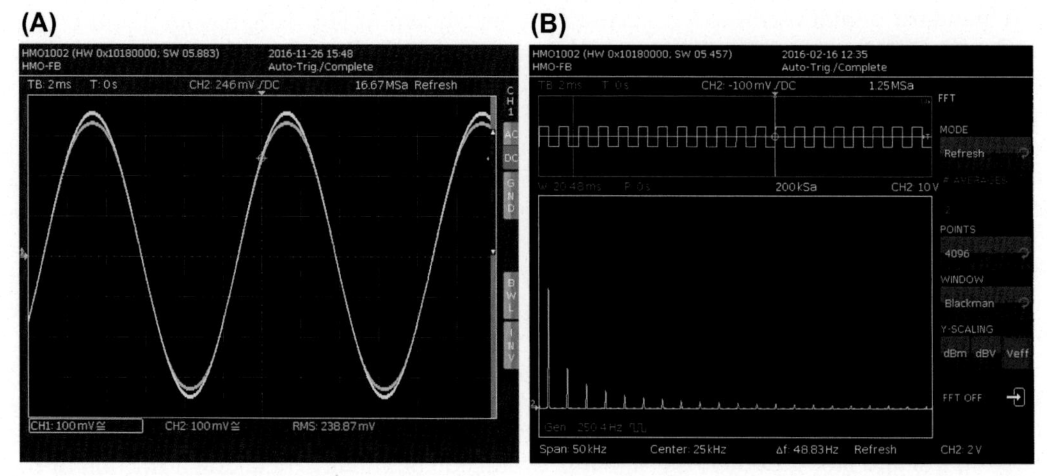

FIGURE 4.24 Oscilloscope traces in yellow and cyan (A) and a spectrum, the result of a fast Fourier transform (FFT) (B).

IMPORTANT PROPERTIES OF OSCILLOSCOPES

1. The bandwidth (BW), i.e., the range of frequencies sampled reliably. Standard oscilloscopes have bandwidth ranging from 0 (DC) to about 20 MHz. For most electrophysiological work, this is more than sufficient.
2. However, since a digital instrument samples the real-world variables in small steps, the user needs to calculate the minimum sample frequency necessary for a reliable record of the sampled signal. Undersampling may lead to erroneous results (see aliasing).
3. Voltage sensitivity. Since the largest signal amplification is performed by preamplifiers, the sensitivity of oscilloscopes does not need to be very high. However, it is useful if the scope has a sensitivity of 1 mV per division.
4. The trigger modes available determine on what aspect of a signal the recording (or averaging) process can be started. Low-pass filtering of the trigger signal can be necessary if it is a bit noisy. Most oscilloscopes offer enough trigger modes to find the best suitable one for any job.

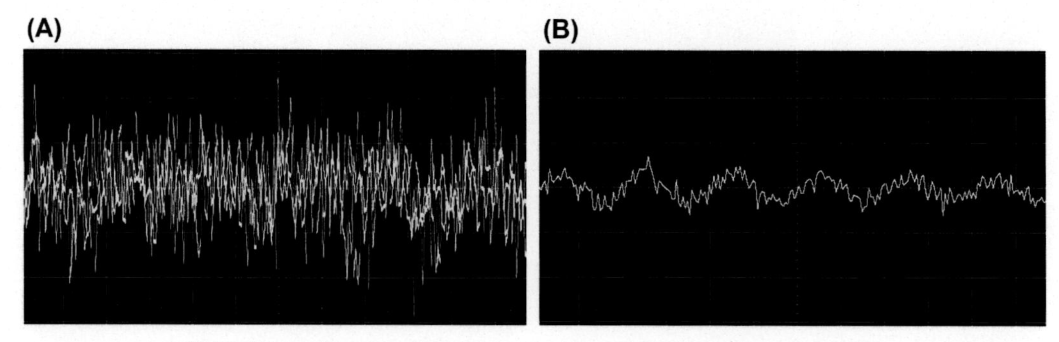

FIGURE 4.25 A sine buried in noise (A) and the result of averaging (B).

5. The possibility of pretrigger recording is important for irregular seldom occurring events. This might be, e.g., occasional spontaneous activity of a neuron or a cardiac arrhythmia.
6. The possibility of XY plotting. Sometimes, two signals can be compared best by plotting the first one on the X-axis and the second on the Y-axis. This holds especially for measuring phases. See Fig. 4.26.
7. Most digital oscilloscopes have one or two extra tools for measuring aspects of a signal, such as a selectable quantity to display in real time or a set of cursors. A selectable quantity can be the DC value of the signal, the amplitude (RMS, positive or negative peak, or peak-to-peak), frequency, or period. Cursors are vertical and/or horizontal lines to select specific points in the waveform by hand (using buttons or dials).
8. Many digital oscilloscopes can perform a fast Fourier transform (FFT) and so double as a spectrum analyzer (Fig. 4.24B, top: signal, bottom: spectrum). This possibility is, however, more limited than in dedicated spectrum analyzers. To get useful amplitude spectra, one has to select the sample rate and sensitivity carefully.

Portable Oscilloscopes

The advent of LCDs made small portable pocket oscilloscopes possible. These little helpers are handy and battery-powered, and so they may be carried easily to any job (see Fig. 4.27). In addition, they go for very affordable prices. However, most types have a single Y/T trace,

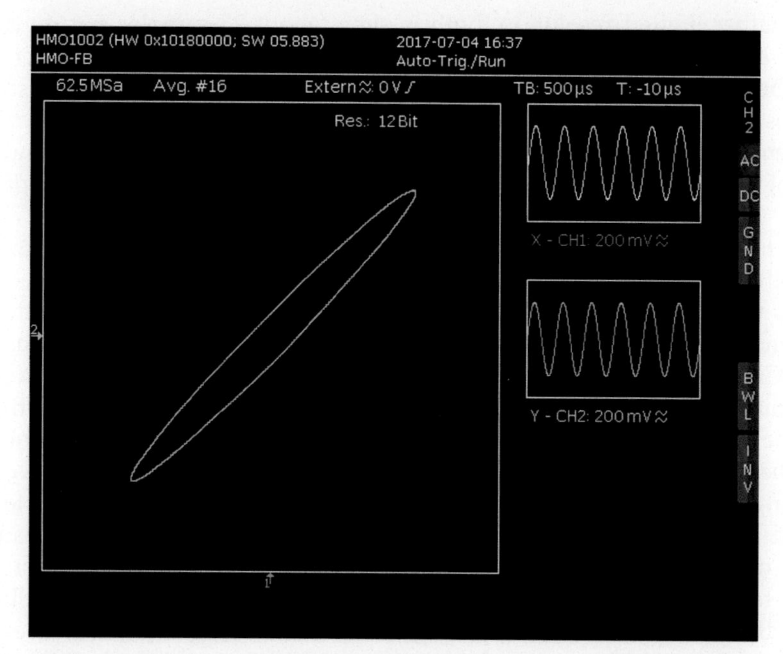

FIGURE 4.26 Oscilloscope XY display. Left: Channel 2 (Y-axis) versus channel 1 (X-axis). Right: The two signals separate.

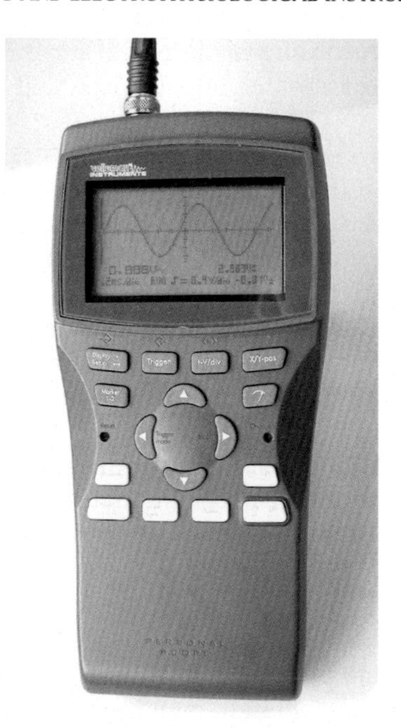

FIGURE 4.27 A portable oscilloscope.

as well as a rather limited resolution, both in time (about 25 samples per division) and in voltage (8 or even 7 bits vertical resolution). Their ease of use has earned them a place in most labs, albeit in addition to the full-blown, higher-resolution dual channel scopes.

Since the oscilloscope is a kind of "sixth sense," or sensory "prosthesis" to people studying electrical phenomena, it is of paramount importance to the validity of measurements and judgments of signals to have a thorough knowledge of its mode of operation and performance.

The most fundamental properties follow directly from the principles discussed above. Scopes are fitted with amplifiers for the conditioning of the input signals, and so they obey all laws we discussed there. The bandwidth is of crucial importance to the success of the measurements for many branches of physics. With ingenious techniques, bandwidth of more than 1 GHz has been obtained. Thus, even people working on television and radar circuits can monitor their signals on an oscilloscope. It will be obvious that for electrophysiological work, the bandwidth must also be sufficient to allow the fastest acts of life, action potentials, channel openings, and so on, to be monitored. However, with even modestly priced modern oscilloscopes, this condition will already be fulfilled. Most customer scopes are designed for television and computer repair and have usually a bandwidth of 10–20 MHz, more than sufficient for electrophysiological signals, at least for the signals that succeed to pass our preamps. Equally important is the sensitivity, usually expressed in volt per scale division (one division being either 1 cm or a half inch). Because we use preamplifiers, we do not need extreme demands on sensitivity: usually 5 or 10 mV/division suffices.

The usual horizontal axis shows the time, the scale of which is expressed in (m)sec/division. Most oscilloscopes support about 1 s/div up to less than 1 μs/div.

Note that most oscilloscopes support uncalibrated intermediate settings by providing potentiometers for all major adjustments. This can cause grossly erroneous readings. So when using an oscilloscope to make measurements of amplitudes, delay times, and so on, it is important to keep the relevant settings in the calibrated position. This is usually either the fully clockwise or the fully anticlockwise position, where the button "snaps in." Digital oscilloscopes have a "reset" or "automatic" function.

Often, one is interested in the synchrony, shape, or any other comparison of two rather than one (vertical) signal. Therefore, most oscilloscopes have two traces.

Some dual-trace oscilloscopes permit the connection of one of the input signals to the horizontal axis, thus creating an X/Y display. For distinction, the normal mode of operation is called Y/T display. The X/Y mode is useful for the comparison of sinusoidal signals and gives rise to the well-known Lissajous figures. These are used for phase measurements (see Fig. 4.26).

The dual-trace oscilloscope is the standard working horse of electrophysiology, where almost always two concurrent signals are studied: stepped voltages and the response of a membrane patch, a stimulus and the response of a sense organ or synapse, two signals from adjacent sense organs or of different pairs of EEG electrodes on the skull, a spike train and a signal derived from it, etc. From this list it will be clear that occasionally, more than two signals will have to be compared. To this end, the dual-trace principle has been extended to combine three or more input signals, yielding three-trace oscilloscopes and so on. Since our ability to analyze several things at once is limited, the task of comparing multiple traces must often be performed off-line.

Obviously, computers are far more powerful and versatile to store, measure, and analyze signals, and so the virtues of the oscilloscope lie more in instantaneous monitoring and control of the experiment, excellent to detect the presence and shape of the wanted signal and to assess the validity of an experiment. Note that measurements with the oscilloscope are of rather limited accuracy. Most digital scopes use 1 byte (8 bits) per voltage value, i.e., a resolution of 1 in 256. Thus, more precise AC or DC voltage measurements must be performed with a digital voltmeter, whereas precision analysis of more complicated waveforms must be left to computers having analog-to-digital converters of the desired precision (speed and bit-depth).

CHART RECORDERS

For many, mostly medical purposes, a direct printed record of signals from subject or patient is mandatory. Especially electrocardiograms and electroencephalograms can be lengthy. The standard letter or A4 paper output of most computers is less well suited for such long continuous records. This is where chart recorders come in handy.

A chart recorder consists of one or more channels—amplifiers and circuits to control one or more pens, writing on a roll of plain paper with ink or with heat on a roll of thermally sensitive paper. Thus, the signal or signals are shown in real time by the momentary pen positions, and the paper shows the entire signal history.

Such a long strip of paper may serve as a kind of log of an entire examination, a surgical operation, or any experiment. EEG recorders may have dozens of pens for the signals from the different electrodes. With chart recorders using mechanical pens, the useful bandwidth is limited by the pen inertia to a few tens of hertz (sufficient for most ECG work). A faster way of writing consists of a long array of microscopic resistors, controlled digitally to heat one such a "hot pixel" at a time, in fast succession, to write on heat-sensitive paper. With this technique, bandwidths of up to a few hundred hertz are possible.

Some chart recorders fold the chart afterward for easier storage.

DIGITAL ELECTRONICS, LOGIC

Contrary to what many people think, digital techniques did not originate in electronics. In fact, we perform "digital" operations, which means "(counting) with your fingers," from ancient times on. Today, the word digital is used to indicate any quantity, operation, or measurement that involves a finite number of states. It is used as antonym of analog, which by definition means any quantity that can vary continuously, possibly between certain limits.

These names might be a bit confusing but can be explained historically. Electrical signals are often used to represent some other physical or chemical quantity. For example, the petrol gauge in a car consists of a potentiometer connected to a float. If the petrol level rises, so does the voltage; if it falls, the voltage falls; etc. Since the electrical signal from the gauge mimics the petrol level, it is called an "analogue" of the petrol level. As a contrast, all things that can be counted are called digital. The car's odometer indicates the mileage in digits, where values intermediate between two numbers are either absent or unintelligible. The odometer may be considered a digital distance meter, in contrast to such things as rulers and tape measures, where in principle any position can be read off.

In fact, the distinction is mostly a practical one, since with sufficient magnification, most quantities we consider to be continuous prove to be quantized, or "grainy." As an example, the length and weight of things are considered analogue quantities, although there is a formal quantization in the length or weight of single atoms.

By the same principle, electrical quantities such as charge and current (charge per time) are quantized by the elementary charge, being about 1.6×10^{-19} C. Indeed, there are cases in which one has to take quanta in account, such as with the reception of light by photoreceptors, but in everyday life, this graininess may be neglected, and voltage, current, and so on are treated as continuous quantities. Digital signals, such as the symbols language is made of, date also from prehistoric times, but in dealing with digital electrical signals, we will use the early, electromechanical telephone exchange as a starting point. This venerable circuit, still in use today, employs so-called relays, having 10 positions that could be reached by feeding it with 1 to 10 pulses, generated by the old telephone dial. By dialing more than one digit (decimal digit), we can theoretically reach a finite but arbitrarily large number of people. In the meantime, back in the 1940s, researchers working on prototypes of the digital computer found that binary relays being either on or off were more versatile as switching element. With the advance of electronics, vacuum tubes and later transistors have replaced the large, slow, and noisy relays.

To analyze the functioning of digital circuits, we may observe that making the signal binary is the alternative solution to the problem we encountered with transistor amplifiers, namely that individual components have different and fluctuating properties. In Chapter 3, we learned that by increasing the gain of an amplifier and by using feedback, we can surmount this problem. Making the signal digital is another way to make reliable circuits with, so to speak, unreliable components. The simplest digital system is binary, i.e., it uses only two alternative states, usually called on and off (electrical), true and false (logical), or zero and one (number system). These circuits use two voltages, often 0 and +5 V, to reflect these two conditions. Contrary to the voltages in amplifiers and comparable transistor circuits, these two voltages are the only states that are allowed.

Fig. 4.28 shows a simple digital logic circuit. It is easy to analyze the functioning: If the input voltage is zero, the transistor is shut off so that the output voltage is virtually equal to the power supply voltage, which in this case is +5 V. If, on the other hand, the input signal is +5 V (a logical 1), the transistor conducts, pulling the output voltage to (almost) zero. Indeed, by the intrinsic properties of transistors, a small voltage is left (about 0.5 V); therefore the operation of binary circuits is defined more precisely. All voltages of +0.8 V or less are taken to signify the logical zero, whereas voltages in excess of +2.4 V are taken as logical one. The range in between is the "forbidden zone." Properly designed digital circuits will only pass this range as quickly as possible (i.e., in a few nanoseconds!) and will never "linger" longer than a fraction of a microsecond in it. Within these specifications, our little circuit of Fig. 4.28 functions as an inverter: The logical number is inverted, which means turned into its opponent state (true becomes false and vice versa).

For such logic circuits, it is customary to describe the function in a so-called truth table, expressed either in logical terms or in the form of numbers:

hence:	input	output	or:	input	output
	true	false		1	0
	false	true		0	1

The functioning may be summarized still shorter by a equation:

$$\text{output} = \text{NOT input}$$

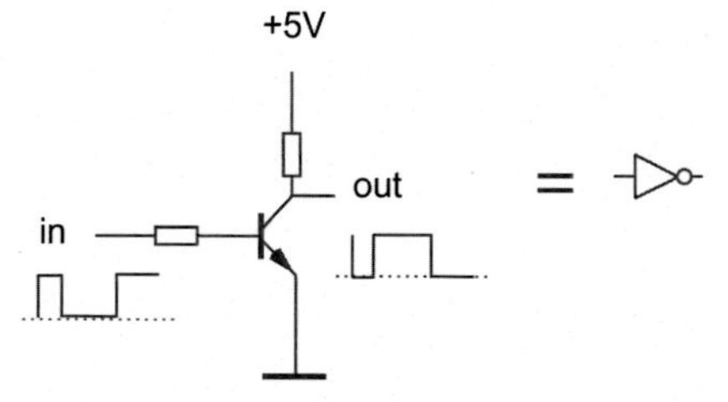

FIGURE 4.28 Simple logic circuit (inverter) and its symbol.

Note that both equation and truth table are exhausting (i.e., sufficient) descriptions of the inverter. Note further that the circuit functions independent of the precise gain factor of the used transistor, provided the gain is higher than a certain minimum value.

Although the inverter is the easiest to understand and is very often used in digital circuitry, its function as such is not very interesting, so let us examine other members of the logic "family." Many logical operations are performed with so-called gates, which can be described as circuits in which the flow of information in one branch is influenced by other branches. The simplest gates have two inputs and one output (the usefulness of more outputs will be dealt with later).

A gate can be made by addition of one resistor to the inverter of Fig. 4.28: see Fig. 4.29. The truth table is as follows:

in1	in2	out
0	0	1
1	0	0
0	1	0
1	1	0

In words, the output is "0" if either input or both inputs are "1." In the jargon of mathematical logic, the corresponding equation is:

$$\text{out} = \text{NOT (in1 OR in2)}$$

Therefore, this circuit is called a NOT OR, or NOR gate for short. By adding the inverter or NOT circuit of Fig. 4.28, the NOR is converted into an OR gate. The plus sign in the symbol indicates the OR operation (which is somewhat similar to algebraic addition). In the same way, AND and NAND gates can be constructed, where the output is 1 or 0, respectively, only if both inputs are 1. A further, important type is called "exclusive OR," abbreviated XOR. The truth table is given below left:

	XOR			XNOR	
in1	in2	out	in1	in2	out
0	0	0	0	0	1
1	0	1	1	0	0
0	1	1	0	1	0
1	1	0	1	1	1

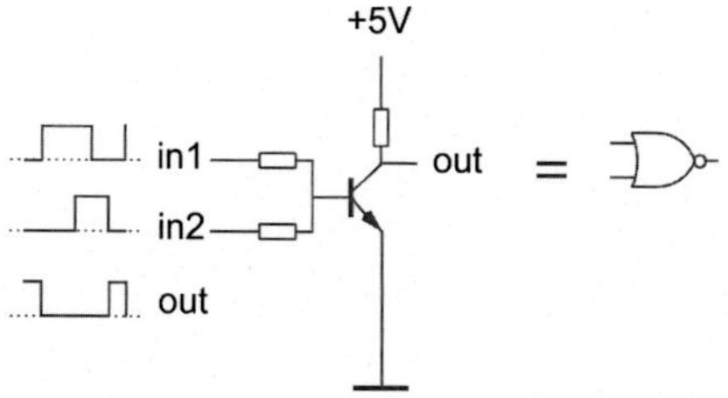

FIGURE 4.29 Two-input NOR gate.

In words, the output is 1 only if the two inputs are different. The negated form, exclusive NOR or XNOR, is shown at right. Here, the output is 1 only if the inputs are equal. This circuit is therefore called binary comparator and plays an important role in digital pattern recognition.

In drawing logic circuits, one is interested in the function, rather than constructive details of the circuit. Therefore, new symbols have been defined to signify the function of gates, inverters, and so on (Fig. 4.30).

In addition to gates, a circuit called flip-flop plays an important part in digital electronics. A flip-flop, officially called a bistable multivibrator, is a simple though powerful circuit built basically with two transistors. This is shown in Fig. 4.31A. In this example, the first transistor is shown in the conducting or saturated state, which causes the second transistor to be shut off. This in turn tends to keep the first transistor saturating so that this condition will last forever. If we could somehow force the second transistor to start conducting ("flip" it), the first one would be shut off and this would again last forever. If we could trickle it to flip again ("flop"), the first condition will be restored. This circuit therefore has a "memory": it simply retains the condition it was forced into and it will hold on forever unless it is flipped again or until the power is shut off. In the latter case, by the way, turning power on results in an unpredictable state: This is why many digital instruments need to be reset or be put into a certain starting condition intentionally, after switching the power on. Most modern instruments, such as counters, frequency meters, computers, and so on, perform a reset automatically.

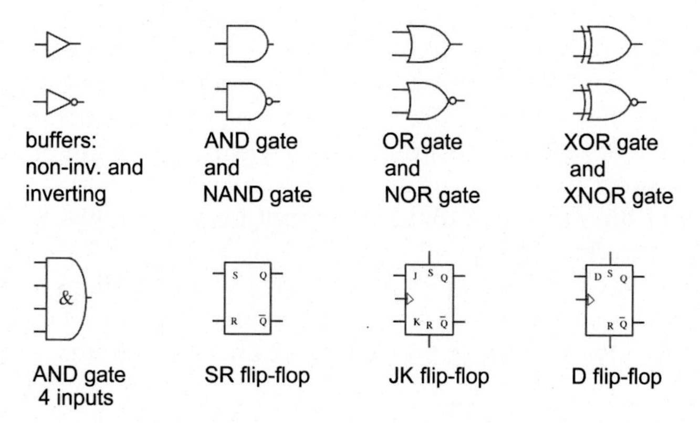

FIGURE 4.30 Symbols for gates and other logic components (for details, see text).

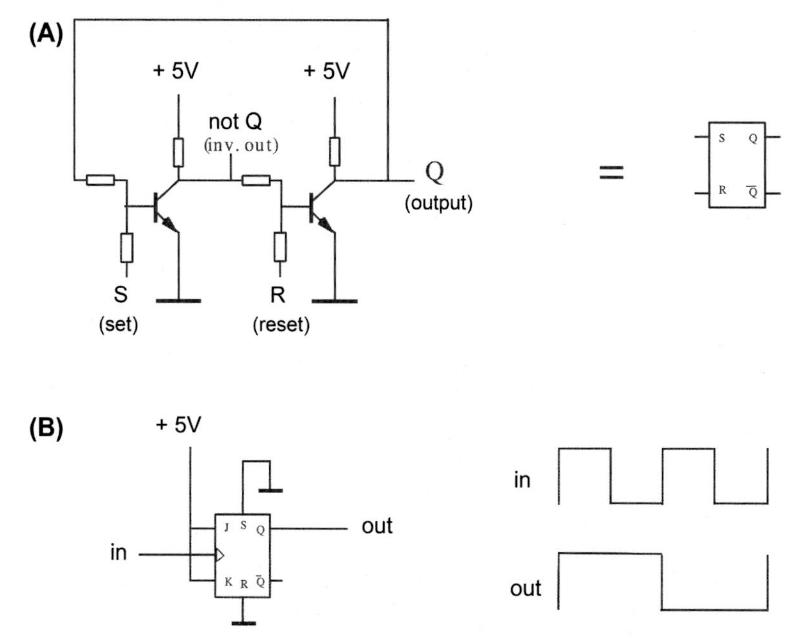

FIGURE 4.31 Principle of basic set/reset (A) and divide-by-two (B) flip-flops.

Note that we may consider each of the two transistors as the output, so a flip-flop has two outputs, in schematics usually designated Q and ~Q (NOT Q), and that both outputs are often used simultaneously to drive several other circuits. A flip-flop has a built-in inverter as it were. In large arrays, however, such as in memory circuits, there is only space to connect one output, if at all.

Indeed, the flip-flop is the basic element of the early forms of digital memory (random-access memory or RAM). It is still used in memory chips called static RAM. The larger and faster dynamic RAM or DRAM used in personal computers use a different principle, where a capacitor is used to retain the 1s and 0s. Flash memory devices, such as USB memory sticks and SSDs in computers, are different again, but outside the scope of this book.

Several kinds of input circuit are added to control the state of a flip-flop, which we will not describe in detail. The most important input signals are set and reset (see RS flip-flop in Fig. 4.31). A type of flip-flop called the JK flip-flop can be wired so that the output flips after each full cycle of the input pulse. Since flipping twice yields one output pulse, such a flip-flop divides the input frequency by a factor of 2 and is known as divide-by-two flip-flop: see Fig. 4.26B. A chain of flip-flops may divide the input frequency by 4, 8, 16, 32, and so on. These are frequency divider circuits.

The circuit of Fig. 4.32 shows a chain of four divide-by-two flip-flops, each output driving a (small) lamp. Therefore, aside from dividing an input frequency by a factor of 16, this circuit may be used to count to 16, since the 16 possible states can be monitored, or read out, by looking at the lamps.

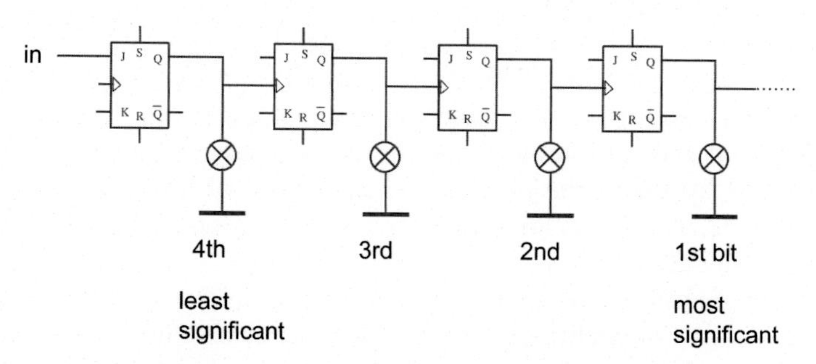

FIGURE 4.32 A chain of flip-flops with signal lamps.

These 16 possible states, running from 0 up to 15, are shown below:

0000	0100	1000	1100
0001	0101	1001	1101
0010	0110	1010	1110
0011	0111	1011	1111

where 0s and 1s stand for lamps in the off or on condition, respectively.

Indeed, we are counting from 0 to 15 in the binary system (often abbreviated as "bin"): Each flip-flop output represents one binary unit of information, or bit. The leftmost column reads, from top to bottom, zero, one, two, and three; the next column from four to seven; and so on. These 4 bits together, reflecting 16 possible states, are occasionally abbreviated as a nibble and are often written as one digit in the hexadecimal system ("hex" for short). And since we have only enough numerals for the decimal system ("dec" for short), the capital letters A through F are used to fill in the missing symbols. Thus, counting from 0 to 15 in the hexadecimal system goes as follows:

0, 1, 2, 3, 4, 5, 6, 7, 8, 9, A, B, C, D, E, F.

After F comes, indeed, 10 in this case is not to be pronounced as "ten" but as either "one, zero" or as "sixteen" (the decimal equivalent). More familiar than the nibble is the composition of two nibbles or 8 bits into 1 byte. Obviously, 1 byte reflects $16 \times 16 = 256$ possible states and is often written as two "hex" digits, the largest one being hex FF or decimal 255.

The preference of digital circuits for powers of two is still found in many other properties of computers, such as the "kilo," actually meaning $1024 = 2^{10}$ rather than the familiar 1000, "Mega," meaning $1024^2 = 1,048,576$ (and neither 1,000,000 nor 1000 kilo), and so on.

Despite this bias to powers of two, we can convert with a few extra gates a 4 bit counter to jump from 1001 (nine) directly to zero. In this way, we can build divide-by-ten (decade divider) circuits and decimal counters. Because the factor of 10 is achieved with binary circuits, this is called binary coded decimal or BCD. A chain of decimal counters composes a digital counter or counter for short (since counting is digital by definition). It is used often in electrophysiology to count action potentials or other events.

Usually, however, one is not interested in total counts but in frequencies or counts per time interval. Therefore, a counter is usually fitted with a gate that is switched on for a predetermined time. Often 1 s is chosen, since this yields a readout of frequency directly in hertz. The 1-s pulses are usually derived from a so-called clock circuit based on a quartz oscillator. The basis of such a square wave generator is an astable multivibrator, a circuit that resembles a flip-flop, but has no stable states at all: It changes states at a high rate. The frequency is stabilized by using the mechanical resonance of a quartz crystal as a "tuning fork." Starting with a 32,768 Hz crystal, one gets 1-s gate pulses (followed by 1-s pauses) using a chain of 15 divider flip-flops.

The full schematic of a digital frequency meter is given in Fig. 4.33. Here, by the way, the indication "digital" is not superfluous, since traditionally frequency meters were analog, accumulating voltage pulses in a capacitor. The input signal, which might be a sine wave such as shown in trace A, is turned into a clean square wave by a circuit called a Schmitt trigger. This is a kind of op-amp circuit resembling the comparator (open loop circuit, see above), but with a little positive feedback added to assure that states intermediate between "0" and "1" (the "forbidden" states) are traversed very quickly. Thus, it helps to make the edges steeper. The clean input signal is shown in trace B. Trace C shows the 1-s gate pulse, and trace D that part of the input signal that passes the gate, and hence is counted. At gate closure, the counter shows the number of pulses counted, and hence the frequency of the original signal. The memory section is added to be able to display a value while the next measurement is being made.

Digital measuring instruments have become increasingly popular. This is in part driven by the technical possibilities to integrate a large number of transistors and accessory components onto a single chip but is based also on the superior properties of digital over analog meters. The most important virtues are reliability and precision. The functioning of digital

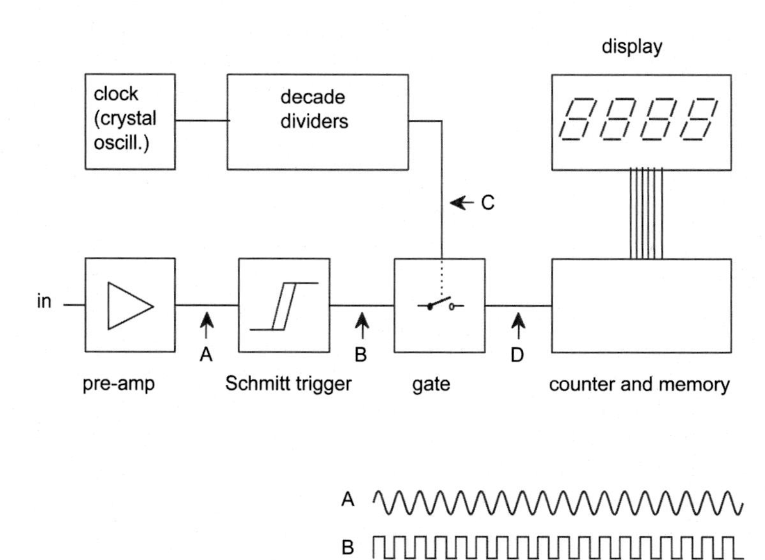

FIGURE 4.33 Schematics of a frequency meter.

instruments is still more independent on the whims of individual components than the op-amp like circuits we described earlier. If we feed exactly 13 pulses to the simple counter described above, the displayed state will each time be a faithful "1101" (hex "D" so decimal 13). Digital operations are exact. This is not to say that digital measurements have infinite precision! The precision of the abovementioned frequency meter depends mainly on the precision of the crystal used as the time base and a bit on the oscillator circuit built around it. Without special precautions, most quartz crystals, such as the ones used in electronic wrist watches, have a basic precision up to one part in 10^5, which is usually sufficient for general laboratory practice. Precision is raised in some instruments by keeping the crystal at a constant temperature (in an "oven").

Another feature that contributes to the reliability of digital instruments lies in the nature of the display. In addition to the abovementioned row of lamps, practical instruments may employ luminous (LED) or black (liquid-crystal) digits that can hardly be misread. This in contrast to the needles of conventional voltmeters can be misread by observing from an oblique angle or by making mistakes in reading the needle position between the different types of scale division lines.

Note that the above statement about the input signal must be true to get the precise answer we expect. This might seem obvious, but it is important to realize that the precision of any digital instrument will collapse if it is fed with a "dirty" signal. This is illustrated in Fig. 4.34, where noise on a low-frequency sine signal causes multiple false triggers that will yield fluctuating and completely erroneous frequency readings. In electrophysiology, one has often signals that are noisy and/or contain spurious interference "spikes" (not to be confused with the wanted action potentials), stemming from refrigerators and hosts of other machines connected to the electrical mains.

Apart from these possible pitfalls, digital instruments are the most reliable and versatile tools available to process electrophysiological and indeed any signals. To prevent the abovementioned pitfalls, the user has always the responsibility to check whether the input signals are suited to be processed digitally. Alas, the often cited phrase "garbage in, garbage out" applies to all signal processing.

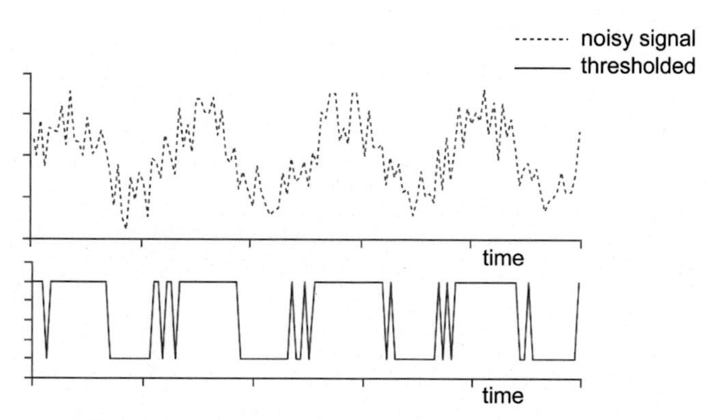

FIGURE 4.34 Erroneous input signal to a frequency meter.

A/D AND D/A CONVERSIONS

The segregation of measuring instruments in analog and digital is not of a categorical kind. Indeed, most real-world quantities, such as weight, length, voltage, and so on, are on an analog scale and must be converted into digital form to be processed and displayed. This conversion step is called analog-to-digital conversion and the circuit doing it is called an analog-to-digital converter or ADC. The inverse circuit, a digital-to-analog converter (DAC), is also often used, especially in combination with computers. A DAC is necessary, for example, if the output of a digital meter must be written onto a paper chart, if a digital instrument is used to control the gain of an amplifier, or when controlling the intensity of a light used as a stimulus.

The frequency meter described above may serve as a vantage point to illustrate the principles of AD and DA conversion. The simplest form of a DAC is shown in Fig. 4.35. Here, we added resistors to the outputs of the four flip-flops from Fig. 4.32 and added the resultant currents with an op-amp. A current through these resistors flows only when the output of the flip-flop is "1."

By choosing the resistance value ratios to correspond with the values ("weights") of the bits, the output voltage consists of 16 steps, from 0 to 15 units, and thus forms an "analogue" representation of the states of the counter.

Practical DACs are a bit more complex; however, the main reason being that the voltage pertaining to a logical one may in reality fluctuate between about 2.5 and 5 V and so would not form a reliable output signal component.

How to convert an analog signal into a digital one? One of the simplest ways to perform an AD conversion is illustrated in Fig. 4.36. First, the voltage to be measured is converted into a proportional frequency and then fed to the digital frequency meter described before. Although this form of ADC is simple to understand, its resulting measurements are not very precise in practice, and there are far more, and better, ways to perform the conversion from analog to digital. Crudely, these fall into two main principles: "successive

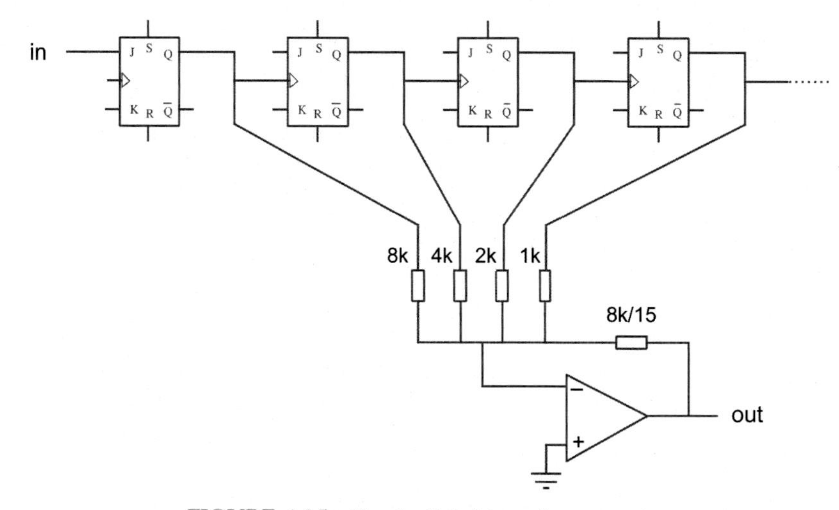

FIGURE 4.35 Simple digital-to-analog converter.

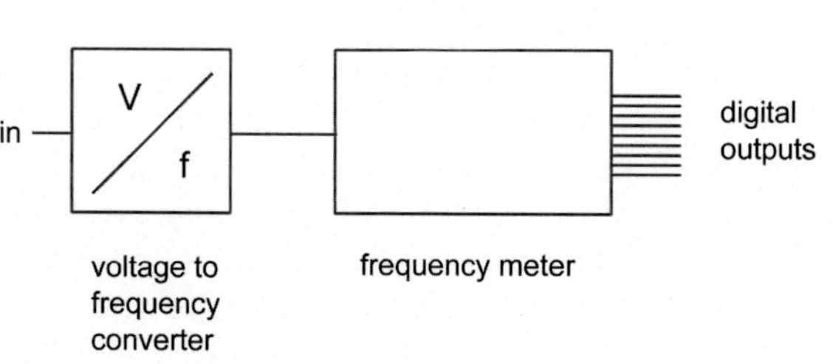

FIGURE 4.36 Simple analog-to-digital converter.

approximation" and "flash." The first type approaches the voltage to convert step by step, every step yielding 1 bit of the final value. These ADCs are relatively slow but can be made very precise (24 bits or more). Flash converters do all necessary comparisons at once, so work fast, but usually with less resolution (say 8 bits). Further details are outside the scope of this book.

Nevertheless, the discussion on ADCs and DACs puts the finger on the weak spots in digital techniques: Any digital instrument is only as good as its AD (and/or DA) converter! And the same holds for transducers.

TRANSDUCERS

All real-world quantities that are not electrical by themselves must be converted into electrical signals, and sometimes back, or into another physical quantity. The examples everyone knows are microphones and loudspeakers. A digital balance must convert the weight to be measured into an electrical signal and then convert that into a digital one with an ADC.

The conversion from any physical or chemical quantity into an electrical one is done by components called transducers. Transducers exist for almost any modality: weight or force to electrical, acceleration to electrical, position to electrical, speed to electrical (can cost you a fine), sound to electrical, and so on.

In Doppler angiography, the speed of blood flow in blood vessels is converted into an electrical signal, analyzed, and stored by a computer but often also made audible by a loudspeaker. In this way, a trained medical person can evaluate the condition of the patient's arteries just by listening.

In general, a transducer must be followed by an amplifier (such as an op-amp) and an output signal and/or a display. In most modern instruments, the results are presented by digitizing and so showing up as numbers. Such digital displays look very exact and can be read out without the errors inherent in analog meters that have a pointer moving above a scale.

But, be warned: Both transducers and amplifiers are analogue components that might suffer from imprecision, temperature dependence, aging, and so on. So do not trust a digital instrument just because the display is digital.

Digital measuring instruments, whether for weight, blood pressure, pH, or automobile speed, must be checked or calibrated every now and then, just like (entirely) analogue apparatus. Note that ion-sensitive electrodes (Chapter 3) are also transducers, converting (the log of) an ion concentration (more correctly: activity) into an electrical voltage.

Transducers are everywhere. In a kitchen scale, for example, weight is determined by measuring the bending of a metal bar using a strain gauge. A strain gauge consists of a thin wire, folded into a small rectangular array that is glued onto the metal bar or of a semiconductor performing the same function. The stretching of the strain gauge by the load changes the resistance of the wire, which can be measured electrically and displayed as numbers. Despite the digital display, strain gauges are notoriously imprecise, making kitchen scales only useful in domestic use.

Statement: To get valid measurements one has to check and double check one's important instruments.

Therefore, DVMs are not inherently precise, and the specifications of many DVMs used in the lab, especially the cheaper ones, are not much better than analogue ones could be. Fortunately, other advantages, such as the reliable readout, the compact and sturdy construction, remain valid, as well as the ease of use: Many DVMs are autopolarity and autorange, which means they sense the polarity and the approximate value of the input signal, respectively, and adjust themselves to give the best display range.

DIGITAL SIGNALS; ALIASING

A series of digital measurements or numbers derived from an analog signal constitute a digital signal. The fact that the continuous real-world quantity is chopped up, or sampled, into separate or discrete parts has consequences that any scientist should know. The process is called sampling or digitizing. One of the pitfalls is shown in Fig. 4.37. The traces in Fig. 4.37A show graphic representations of a sine wave (line) and its digitized counterpart (dots). For the human eye at least, the digital signal is a fair representation of the original, analog signal.

Fig. 4.37B, however, shows a case in which something went wrong: The high frequency sine is sampled too infrequently, which results in a digital representation that reflects a nonexistent, low-frequency sine wave. This phenomenon is called aliasing (from alias—a false name or identity).

It can, and must be, prevented by taking care that the sampling frequency fulfills the Nyquist criterion, which states that any waveform of interest must be sampled with a frequency of more than two times the (highest) frequency of interest (hence called the Nyquist frequency). For electrophysiological measurements, this implies that we must either choose a sampling frequency so high that it is without question higher than twice the highest signal frequency, or we have to limit the bandwidth of the preamp to fulfill the Nyquist criterion. To this end, some digital instruments have a special antialias filter built in. This is often a higher-order filter that allows signal frequencies from zero to slightly under the Nyquist frequency to pass unattenuated but prevents aliasing by cutting off higher frequencies very sharply, up to −100 dB/octave. Nevertheless, one must always keep in mind that a digital signal is undetermined between two samples, i.e., that one is not informed about the real world during the time between two samples. Thus, the digital signal shown in Fig. 4.37A might stem from the signal shown in Fig. 4.37C: The fast sine is lost in digitization.

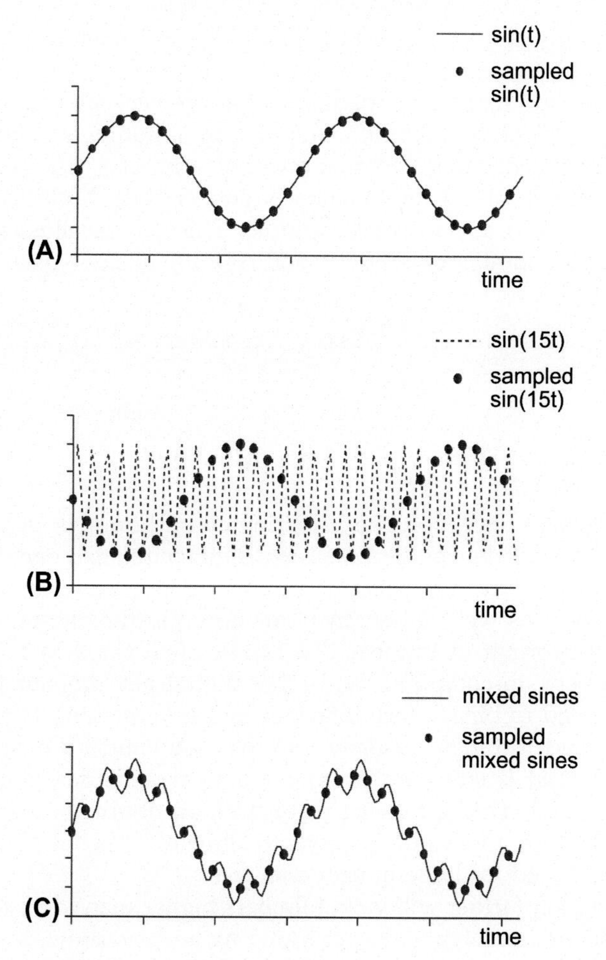

FIGURE 4.37 (A) Correct sampling. (B) Undersampling causing problems of waveform digitization. (C) Rapid signals go unnoticed after digitization.

COMPUTERS AND PROGRAMMING

The most complex, versatile, and certainly the best known type of digital apparatus is the computer, the operation of which can be derived from the simpler digital circuits we dealt with above. Modern personal and office computers, including the ones in electrophysiological labs, are very complex machines that are hard to fathom. For the users, much depends on the operating system (OS), such as Microsoft's Windows, Apple's Mac OS of the many versions of Unix, and Linux. Many good books to learn to control these OSs exist. In physiology, however, computers must be extended to handle electrical signals. Some turnkey systems do exist, but most researchers have to bolt their electrical apparatus to the computers themselves. In these cases, some insight in the architecture might help to understand the possibilities and problems of data acquisition and analysis. Therefore, although this is not a book on computer technique, a few basic operating principles will be treated below.

Apart from counting and measuring, digital circuits are very well suited to do mathematical computations and indeed any kind of manipulation of numbers in general (the numbers may of course also represent letters, words, or other symbols). Adding two numbers, for instance, can be performed with a combination of gate circuits. Multiplying and dividing a number by a power of two is particularly simple. This can be illustrated with our 4-bit counter: take the decimal number 6 (bin 0110) and shift the bits one position right or one position left.

decimal:	12		6		3
		<- left		right ->	
binary:	1100		0110		0011

Obviously, shifting right once more, i.e., dividing an odd number by two, will yield fractional numbers, but these can be handled easily by digital circuits if one makes the necessary conventions (0001.1 would have to mean dec 1.5). In practice, real numbers (i.e., integers as well as fractions, positive as well as negative) are stored in floating-point format. It is also known as the scientific format. A number like 12,345.6 is notated as $1.23{,}456 \times 10^5$ so as a fraction and an exponent. In computers and other digital instruments, this means that the fraction part is converted to binary form and put in a few memory bytes, followed by the exponent, also converted to binary, usually in 1 byte. Of course the sign is also included. This 4-byte (32-bit) format is still used today (as data type, it is called "float"), but often more resolution is needed. This led to the 64-bit format, double-precision or "double" for short. This is the main format for numbers, next to "integer." The latter form uses less memory per number and is used for large integer data sets.

Conversely, shifting left farther will yield numbers higher than (dec) 16 (11,000 = dec 24). When bounded to the original 4 bits, we get rounding and overflow, respectively:

$12 \times 2 =$ overflow	$3/2 = 1$ (rounded)
*1000	0001

where the asterisk stands for the signaling of the overflow condition. In larger binary chains, this is simply the signal to set the next, more significant bit, but any digital instrument should signal overflow of the most significant bit to the user (e.g., an "overflow" LED on a counter, or an error condition or warning in a computer program).

In the early days of digital electronic instruments, the problem with digital circuits was that different operations demand different forms of wiring of the circuits performing the calculations, and that it had become unwieldy to change the wiring by patch panels or by flipping switches all the time. This was solved by the principle of the computer, where the circuits are made flexible, and where "changing the wiring" is performed in a highly mechanized or electronic way and made sensitive to electronic command signals called instructions.

Therefore, the single, most essential property of a computer is that it is a digital circuit, the function of which is determined in rapid succession by a set of input signals, the instructions, rather than by the wiring. The sequence of instructions is called the program. In principle, these instructions could be fed in on demand as an input signal, but in practice, a number of instructions are stored at once, in digital memory.

In principle, a memory is a large array of flip-flops, each holding 1 bit of information. One can write into memory by setting the appropriate flip-flops to 0 or to 1 at will, and read from memory by connecting the outputs of the relevant flip-flops to the desired output wires. This is static memory. A dynamic memory is a large array of capacitors, each holding 1 bit of information. One can write into memory by (dis-)charging these capacitors at will, and read from it by connecting them to the desired output wires. Since capacitors would lose their charge slowly, the original charge level (either 0 or 1) must be restored on a regular basis (refresh). The advantage of dynamic memory is its speed and simplicity, making huge and fast memory, such as the gigabytes in our PCs, feasible.

The stored-program feature is the second important characteristic of the computer.

Thus, the main functional parts of a computer are the processor, which performs most operations and the memory, which holds the program as well as the data, i.e., the numbers that have to be processed. The processor may be a single chip but is often a group of related, partly specialized chips, including a math coprocessor and a memory management unit, that reside either on the same chip or as separate chips.

The speed of computer functioning is largely dependent on the clock frequency and hence speed of the processor. This is the speed mentioned in advertisements, bench tests, etc. However, the overall speed available to the user depends to a large extent on other architectural details and on the speed of other components like the hard disks and other storage media and of course on the network connections, if available.

A third, indispensable computer characteristic is the bus structure of the connections between these different parts. A bus is a set of wires that are connected to all component circuits. Together, the bus wires code for as many bits of information as the number of wires, often 16, 32, or more. Today, most personal computers use 64-bit wide busses. In general, three types of busses can be distinguished that provide for the flow of all signals in a (simple) computer.

In the first place come the data bus that carries the instructions as well as the numbers that are processed and the address bus that specifies which memory cell is to be read from or written into. A third set of common wires, occasionally called the control bus, switches the right parts on and off, so that the screen gets the information to be displayed and the printer gets the information that must be printed. The bus speed, rather than the processor speed, is the bottle neck in the maximum overall speed that can be attained for most types of personal computer. Therefore, cache memory is added to speed up most operations. This is a small RAM chip connected to the processor with a special, faster type of bus. How this speeds up the work can be understood by the fact that, if the computer fetches the contents (byte) of a memory location, most often the next series of addresses will be needed a short time later. Therefore, if a certain address is asked for, the computer's OS puts the contents of a whole block of main memory into this cache memory, so that subsequent memory calls are handled faster. Many computers even have more levels of cache memory: one on the chip, the others as close as possible to it (backside and motherboard caches).

The busses together interconnect not only all internal parts, such as processor, memory, hard and floppy disk drives, and so forth, but also external parts (peripherals) such as keyboard, mouse, printers, and so on. A simplified block diagram is shown in Fig. 4.38.

It is obvious that it must have a user interface to work with a computer, i.e., a set of devices to control it. This usually comprises of a monitor screen together with a keyboard and a mouse or trackpad. Most often, the screen is an LCD.

In addition to the mentioned peripherals, most computers have extra, free connectors or slots, to connect a host of other instruments, such as the AD and DA converters necessary to handle analog signals from the real-world, modems to send data through a data line, extra printers or monitors, etc.

The abovementioned parts are collectively called the hardware of the computer, in contrast with the software, which is the familiar description of the set of all possible or available programs, routines, or instructions. The software is "soft" in the sense that it is fundamentally a collection of mathematical and verbal ideas about how to process data and thus exists only in the brains of people. To feed software into a computer, some form of carrier of these ideas is necessary. Formerly, the carrier was a punched paper tape or a deck of punched cards, nowadays it is mostly a hard disk, USB stick, or other medium. All kinds of data carriers together are called the media.

The principle of the computer and the first working prototypes were designed before and independent of the development of electronics. The first mechanical devices resembling computers date back to the 1820s when Charles Babbage worked on his (unfinished) "analytical engines." Several of such mechanical calculators were built in the 1940s using electromechanical relays comparable to those in telephone exchanges. Not much later, however, the usefulness of vacuum tubes as electronic "switches" was recognized: They needed less current, could switch immensely faster, and could be built smaller than relays. This led to the construction of the first electronic computers, such as the "electronic numerical integrator and computer," ENIAC, in 1943–46. The vacuum-tube computer grew up to the very limits of construction: the more components, the more frequently an occasional failure would hamper

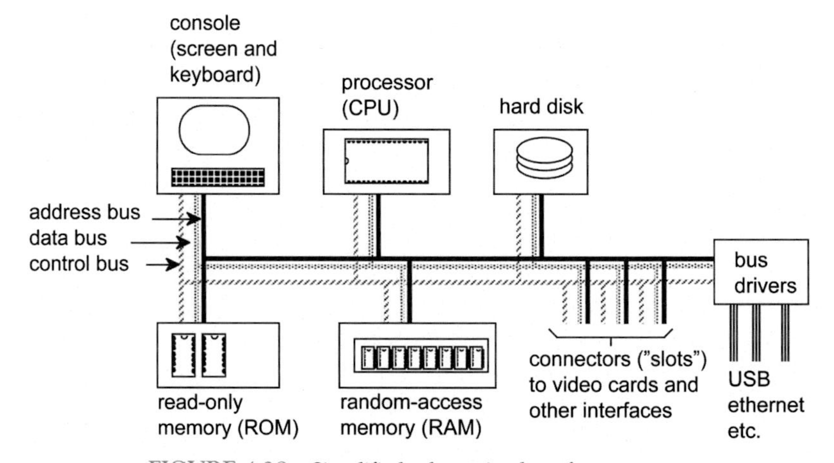

FIGURE 4.38 Simplified schematic plan of a computer.

functioning of the whole apparatus. Remember the chain of flip-flops we discussed as a counter: Counting would stop or become erroneous if any of the participating flip-flops would fail. So, a computer with more than about 50,000 tubes would hardly function, if at all. The advent of transistors, being not only much smaller but also far more reliable components, got the principle of the computer where it is now: millions (or even billions) of transistors, joined into large-scale ICs, performing billions of instructions per second.

The first generations of (large) computers employed small magnetic rings, or cores, as 1-bit memory cell. 0s and 1s were stored as the direction of the magnetic field, clockwise or anticlockwise. The term core memory, still heard today, stems from these machines.

Larger data sets than could fit in memory were stored on magnetic tape. The disadvantage is that a tape is one long chain of data, so that memory locations could only be reached sequentially.

Today, computer memory consists of semiconductor circuits, and with the advent of large-scale MOS memory chips, the prices came down by a factor of more than 1000. These chips are called RAM because each cell can be accessed at will, in contrast to other data storage media such as tapes and disks. This, together with the construction of the single-chip microprocessor, led in the late 1970s to a stripped down version of the large computer, the "microcomputer," that was affordable for small companies and even family households. This microcomputer evolved into today's personal computer, a device that does not need any introduction at all.

However, for a better understanding of the functioning of these everyday machines and to see what is involved in the use of computers in electrophysiological research, we will discuss the main principles and the gross construction of the average PC, together with examples of data manipulation.

The PC is evolving very rapidly, and so giving details about specifications of PCs is a hopeless task, since this kind of information is often obsolete when it appears in print. Therefore, we will stick to a simple, hypothetical type, not unlike the first microcomputers, the working principle of which is most easily grasped.

A "new born" (just built) computer would consist of all the necessary parts but, having no program, would not be able to perform any tasks at all. Therefore, the most basic pieces of program, or routines, that are necessary to make a computer able to process commands from the keyboard or to read something from a disk is built into a fixed-content memory, or read-only memory (ROM). The collection of programs to access the disks, listen to the keyboard, put text or graphics on the screen, and so on forms the OS or, since the disk plays the central part, a disk operating system (DOS). Although the smallest of computers may have their entire OS in ROM, most PCs have only the most basic routines, the basic input and output system (BIOS) in ROM. This BIOS is factory-programmed to read the (rest of the) OS from disk into RAM at start-up or rather to read only that part of the OS needed at start-up. A big disadvantage of ROM is that any inconvenience or bug in it cannot be repaired. So, modern ROM chips are not entirely fixed but can be changed incidentally (EAROM, electrically alterable ROM). The EAROM chips are collectively known as the firmware of a computer, i.e., the relatively fixed part of the software. Essential is that the contents of these memories are retained when the computer is switched off.

All other instructions to a computer, be it parts of the OS or user applications, are called software and are fed to the computer on demand, usually either from the built-in hard disk or from any kind of removable media. The instructions must be put in a form the computer understands, in a programming language (or computer language), such as Basic, C, or Java.

Computers, however, do not understand "languages" directly. The statements from these languages must be translated first, often by the same computer, into machine language. The type of machine language used depends on the processor and differs widely among different brands of chips. But strictly speaking, the only "language" a computer understands consists of combinations of voltages (about 0 V and about 5 V, signifying 0s and 1s, respectively) at certain times at certain inputs (see digital electronics and logic discussed above).

To make these bunches of binary voltages readable for humans, 4 bits are bundled to yield a hexadecimal (or "hex") number (0...15, notated as 0, 1, 2, 3, 4, 5, 6, 7, 8, 9, A, B, C, D, E, and F). Two hex digits together describe the value of 1 byte. So hexadecimal "FB" means $15 \times 16 + 11$, or 251.

In the latter form, a piece of machine language for some processor would look like this: 2C 07 02 46 00 0F E8 4F 06 46 00 30...

Some of these numbers code for instructions, others for addresses where to get or store a number, and some are numbers that are to be used, such as constants or the momentary values of variables. Obviously, this is still hard to read and led to the development of assembly languages, by giving instructions, processor registers, and actions short names. These abbreviations, mnemonics, help to understand what the code performs. In assembly language, the above program fragment would be notated as:

```
move.l  d7,d6
andi.w  #$f,d6
lsr.w   #4,d7
addi.w  #$30,d6
```

Here, a programmer will be able to recognize the instructions easily as "move a longword," "immediate AND," "left shift," and "immediate add." By the way, in the daily computer life, assembly language is often called "machine language," in contrast with the higher programming languages mentioned earlier.

Writing complex computing or data manipulation programs entirely in assembly would be a tedious task, since even the simplest calculation would cost hundreds of lines of code.

Note that the assembly language example given above is only a minute part of a decimal-to-hex conversion routine. Therefore, most programmers use a higher language to write their programs in, such as Basic, C, or Java. These languages look more like "algebra stated in English" and are both easier to read and remember and far more compact. The simple statement "$c = a + b$" would need many assembly-level instructions to fetch numbers, put them in registers, convert them to binary form, add them byte by byte (or send them to a floating-point processor), convert them back to decimal, and so on.

As an illustration, the listings below show similarities and differences between three programming languages: FORTRAN, Basic, and C. The output of all three miniprograms is the same: a table of the first 100 natural numbers and their squares.

```
C IN FORTRAN        C means comment
INTEGER I, IMAX     C declare loop counter and max
PROGRAM             C start of program
IMAX = 100;         C set loop maximum
```

```
DO 100 I=1,IMAX                C start of loop
WRITE(6,200) I, I**2           C main instructions, yield the output
100 CONTINUE                   C end of loop
200 FORMAT (I8, 1X, I8)        C specification of printing format
END                            C end of program
```

```
5 REM IN BASIC                 REM means remark (start of program is implicit)
10 IMAX=100                    REM setting loop maximum
                               REM (declaration is implicit)
20 FOR I=1 TO IMAX             REM start of loop
30 PRINT I, I∧2                REM main instructions
40 NEXT I                      REM end of loop
50 END                         REM end of program
```

```
*/ IN C                        "*/" means remark
#include <stdio.h>             */ invoke library of standard input and output
                               */ (stdio)
main() {                       */ start of program
int imax=100;                  */ declaring and setting the loop maximum
int i;                         */ declaring the loop counter
for (i=1; i <= 100; i++)       */ start of loop
printf ("%8d %8d", i, i*i);    */ main instructions
}                              */ end of program
```

FORTRAN ("formula translator") was the first higher language that became widely used in the 1950s. It was extended and modernized over the decades and is still in use today, mainly in physics and technology. BASIC is said to mean "beginner's all-purpose symbolic instruction code," but it is almost certain that the acronym preceded its explanation. It was invented in the 1970s, mainly for didactic purposes, but became very popular around 1980 because a basic interpreter could fit in the first generation of personal computers making computer programming available to the masses. However, today Basic is replaced largely by so-called "higher" programming languages.

In those modern languages, more complex manipulations can be coded with less lines of instructions. See examples later on.

Any program written in a higher-level language must be translated into the mentioned machine code by either of two methods: compiling or interpreting. This is performed by computer programs, known as compiler and interpreter, respectively.

An interpreter translates the instructions (program lines) at runtime and then executes them. Thus, a Basic interpreter would take line 10 of our Basic example and do the following: reserve a few bytes of memory to hold the variable IMAX, attach this name to the address of this memory register; convert the characters "100" into binary representation (00000000 01100100), and put it in the reserved memory locations. Then it would read the next line.

A compiler, to the contrary, would take the complete program text, the source code, and convert all instructions into machine-language form. Thus, the output of a compiler, called object code or runtime code, is a complete, independently working program (stand-alone program), which may be given (or sold) to people that use the same type of computer but do not need to own the compiler used. Large stand-alone programs are known as executable (EXE) applications (program, App) or simply program. This is in contrast with an interpreter text (often called a script), which needs the interpreter each time it is executed (i.e., at runtime).

However, there is a price to the added convenience of higher-level languages: Most of the codes generated by these compilers or interpreters do not run so fast as when the same task would be put directly in machine language form. In this respect, the C language generates the code running fastest, which is said to be the most efficient code. Therefore, most OSs are written in C as are user applications such as word processors, spreadsheets, and image analysis programs. Even then, assembly language is often still used for the most time-critical parts of the programs.

With languages that use compilation, such as C (and variants such as C++, C#) and Java, one writes programs (series of instructions) in a human-readable form, in a kind of text processor called the programming environment. The result is then turned into executable machine code by the compiler, which is itself a computer program. The entire user program is compiled into a computer-executable series of instructions, so the end result is a file ending in "exe" or "app" that can be executed by people not having the programming environment. Compiling yields the fastest executing programs but is less friendly to the programmer. First, a full program must be typed and saved, then compiled, and finally run to find out whether it works properly.

With an interpreter, the user also writes human-readable code, but this code is only converted during execution, line by line. So to run such a program, the user must have the corresponding "programming environment" installed. Interpreted languages yield slower running code in principle, but programs are easier to understand and develop. Each written chunk of code can be run (i.e., executed) to see whether it works correctly.

Popular interpreted languages are MatLab, Mathematica, and R. The latter is free, open source and maintained by hundreds of people worldwide.

Most modern computer languages, such as the ones just mentioned, are vector-oriented. A vector in this respect is a row of numbers.

In earlier languages, if one needed, e.g., the numbers from 1 to 100, a loop must be written such as (in Basic):

```
dim x(100)
for i=1 to 100
x[i] = i
next i
```

In vector-oriented languages, this is simpler (example in R[5]):

```
x <- 1:100
```

Read this as: "the variable x gets the numbers of 1 through 100."

The double symbol <- is the assignment operator.

So a variable may hold many numeric values that can nevertheless be reached or changed individually by indexing, so x[1] through x[100]. The above square-printing routine could be programmed in R as:

```
x <- 1:100
print(paste(x, x^2))
```

Note that the loop is implicit because x is a vector. Vectors are also called arrays. In most languages, an array can have more than one dimension. A 2D array is a matrix or table, a 3D array a stack (e.g., a stack of images rendering a volume).

Defining a matrix in R is simple:

```
matrix(data = 0, nrow = 12, ncol = 4)
```

This defines a matrix with 4 columns and 12 rows, all filled with 0s.

Of course, the data can be more interesting, such as the contents of a loaded data file. Alternatively, a number of equally long vectors (say A, B, C, and D) in R can be joined into a matrix:

```
mymat <- cbind(A, B, C, D)
```

If one or more of the columns to join is not numeric, the result is called a data frame. This is the equivalent of an Excel spreadsheet in which some columns may contain names, others numbers:

```
mydf <- data.frame(name, address, age, salary)
```

The R language is used worldwide in many branches of science, such as economics, DNA analysis, and signal analysis. Packages are sets of functions that can be added to the standard R implementation, for instance to add digital filters, handling wave files, etc.

Finally, LabView is a graphical programming language specifically designed for the direct control of AD cards and other laboratory instruments. It works by drawing lines ("wires") on screen to connect components like signal sources, amplifiers, filters, and measuring instruments such as on-screen oscilloscopes.

References

1. Li H, Liu X, Li L, Mu X, Genov R, Mason AJ. CMOS electrochemical instrumentation for biosensor microsystems: a review. *Sensors (Basel)* 2016;**17**(1).
2. Ceriotti L, Kob A, Drechsler S, et al. Online monitoring of BALB/3T3 metabolism and adhesion with multiparametric chip-based system. *Anal Biochem* 2007;**371**(1):92–104.
3. Sohn K, Oh S, Kim E, et al. A unified potentiostat for electrochemical glucose sensors. *Trans Electr Electron Mater* 2013;**14**(5):273–7.
4. Ahmadi MM, Jullien GA. Current-mirror-based potentiostats for three-electrode amperometric electrochemical sensors. *IEEE Trans Circ Syst* 2000;**56**(7):1339–48.
5. The_R_Core_Team. *R: a language and environment for statistical computing*. 2017. R, https://www.R-project.org.

CHAPTER

5

Electro(de) Chemistry

OUTLINE

Electrical processes in living organisms take place in water solutions containing salts, proteins, carbohydrates, and a host of other organic and inorganic substances. The electrical processes are dominated to a large extent by the various salts. Therefore, we will need a good understanding of the properties of electrolyte solutions and of the processes associated with them. In addition, most methods to get measurements from the wet medium take place with electronic instruments that must be connected somehow to the process studied. We are therefore also interested in the processes at the electrodes used for measurement and stimulation. More precisely, the point of focus is the boundary between the electrode and the solution or metal–electrolyte interface (conducting solutions are called "electrolyte"). Insulators

Introduction to Electrophysiological Methods and Instrumentation, Second Edition
https://doi.org/10.1016/B978-0-12-814210-3.00005-0

do not support an electric current of course but may show electrical influences when brought in contact with electrolyte solutions. This is caused by the fact that the surfaces of many insulators consist of charged particles. Finally, electrical processes take place at the boundaries between different solutions. Thus, we will have to deal with the following boundaries:

1	Metal	|	Electrolyte
2	Insulator	|	Electrolyte
3	Electrolyte 1	|	Electrolyte 2

We will treat the metal–electrolyte interface in greater detail below. Boundaries at insulators such as glass and plastics give rise to electrokinetic effects, which will be described briefly.

Finally, the third case is called a liquid junction or diffusion boundary. It arises where two different solutions meet in situations where convection or mixing is prevented, such as in gels, in porous substances, and in capillaries. This causes the dreaded diffusion potential or liquid junction potential (LJP) that forms a threat to the validity of many electrophysiological DC measurements. Knowledge of these processes can be extended easily to include the electrical phenomena at the cell membrane, where two solutions are separated by a selectively permeable membrane. Thus, we will add the following system:

electrolyte|membrane|electrolyte

The processes involving membrane potentials and conductances are important in neurophysiology. In addition, potentials across artificial membranes are used to measure ion activities, such as pH.

ELECTROLYTES

Electrolytic solutions arise, according to their name, by the fact that a number of substances, electrically neutral in dry condition, can dissociate or split into charged particles or ions when they are dissolved in a suitable solute (usually water). In principle, all dissolved salts are dissociated fully into their constituent ions, but not all salts dissolve readily.

The degree to which a salt dissolves and ionizes depends on the solubility product, which is the product of the concentrations of the constituent ions. For silver chloride, AgCl, an important substance in electrophysiology, the solubility product, i.e., $[Ag^+] \times [Cl^-]$, is about 10^{-10}. Dissolved in pure water, the solid AgCl dissolves until both ion concentrations are 10^{-5}. When other salts are present, things may change. In a physiological saline environment, for example, the chloride concentration is in the order of 0.1 M. Therefore, solid silver chloride will only dissolve up to a silver ion concentration of 10^{-9}.

Some salts, such as rock minerals, are so insoluble that their dissolution may take lots of water during millions of years (or takes place by wearing down mechanically).

Acids and bases are a bit different. In principle, acids split into hydrogen ions (H^+) and an anion that characterizes the particular acid. In the same way, bases split into a characteristic cation and the negative hydroxyl ion (OH). Some acids and bases, called strong, are almost fully dissociated, but many acids and bases are only partially dissociated, and hence called weak. The strength of acids and bases is characterized by their respective dissociation constants. Examples of strong acids are hydrochloric acid and nitric acid, of weak acids carbonic acid (H_2CO_3) and hydrogen sulfide (H_2S). Note that in solution, weak acids and bases are for the greater part undissociated, and so do not take part in reactions. Therefore, weak acids and bases are very useful to make buffer solutions: If some H^+ (or OH^-) is "used up" by the substance the buffer is added to, a bit of extra acid (or base) is split, so that the H^+ (or OH^-) concentration is kept approximately constant.

In electrophysiology, several salts play important parts. A simple electrolyte, such as sodium chloride (NaCl), splits in one cation (Na^+) and one anion (Cl^-). In general, hydrogen and metals are positively charged, whereas the other half of the original salt carries a negative charge. This was found by early electrochemists by forcing an electric current through a salt solution. The metals, being positive, accumulated at the cathode (negative electrode) and hence are called cations, whereas the negative ions are called anions because they migrate toward the anode (positive electrode).

The ions can move through the water, and so they are able to transport charge, just like electrons in a vacuum or the electrons and holes in a semiconductor.

We will summarize the properties of electrolytes with a simple example, such as NaCl. Unfortunately, simple electrolytes do not show simple behavior, as will become clear later on. In addition, the solute itself, water (H_2O), shows remarkable features. First, the hydrogen atoms in the water molecule are distributed asymmetrically, causing the water molecule to have a positive side and a negative side: It is said to be polar. This is the cause of the very high relative dielectric constant of pure water: about 84 (most conventional dielectrics have a dielectric constant of about 2–6). This feature is of direct consequence to the behavior of metal electrodes put in an electrolyte solution, explained further on. Water would be the ideal dielectric for capacitors if it would be a better insulator. In fact, water conducts slightly because of its partially ionization into H^+ and OH^-. Under normal conditions (room temperature and atmospheric pressure), a fraction of about 10^{-7} is split. Therefore, the pH of neutral solutions at room temperature is 7. At higher temperatures, more water is split, until at the boiling point (100°C), the pH is about 6.

The presence of ions, being mobile charges, makes salt solutions fairly good conductors of electricity. How well they conduct is usually expressed as the resistivity, abbreviated ρ (Greek rho), and often expressed in Ωcm (cgs system) or Ωm (SI). The inverse of resistivity, conductivity and abbreviated g, is also used frequently. The unit of conductivity is the siemens (S, see Chapter 1). Note that resistivity is a characteristic of a substance rather than of an object. The resistivity, expressed in Ωcm, is the resistance of a column or tray with a length of 1 cm and a contact surface area of 1 cm^2: in other words, the resistance of a cube of 1 cm edge. In the case of other forms of column, the resistance is proportional to the length and inversely proportional to the contact area. As a consequence, the resistivity is expressed in $\Omega\ cm^2/cm$, which simplifies to Ωcm. This is easy to understand, since a longer column can be thought of as composed of resistors connected in series, whereas a wider column can be considered as a circuit of resistors in parallel.

Conduction of electrolytes is not as good as in metals, wherein the free electrons may cause resistivity values as low as 10^{-6} Ωcm. Yet, strong electrolyte solutions, such as 3 M KCl, have a resistivity of no more than 5 to 10 Ωcm. Seawater, having ion concentrations of about 0.5 M, has also a rather low resistivity. Other values can be taken from the following table.

Approximate Resistivities of Common Electrolyte Solutions at 18 °C

Copper (for comparison)	1.7	μΩcm
3 M KCl	5	Ωcm
Seawater (mid-ocean)	22	Ωcm
Seawater (coastal areas)	25	Ωcm
Physiological saline (about)	70	Ωcm
0.1 M KCl	90.9	Ωcm
0.01 M KCl	820	Ωcm
Freshwater	0.2 to 10	KΩcm
Distilled water	1 to 5	MΩcm
Pure water (theoretically)	23	MΩcm

The value given for pure water at a pH of 7 can be approached closely if boiled (i.e., degassed) water is deionized in a mixture of ion exchange resins. Distilled water and deionized water that are allowed to equilibrate with air absorb carbon dioxide. It is partially dissociated into H^+ and HCO_3^- and so causes the lower resistivities given. The two concentrations of KCl given are used to calibrate resistivity meters.

Many of the stated resistivity values have direct consequences for electrophysiological measurements. For instance, the resistance of glass capillary electrodes depends on the resistivity of the filling fluid. Strong salt solutions such as 3–4 M KCl or KAc (acetate) are used to keep the electrode resistance to a minimum. When dealing with the skin and skin sense organs of aquatic animals, the resistivity of the environment, being either freshwater or seawater, may play a role. Measurement of small currents in seawater is difficult because they yield far lower voltages than in freshwater. The composition of physiological salines or "Ringer's fluids" is chosen to be similar to the natural body fluid of the experimental animal species in both chemical composition and resistivity.

The resistivity of simple solutions depends on the salt concentration but not in a completely linear way. The ions are relatively independent in very dilute solutions and behave, despite the presence of water molecules, approximately like atoms in a gas. In this case, a doubling of the concentration causes the conductivity to be doubled, and so the resistivity to be halved. Hence, resistivity is inversely proportional to concentration. In stronger solutions, however, the ions interact with each other in a rather complicated way, described among others in the famous works of Debye, Hückel, and Onsager. Roughly speaking, the effect of interactions can be described as if ions are screening each other, reducing their electric "visibility" in the solution.

The main effect of these interactions is that the electrically active concentration, or activity for short, of ions in concentrated solutions is less than that expected from the true concentration. Thus, the ion activity can be considered to be the electrically *effective concentration*. To be able to distinguish this quantity from the true concentration, i.e., the total amount of substance present, this latter quantity is called, somewhat superfluously, the stoichiometric concentration. This means "by weight," that is, the amount found by chemical analysis, by drying and weighing, etc., or simply the amount "put into it." Interactions of ions take place mainly at high concentrations. At concentrations below 0.1 mM, the activity is virtually the same as the concentration. At concentrations more than 1 mM, however, the screening becomes apparent. As an example, a 100 mM KCl solution behaves electrically as if the concentration were 76 mM. Therefore, the ion activity is said to be 76 mM. The conversion factor is called the activity coefficient, f:

$$f = a/c \quad \text{or} \quad a = c \cdot f$$

where c is the concentration and a, the activity of an ion.

At higher concentrations, the activity coefficient can become as low as 0.55 so that a 3 M KCl solution behaves electrically as if it were 1.65 M. There is a complication, however. Activity coefficients of individual ion species cannot be determined directly. The activity of ions is reflected most prominently in the conductivity of salt solutions, but these consist necessarily of cations and anions so that a resultant or mean activity coefficient is obtained. Fig. 5.1 shows the mean activity coefficients of KCl, NaCl, and $CaCl_2$ as a function of concentration, derived from several compilations of measurements.[1–3]

Under certain assumptions, the activity coefficients of salts having a common anion can be used to estimate the activity coefficients for individual ion species. The result of such computations is shown in Fig. 5.2.

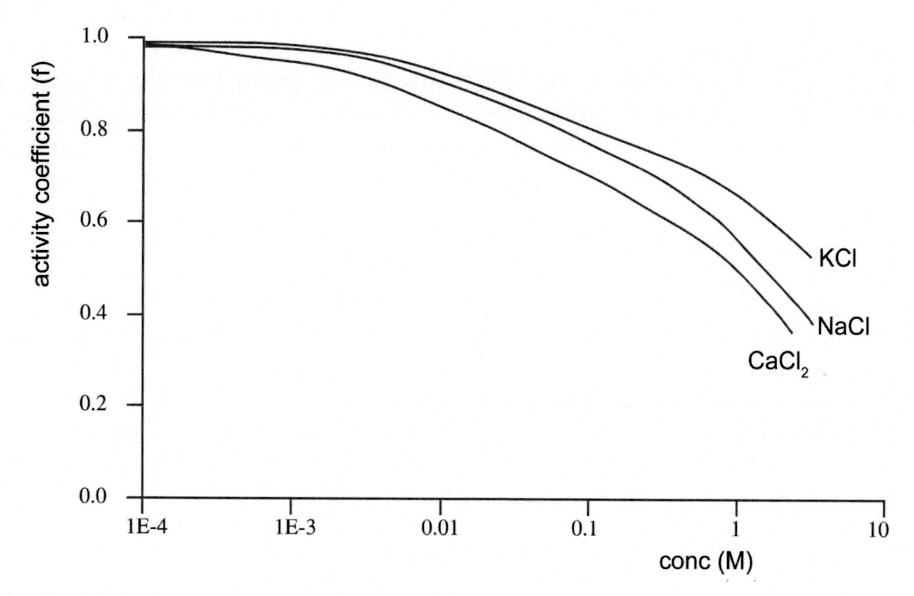

FIGURE 5.1 Mean activity coefficients of salts versus concentration.

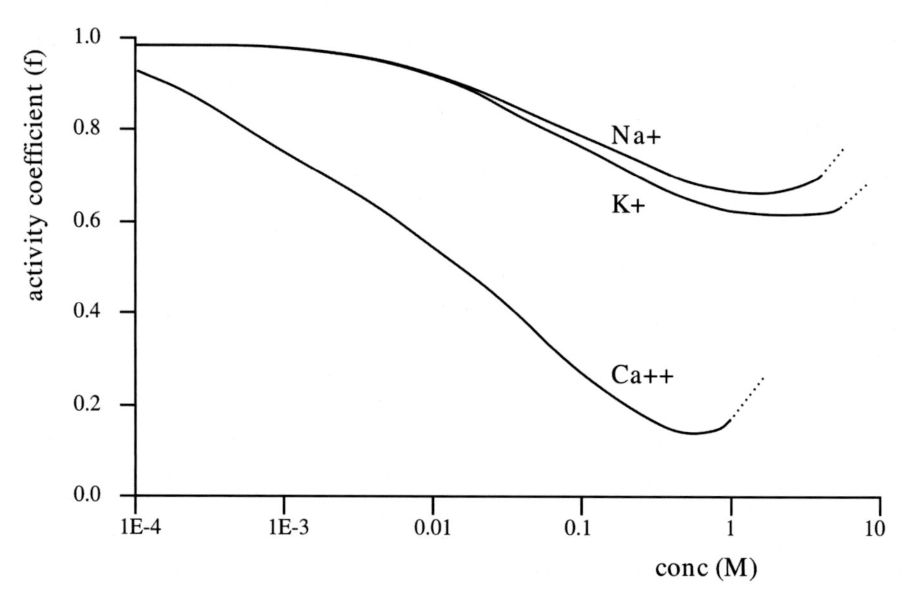

FIGURE 5.2 Estimated activity coefficients of cations as a function of concentration.

Note that, especially for divalent ions such as calcium, the reduction of the activity coefficient is very prominent: At 1 mM, the activity is already reduced to 0.9, and at 0.5 M, the activity coefficient is reduced to a mere 0.2. The activity coefficients are hard to determine at concentrations of 1 M and higher and are usually estimated by compiling the results of different methods. Some of these measurements suggest that at very high concentrations, the activity coefficient rises again, to reach values far higher than 1 (dotted lines in Fig. 5.2). This would mean that the negative interactions found at lower concentrations are changed into positive, or reinforcing interactions. Although the mechanism hereof is not understood fully yet, the effects are large and must be taken into account whenever strong salt solutions are used, e.g., as salt bridges.

It is important to note that in all electrochemical calculations, ion activities rather than concentrations must be used.

Because of the interactions mentioned, the electrical resistivity of salt solutions depends on the activities of all the constituent ions. In addition, a quantity called the ion mobility plays an important part. For an ion to transport electricity, it must be mobile, i.e., it must be able to move with respect to the solute. The mobility of the familiar ions varies widely. It seems to depend on the radius of the ions, but many ions are slower than expected. Therefore, one has to take into account that some cation species drag water molecules with them, and this hydration mantle is strongest in small cations. Mobility is abbreviated u, and expressed in $10^{-8}cm^2/s\ V$. The table below shows the mobility of a number of common ion species at room temperature.

Cations	u	Anions	u
H^+	36.3	OH^-	20.5
Li^+	4.0	Cl^-	7.9
K^+	7.6	NO_3^-	7.4
Na^+	5.2	HCO_3^-	4.6
NH_4^+	7.6	Ac^- (acetate)	4.2
Ag^+	6.4	1/2 SO_4^-	8.3
1/2 Ca^{++}	6.2	1/2 $H_2PO_4^-$	3.7
1/2 Mg^{++}	5.5	1/2 HPO_4^-	5.9

From the table, hydrogen seems to be an exception, having a mobility far greater than all other metal, metalloid, and composite ions. Therefore, it is suggested that hydrogen ions move as free protons. Note that the mobility of a single ion species cannot be determined directly so that the values in the table are computed from many experiments involving different salts with either a common cation or a common anion. The mobility of an ion contributes directly to the conductivity of the solution. The conversion factor from mobility u to conductivity g follows from:

$$g_i = Fu_i$$

where g_i is the *contribution* of one equivalent (i.e., 1 mol of a univalent ion, half a mole of a divalent ion, etc.) of an ion species *to the conductance* of a solution, and u_i is the mobility of that ion. The proportionality constant F is called Faraday's constant. This is the amount of charge carried by one equivalent and has the value of 96,500 (coulomb per equivalent). Faraday's constant is found very often in electrochemical relations, such as the equations of Nernst dealt with later on.

When dealing with composite electrolytic solutions, being mixtures of several salts at different concentrations, the notion of concentration can be replaced by the total ionic strength. The ionic strength S is defined by:

$$S = \frac{1}{2}\sum c_i z_i^2$$

where c_i is the concentration and z_i, the valence of each ion species. The square arises because the electrostatic force between two point charges is proportional to the product of the two charges. Thus, the ionic strength of a 10 mM KCl solution is 10 mM and that of a 10 mM $CaCl_2$ solution is 30 mM.

Although the ionic strength is a useful indication of the approximate electrical properties that can be expected, the detailed behavior of ions may still depend on the precise composition of the solution, that is, on the individual activities and mobilities. Because of interactions between ions of different species, the activity coefficients shown above depend on ionic strength rather than on the concentration of only one ion species. To be more precise, ionic strength, rather than concentration, is the quantity shown along the horizontal axis in Figs. 5.1 and 5.2.

It is important to note that the exact composition of many solutions, such as the natural body fluids, is not known. Therefore, many electrochemical quantities *can only be approximated*.

In subsequent chapters, we will need the following constants frequently:

Molar gas constant	R	8.314 J/mol K
Avogadro's constant	N	6.03×10^{23} mol^{-1}
Faraday's constant	F	96485 C/equiv
Elementary charge	e	1.6022×10^{-19} C

THE METAL–ELECTROLYTE INTERFACE

At a metal–solution interface, three things can happen:

- "Nothing"—the boundary behaves as capacitance
- Faradaic processes—the well-known redox reactions
- Non-Faradaic processes—adsorption and other complications

What will happen depends on the compositions of the metal and of the solution and on the voltage applied.

A major complication concerning all electrochemistry is that the processes at a metal–solution boundary, called an electrochemical half-cell, cannot be studied in isolation. They can be studied only by joining two half-cells into a full electrochemical cell. This implies that any electrochemical measurement is in fact the difference of two components that can be separated only by the careful design of several experiments and by choosing an appropriate standard.

The standard, adopted universally in electrochemical research, is called the normal hydrogen electrode. It has been chosen carefully to be reproducible within very narrow tolerances in the laboratory and consists of a porous platinum electrode saturated with hydrogen gas at 1 atm pressure in a solution containing 1 mol of hydrogen per liter (1 normal or N). Although it may seem strange that hydrogen is used as a metal, this is found to be the best way to build a standard half-cell and has been used for more than a century. In electrophysiology, however, potential measurements are almost always relative and so may be taken against any stable but arbitrary standard. In the following discussion, all stated potential *differences* are meant to be the potential difference across a whole electrochemical cell, whereas all stated *potentials* are half-cell potentials taken with respect to the abovementioned standard.

The best known examples of electrode reactions pertain to the combination of a metal in a solution of a salt of that metal, such as copper in a copper sulfate solution or zinc in zinc chloride. Since most metal salts are poisonous, this type of electrode cannot be used very often. An example is a pair of copper electrodes in porous pots filled with copper sulfate that is used in the earth sciences to measure geoelectric fields. In electrophysiology, we need nonpolarized electrodes in combination with physiological saline and, in general, solutions that are not poisonous to living tissues. In these cases, the solution does not contain the electrode metal in significant concentrations. This has severe consequences for the processes that take place to move charge across the boundary.

As an example, we will take the half-cell composed of a mercury electrode and a 1 M KCl solution. At potentials greater than +268 mV, mercury is oxidized by combining with chloride ions:

$$2Hg + 2Cl^- \rightarrow HgCl_2 + 2e^-$$

At strongly negative potentials, at least 2.9 V, potassium ions are reduced (and form an alloy with the mercury):

$$K^+ + e^- \rightarrow K(Hg)$$

CAPACITANCE OF POLARIZED ELECTRODES

In the range of potentials between the two voltages mentioned above, no electrode reactions occur, and so no steady (DC) current flows whatsoever. This can be seen from a graph of the current flow (I) as a function of potential (E), a so-called I/V characteristic (or better I/E). This is illustrated in Fig. 5.3 (left). Note that the I/V characteristic of a pure resistance would be a straight line, the slope of which depending on the value of the resistance. The I/V characteristic of the metal electrode is flat in the potential range mentioned, i.e., no current flows.

In other words, the resistance is infinite. In this range, the electrode is called a polarized electrode. A polarized electrode behaves as a rather large capacitance. This can be understood by the following argument. An electrode can be considered as a border between two conducting substances (the metal and the solution). Since no charge can cross that border, it behaves as an insulator and so the whole acts as a capacitance. It can also be seen that the capacitances formed have very high values. If the boundary would be sharp, the "insulator" would be infinitely thin, and so the capacitance would be infinite. Since in reality the

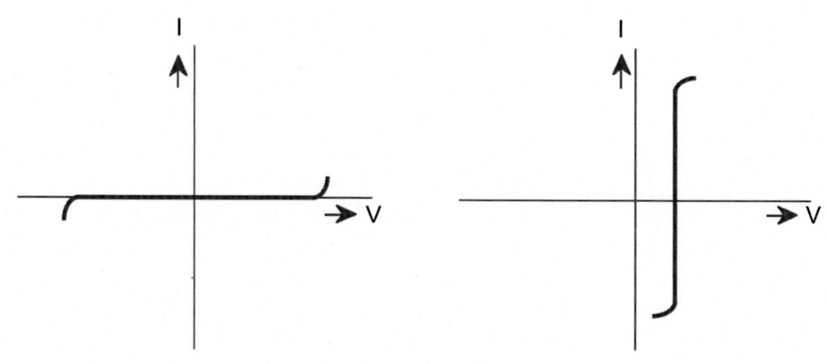

FIGURE 5.3 I/V characteristics of ideal polarized (left) and nonpolarized (right) electrodes.

capacitance is finite, there must be a finite gap between the charges on opposite sides. In the 19th century, Helmholtz formulated a theory about this electrical double layer.

One side of the double layer consists of the electrons in the metal. Because of the high conductivity of metals, this layer is very thin, far less than a nanometer. The other side, the charge in the solution, is more complex. Several processes determine the effective thickness of this layer. In the first place, ions have finite diameters and can only approach the boundary down to the ion radius. In the second place, some ions adhere to the metal surface, thus keeping others from getting closer. These ions have hardly any mobility. This inner layer is called the Helmholtz layer and determines the properties of an electrode to a great extent. The ions farther out in the solution are more and more mobile, approaching the mobility in the bulk solution.

The profile resembles that of an "atmosphere," i.e., it assumes the same shape as the air pressure at different altitudes in the earth's atmosphere. In this way, the potential changes gradually from the value in the metal to the value in the solution. This is shown in Fig. 5.4.

The size of the double layer depends on the ion species involved, their concentrations, and the potential and may vary between less than 1 nm in concentrated solutions to tens of nanometers in very dilute solutions. Since the capacitance is dependent on the thickness of the double layer, it will depend on the same factors. Therefore, metal electrodes, very popular in electrophysiology, have complicated properties that must be assessed or calibrated in any practical situation.

FARADAIC PROCESSES: THE REDOX COUPLE

In the last year of the 18th century, Alessandro Volta had been inspired to make the first battery based on zinc and copper electrodes separated by an electrolyte containing sulfuric

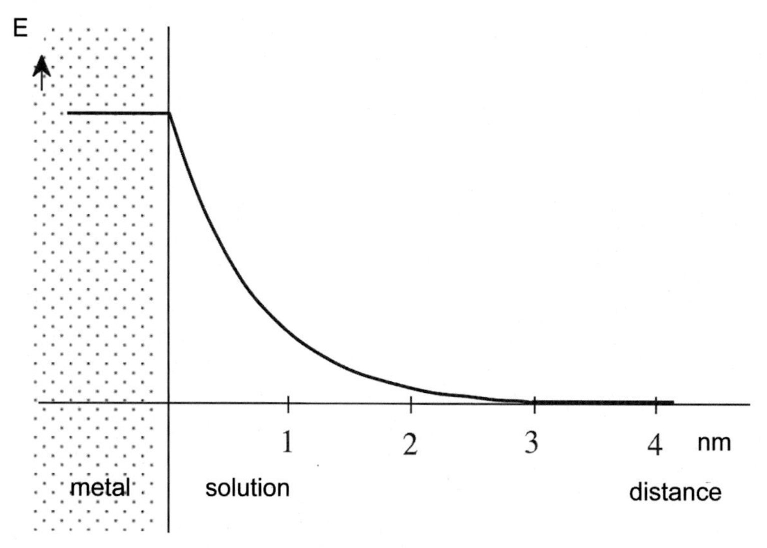

FIGURE 5.4 Potential distribution across a metal–electrolyte interface.

acid. His inspiration came from Luigi Galvani's observation that a frog's leg contracts when touched with two different metals. When current was drawn from Volta's battery, the zinc electrode was slowly consumed by its transformation into Zn^{2+} ions, while H^+ ions combined to form H_2 gas at the copper electrode. Since the process was irreversible, mostly due to the H_2 gas escaping, his battery could not be recharged.

The principle of a Zn/Cu battery in a reversible form is shown in Fig. 5.5. It is made of two half-cells connected by a salt bridge (KCl in agar).

At the anode, made of zinc, ionization of metallic Zn takes place as in the original Volta pile:

$$Zn \rightarrow Zn^{2+} + 2e^-$$

while copper is deposited on the cathode in the other half-cell:

$$Cu^{2+} + 2^{e^-} \rightarrow Cu$$

A copper salt, such as $CuSO_4$, has to be present in the electrolyte for the latter reaction to work. A positive potential of 1.1 V is established between the two electrodes in this configuration, i.e., 1 M KCl and 25°C. The flow of electrons from the Zn electrode to the Cu electrode is relayed by the flux of Cl^- ions out of the salt bridge into the half-cell containing the zinc electrode and the influx of K^+ ions into the half-cell containing the copper electrode. A current flows from the copper electrode to the zinc electrode if the two electrodes are connected by a device that consumes electric energy, such as a small light bulb. In order to reverse the chemical reactions and hence recharge the battery, a tension of more than the equilibrium potential of 1.1 V has to be applied. In that case, metallic Cu becomes ionized and Zn is deposited on the Zn electrode. The two ionization reactions are two halves of a single reaction,

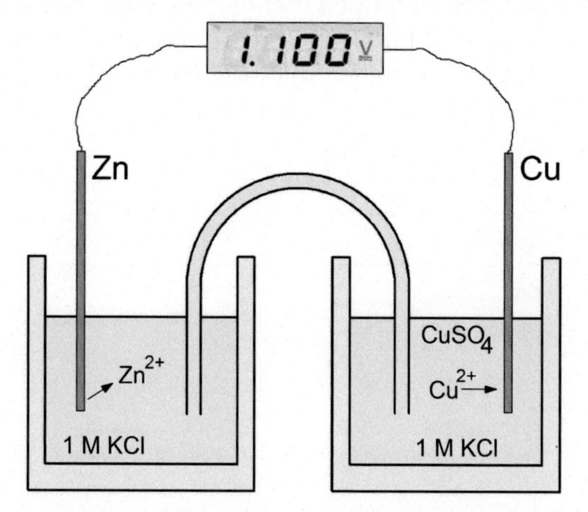

FIGURE 5.5 Two half-cells containing a zinc electrode and a copper electrode in a solution of KCl coupled by an agar bridge.

more generally known as a reduction–oxidation reaction or a redox reaction in short, which in this example resumes to:

$$Zn + Cu^{2+} \leftrightarrow Zn^{2+} + Cu$$

The term reduction–oxidation used to be limited historically to reactions engaging oxygen as in the burning of wood. However, as the principle involved, i.e., the exchange of electrons, is much more general, the definition has been enlarged since.

The current definition of oxidation is: a process in which a positive charge is accepted by a chemical substance (in the example above, Zn). Reduction, its reverse, that is: a process whereby a substance loses a positive charge (Cu in the example).

Now let us consider only one of the half-cells, the one housing the Zn electrode.

As we have seen, if the electrode is made sufficiently positive, zinc atoms can be oxidized into zinc ions and at sufficiently negative electrode potentials, zinc ions are reduced. The metallic zinc precipitates onto the surface of the electrode.

In such a half-cell, the electrode assumes a stable equilibrium potential, virtually independent of current strength. The potential has been shown to depend on the concentration of the salt and on a theoretical property of the metal called the metal's solution pressure, or tendency to dissolve. Nernst established the relation between these two as:

$$E = \frac{RT}{zF} \ln\left(\frac{c}{K}\right)$$

where E is the electrode potential, T is the absolute temperature, z is the valence, c is the concentration of the metal ion, and K is the mentioned solution pressure. Although the latter is a hypothetical quantity and is even criticized as to its physical meaning, the differences between metals are very real. Some metals, such as silver, gold, and platinum, are very hard to dissolve (i.e., oxidize) and are called precious metals. Others, such as aluminum, iron, and zinc, are oxidized very easily. Thus, all metals can be ordered according to their relative solution pressures. Hydrogen, although not a metal in the strict sense, fits in this electrochemical series perfectly. Its use as a standard was already mentioned.

The list below shows electrode (i.e., half-cell) potentials for a number of reactions, all taken at "normal" (one equivalent per liter) concentrations of the salt. These are called "standard electrode potentials" and are abbreviated E_0.

System	**E_0 (Volt; 25°C)**
Zn^{++}/Zn	−0.761
Fe^{++}/Fe	−0.440
Pb^{++}/Pb	−0.126
H^+/H_2	*0 by definition*

System	E_0 (Volt; 25°C)
$AgCl/Ag^+$	+0.222
Calomel	+0.281
Cu^{++}/Cu	+0.337
Hg^{++}/Hg	+0.789
Ag^+/Ag	+0.799
Au^{+++}/Au	+1.50

Note that the standard hydrogen electrode half-cell has a pH of 0. Since in biochemical systems, pH values are about pH = 7 rather that at 0, biochemists often use electrode potentials referred to a hydrogen electrode at pH = 7.0. These are indicated as E'_0 and can be found by subtracting 406 mV from the values in the table (why?). Standard potentials of other redox systems, such as metabolic systems, can be expressed in the same way and are called standard redox potentials. These are almost always expressed as E'_0 values.

As a result of Faradaic processes, these electrode half-cells conduct electric current with very little change of the potential. Therefore, a metal in a solution of one of its salts is known as a nonpolarized electrode. The I/V characteristic of an ideal nonpolarized electrode is shown in Fig. 5.3, right. The electrode voltage is constant over a large range of current strengths. In that respect, the I/V characteristic of nonpolarized electrodes is the opposite of the characteristics of polarized electrodes, where the current is zero over a large voltage range.

PRACTICAL ELECTRODES

Practical electrodes do not follow the I/V characteristics of the ideal polarized or nonpolarized electrodes exactly. This is shown in Fig. 5.6.

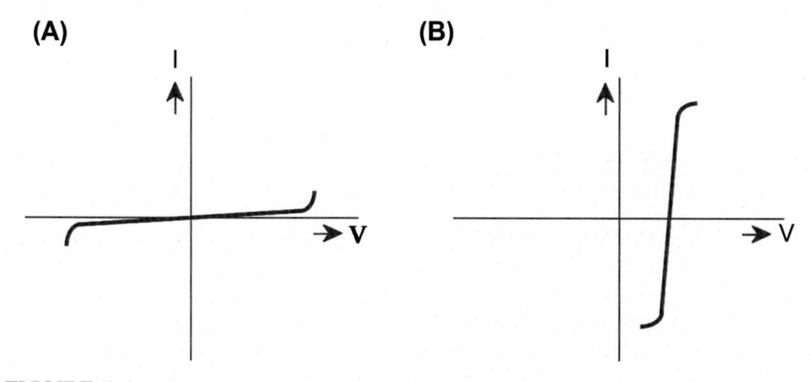

FIGURE 5.6 I/V behavior of practical polarized (A) and nonpolarized (B) electrodes.

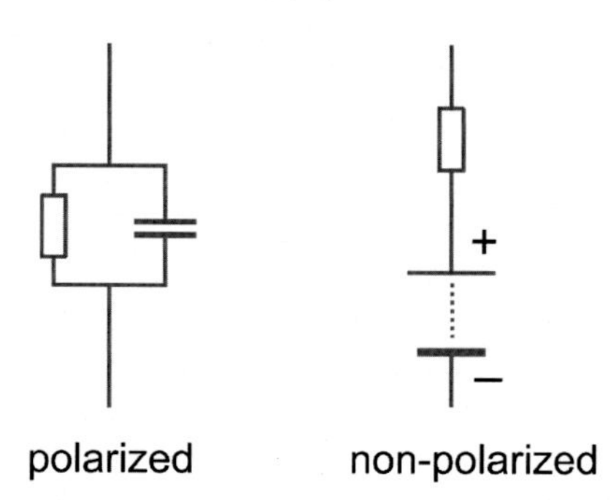

FIGURE 5.7 Equivalent circuits of polarized and nonpolarized electrodes.

In the case of a polarized electrode, which as we saw shows mainly capacitance, a small

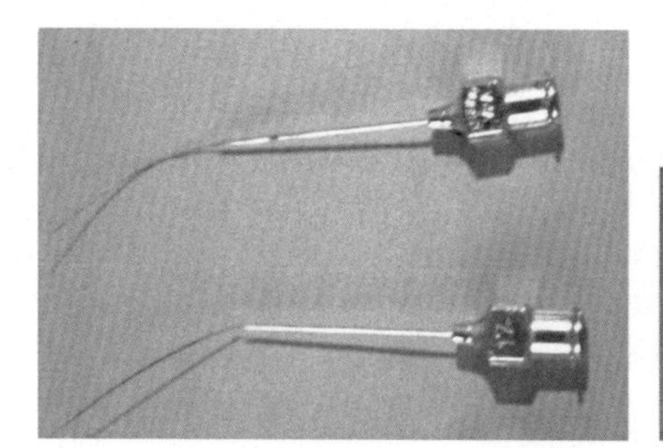

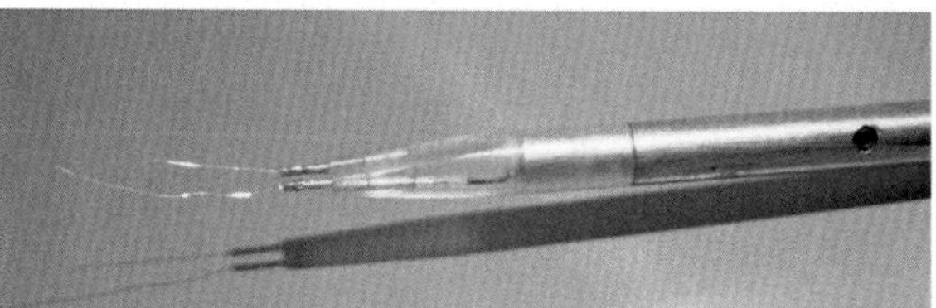

FIGURE 5.8 Left: Practical electrodes: pointed and coated tungsten wire electrodes, mounted in all-metal hypodermic needles. Right: Thin silver wire electrodes in a holder.

current can nevertheless traverse the electrode surface because a tiny amount of water can be ionized. Thus, the I/V characteristic is not exactly horizontal. In fact, a polarized electrode can be represented by a leaky capacitance, i.e., a capacitance shunted - by a (relatively high) resistance. In the same way, a nonpolarized electrode cannot conduct current without any deviations of the potential. Hence, a nonpolarized electrode may be represented by a voltage source (the electrode potential) in series with a (relatively low) resistance. The voltage is often referred to as electromotive force (emf, although it is not a force in the usual physical sense) and abbreviated as E. The two equivalent circuits are shown in Fig. 5.7.

Practical forms of electrodes are shown in Fig. 5.8.

ELECTROCHEMICAL CELLS, MEASURING ELECTRODES

The aforementioned processes pertain to one electrode dipped into a salt solution. Any practical case, however, will consist of a full electrochemical cell made up of two electrodes in one solution or alternatively of two joined half-cells. The latter case comprises a liquid junction that will be dealt with later. In general, two forms of electrochemical cells are used. For the first one, one picks metals as far apart in the list of electrode potentials as possible such that the resulting cell voltage is as high as possible. This approach is used for the design of electrochemical cells as energy sources, i.e., batteries and accumulators. The other extreme is used in electrophysiology: Two electrodes are made from precisely the same metal and dipped into a solution that is homogeneous throughout such that the potential difference should be zero. Small differences between electrode compositions and wear and tear cause the potential to be slightly different from zero. This potential is called "spontaneous" polarization because it occurs in the absence of current fed through the electrode pair. Both amplitude and polarity of spontaneous polarization are impossible to predict. It is a nuisance when one wishes to record small electrical signals. Since the polarization in this case is a random variable with zero mean, both amplitude and polarity are hard to predict and can interfere with electrophysiological measurements unless one takes appropriate precautions.

THE AG/AGCL ELECTRODE

In electrochemistry, the reliable, nonpolarized standard electrode used for more than a century is the Hg/HgCl, or calomel electrode. In can be made stable and reproducible to very narrow tolerances, necessary for the precision of fundamental, electrochemical research. However, since calomel electrodes contain liquid mercury, they must be handled with care and kept in the upright position. In addition, electrophysiological measurements seldom need the precision mentioned above. Therefore, the similar nonpolarized Ag/AgCl electrode is used far more often. Ag/AgCl electrodes are made by inserting a silver cathode and a platinum anode in a solution of HCl and applying a voltage difference between the electrodes (Fig. 5.9).

At the cathode, a deposit of the (almost) insoluble AgCl on the silver wire takes place:

$$Ag + Cl^- \leftrightarrow AgCl + e^-$$

while at the anode the formation of hydrogen occurs:

$$H^+ + e^- \leftrightarrow 1/2H_2$$

As mentioned earlier, the redox potential of the couple $2H^+ \leftrightarrow H_2$ is per definition 0 mV at 1 M H^+ at all temperatures. The redox potential of the first equation is 222 mV at 1 M HCl. The redox potential difference between the two equations remains 222 mV with changing HCl concentration because the redox potentials of both equations vary identically with the HCl concentration:

$$dV\ \text{cathode} = 0.222 - RT/zF \times \ln[Cl^-]$$

and

$$dV\ \text{anode} = 0 - RT/zF \times \ln[H^+]$$

Hence, AgCl is deposited at the cathode at a potential difference between the electrodes larger than 222 mV irrespective of the HCl concentration.

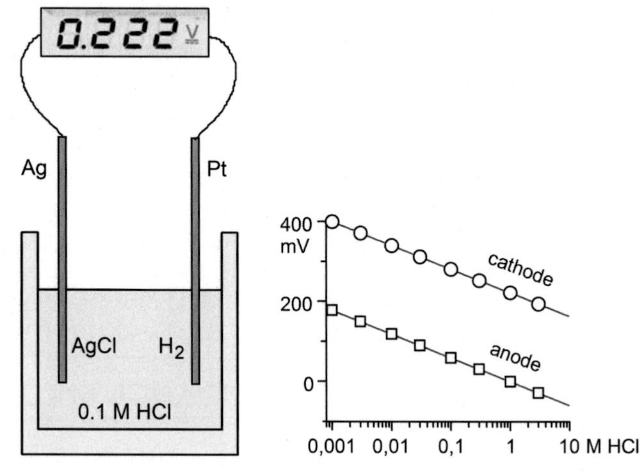

FIGURE 5.9 Left: The Ag/AgCl versus platinum electrodes. Right: Potentials of the two electrodes as a function of HCl concentration.

Often Ag/AgCl electrodes are created by passing a current between Pt and Ag electrodes dipped in 1 M KCl at neutral pH. Since the H^+ concentration is now 10^{-7} M instead of 1 M, an extra voltage difference of $RT/zF \times \ln[10^{-7}] = 406$ mV in excess of the 222 mV needs to be applied in order to deposit AgCl on the silver surface.

An AgCl deposit can also be created on a silver wire by oxidation of silver with sodium hypochlorite (bleach, NaOCl):

$$Ag + NaOCl + 1/2H_2O \leftrightarrow AgCl + NaOH$$

Ag/AgCl electrodes made this way tend to be off-white due to the presence of AgOCl in the AgCl pellet/deposit. In electrophysiology, Ag/AgCl is used almost exclusively for the recording of DC and low-frequency potentials.

After having discussed the Faradaic processes at a metal electrode dipped in one of its own salts, it may seem strange that a metal covered with a solid, insoluble salt is used as a non-polarized electrode. So let us first see:

- why Ag/AgCl is an electrode at all;
- why it is a nonpolarized one.

The mode of operation relies on two facts, which answer these questions:

1. In the AgCl crust, silver atoms have a low but useful mobility.
2. The Cl atoms, in contact with a solution containing chloride, can be exchanged just like metal atoms in contact with a salt of that metal.

Thus, the Ag/AgCl electrode can be considered as "a chlorine electrode in a salt of chlorine." Here, the anion rather than the cation is the reactive substance that supports an electric current at a virtually constant potential. The potential is determined by the concentration of Cl^- in the solution and can be assessed with Nernst equation, yielding the familiar 58 mV per decade. The stability of Ag/AgCl electrodes is served further by the fact that physiological salt solutions have a high and stable chloride content of about 150 mM in most animal species. By the low solubility of AgCl, the concentration of silver ions near an Ag/AgCl electrode is very low, usually about 1 nM. However, other anions in living tissue that can form insoluble salts of silver, such as sulfide, can "poison" an Ag/AgCl electrode. The poisoning effect of sulfide, for instance, can be understood by recalling that the silver sulfide that is formed would try to establish a Nernst equilibrium with sulfide ions (more precisely HS^- or S^{--}). Since sulfide concentrations are very variable, the potential of an involuntary Ag/Ag_2S electrode would be unstable. To avoid contamination, the electrode proper is usually screened from body fluids, water, etc. by an electrode holder filled with a clean, filtered KCl solution (called a salt bridge, and dealt with below).

When Ag/AgCl electrodes are not carefully prepared or wear out, parts of the metallic silver surface may become exposed. Oxidation of these surfaces forms Ag_2O and AgO oxides. Bringing these oxidized surfaces subsequently in contact with reducing agents, among which biomolecules containing free cysteine groups may lead to large (tens of millivolts) changes in electrode offset potentials during an experiment, could be misinterpreted as electric activity.[4]

Non-Faradaic Processes

These processes include adsorption and desorption of ions at the electrode surface as well as structural changes of the substances involved in the electrode reactions. Because of these processes, electrochemical measurements, including electrophysiological ones, depend on *approximations* to the theories presented. Therefore, measured values of electrode potentials, impedances, etc. will often deviate more or less from the predicted values. This situation aggravates when intracellular or extracellular body fluids are involved, especially since their complex ion composition may include proteins and other molecules with intricate chemical properties.

Electrokinetic Processes

The electrical double layer at the metal–electrolyte interface, described above, has a counterpart at the boundary between an insulator such as glass and an electrolyte solution. Although it may seem less obvious, the separation of charges at these boundaries lead to several electrical phenomena found and described in the 19th century. They are collectively called electrokinetic processes. These phenomena can show up unintentionally in electrophysiological measurements. The existence of an electrical double layer at an insulating surface was concluded from the voltage that develops across a porous plug where an electrolyte solution is forced through it; a process known as a streaming potential.

It is best illustrated with a glass capillary filled with a salt solution and with measuring electrodes at the ends (Fig. 5.10).

A sustained flow of liquid is indicated by the arrows. It was found that by the nature of the wall and the ions in the solution, one of the ion species is adsorbed to the surface, leaving an "atmosphere" of counterions in a narrow zone along the wall. Together, they form an electrical double layer that is analogous to the Helmholtz double layer described for the metal–electrolyte boundary. Movement of the liquid drags the ion "atmosphere" in the direction of

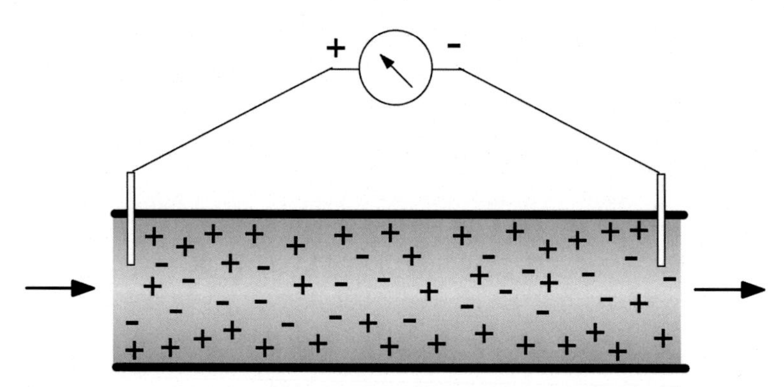

FIGURE 5.10 Streaming potential in a salt-filled glass capillary tube.

the flow, leaving a small charge of opposite polarity behind. This shows up as a streaming potential that can be measured with a pair of electrodes at the ends. Strength and polarity depend on which ion species is absorbed, how strong, and hence on both the wall material and the composition of the fluid. Given these circumstances, the potential difference depends entirely on the speed of flow. Often positive ions adhere to the surface, whereas the negative charge is dragged toward the end of the tube. This is the case shown in the figure. Porous substances, such as cotton wool, paper, and some ceramics, can be thought to consist of numerous capillary spaces and so show the same phenomena. In natural waters, the same process, here called filtration potential, causes voltage gradients to develop across the sand or clay bottom. The currents associated with these voltage gradients can be perceived by some species of fish that have special, so-called ampullary electric sense organs.

LIQUID JUNCTION POTENTIALS

Often, electrophysiological measurements involve one or more transitions from one solution to another. The most familiar one, known from the chemistry lab, is the transition from the filling fluid of a pH reference electrode to the external medium. In electrophysiology, one has to deal with the transitions from the microelectrode filling fluid to either the cytosol of an impaled cell or the extracellular body fluid or saline. In all these cases, the transition between two electrolyte solutions with different compositions is known as a liquid junction. Here, LJP (diffusion potential)[1] develops, despite the fact that the different ion species are allowed to diffuse freely across the boundary. The potentials arise because different ion species move at different speeds through the solute, thereby creating a slight separation of charges. The ion mobility, stated in Chapter 1, is the quantity that determines the magnitude of the junction potential. It can be noticed from the table that K^+ and Cl have closely matched ion mobilities. This is one of the main reasons why KCl is so important in electrochemical measurements.

[1]Note that some authors call a membrane potential that arises by passive ion permeabilities rather than by ion pump activity a diffusion potential. Therefore, the former name is less ambiguous.

The simplest liquid junction consists of a single salt at two concentrations. In this case, LJP E_l follows from the following equation, derived by Nernst:

$$E_l = \frac{RT}{F}\left[\frac{u^+ - u^-}{u^+ + u^-}\right]\ln\frac{c_1}{c_2}$$

As an example, the LJP between solutions of 0.1 M KCl and 0.01 M KCl is 1.12 mV. NaCl solutions cause larger LJPs because the mobilities of Na^+ and Cl ions are more different from one another: taking 0.1 M versus 0.01 M again, $E_l = 12$ mV.

For more complex junctions, the LJP can be computed with the aid of Henderson's equation:

$$E_l = \frac{RT}{F}\left[\frac{\sum_i \left\{\frac{u_i}{z_i}(c_{2i} - c_{1i})\right\}}{\sum_i \{u_i(c_{2i} - c_{1i})\}}\right]\ln\frac{\sum_i u_i c_{1i}}{\sum_i u_i c_{2i}}$$

The dissociation of water in H^+ and OH^- must be taken into account to approximate the true LJPs as closely as possible. Note that the pH of demineralized water and of unbuffered, neutral solutions is about 5 rather than 7 because CO_2 from the air is dissociated partially into hydrogen (H^+) and bicarbonate (HCO_3^-) ions.

The table below states LJPs of a number of junctions that can show up in the electrophysiological practice. The abovementioned factors have been taken into account.

It must be kept in mind that the computed values are approximations again, mainly because the mobilities are known for very dilute solutions only, whereas in electrophysiology, one uses concentrations of 0.1 M and higher.

From the table, it is clear that, in general, LJPs involving 3 M KCl are relatively low, whereas lower concentrations of KCl cause relatively higher potentials (the situation given for 1 mM $CaCl_2$ is only an occasional exception: for other concentrations of $CaCl_2$, the situation might be reversed again).

Solution 1		**Solution 2**	**E_l ($E_1 - E_2$)**
3 M KCl	|	0.1 M KCl	1.66 mV
3 M KCl	|	0.15 M NaCl	0.89 mV
0.1 M KCl	|	0.15 M NaCl	−5.4 mV
3 M KCl	|	1 mM $CaCl_2$	3.68 mV
0.1 M KCl	|	1 mM $CaCl_2$	1.21 mV
3 M KCl	|	Demineralized water	5.69 mV
0.1 M KCl	|	Demineralized water	40.3 mV
0.15 M NaCl	|	Demineralized water	44.1 mV

The beneficial influence of high KCl concentrations can be explained by analyzing Henderson's equation. The middle factor, involving the sum of concentration differences multiplied by the mobility, causes the salt with the highest concentration to dominate the LJP. In other words, choosing a high concentration of a salt where cation and anion have almost equal mobilities helps to keep LJPs low. This is the main reason that KCl solutions of 3 M or higher are used so often in electrochemical chains containing liquid junctions. By the same argument, KCl is also used as the main filling fluid of glass microelectrode pipettes. Here, a second reason to use concentrations of at least 3 M is that it helps keeping the tip resistance as low as possible. Occasionally, however, the leakage of potassium or chloride from the tip would have an unwanted influence on the cell one is recording from. In these cases, other salts can be used, such as LiCl if potassium must be avoided, or KAc (acetate) if chloride must be avoided. Unfortunately, these filling fluids cause higher LJPs. The mobilities of NH^{4+} and NO_3^- are matched better but are too poisonous for many tissues.

MEMBRANE POTENTIALS

In general, the object of electrophysiological experiments consists of membrane potentials and changes thereof. More precisely stated, a membrane potential is a potential difference, between the inside and the outside of a biological cell.

The magnitude of membrane potentials can be derived from the case of the liquid junction. For the simple case of one permeant ion, the equation of Nernst stated above can be used, taking either u^+ or u^- as zero. The equation then reduces to the familiar one:

$$E_{eq} = -\frac{RT}{zF} \ln \frac{c_{in}}{c_{out}}$$

where the potential E_{eq} may be called an equilibrium potential because the situation reaches a thermodynamic equilibrium. In addition to the familiar constants, c_1 and c_2 are the concentrations, more precisely the activities, of the permeant ion at the inside and outside of the membrane, respectively. For single-valued ions ($z = 1$) such as K^+ and Na^+, the Nernst equation amounts to the well-known 58 mV per decade of concentration ratio at room temperature. At the body temperature of warm-blooded animals, the value is 61.5 mV/decade.

DERIVATION OF THE EQUILIBRIUM POTENTIAL

Simple diffusion of an (uncharged) molecular species in one dimension is given by Fick's law:

$$J = -D\frac{\partial C}{\partial x}$$

where J is flux; C, concentration; x, distance; and D, the diffusion constant.

Electrodiffusion, which describes the diffusion of charged particles like ions in an electrical potential gradient, takes an extra voltage-dependent term. With V, the electrical potential; u, the mobility of the particle; and z, its charge:

$$J = -D\frac{\partial C}{\partial x} - u \cdot z \cdot C\frac{\partial V}{\partial x} \tag{5.1}$$

According to the Stokes–Einstein equation, D and u are related by:

$$D = u\frac{RT}{F}$$

with T, the absolute temperature; R, the gas constant; and F, Faraday's constant. Combining the latter two equations gives the Nernst–Planck equation of electrodiffusion:

$$J = -u\left(\frac{RT}{F} \cdot \frac{\partial C}{\partial x} + zC \cdot \frac{\partial V}{\partial x}\right) \tag{5.2}$$

In equilibrium the net flux, J, equals 0 and therefore:

$$\frac{\partial V}{\partial x} = -\frac{RT}{zF} \cdot \frac{\partial C}{C\partial x} = -\frac{RT}{zF} \cdot \frac{\partial \ln C}{\partial x}$$

Integration yields the Nernst equilibrium equation:

$$V_1 - V_2 = \frac{RT}{zF} \cdot \ln\left(\frac{C_2}{C_1}\right)$$

$V_1 - V_2$ is the transmembrane equilibrium potential, given the intra- and extracellular concentrations of an ion species, C_1 and C_2, respectively.

THE REVERSAL POTENTIAL

Eq. (5.2) is difficult to solve, if at all, without simplifying it to some extent. Goldman, Hodgkin, and Katz have done so by assuming a constant electrical field, i.e., V changes linearly across a slab of width d that is voltage-clamped at the borders.

Hence, $\frac{\partial V}{\partial x} = \frac{V_1 - V_2}{d} = \frac{V_m}{d}$ for all x within the slab. Then Eq. (5.2) becomes:

$$J = -u\frac{RT}{F} \cdot \frac{\partial C}{\partial x} - u\frac{zCV_m}{d}$$

Rewriting gives:

$$\frac{\partial C}{\frac{-JF}{uRT} - \frac{zFCV_m}{dRT}} = \partial x$$

which yields after integration:

$$\frac{-dRT}{zFV_m} \cdot \ln\left(\frac{\frac{zFV_mC_1}{dRT} + \frac{JF}{uRT}}{\frac{zFV_mC_2}{dRT} + \frac{JF}{uRT}}\right) = d$$

with C_1 and C_2, the concentrations at the borders of the slab. Exponentiation gives:

$$\frac{\frac{zFV_mC_1}{dRT} + \frac{JF}{uRT}}{\frac{zFV_mC_2}{dRT} + \frac{JF}{uRT}} = \exp\left(-\frac{zFV_m}{uRT}\right)$$

After a lot of rewriting, the result representing the flux of a single ion species across a voltage- and concentration-clamped slab is:

$$J = -\frac{uzFV_m}{d} \cdot \frac{C_1 - C_2 \exp\left(-\frac{zF}{RT}V_m\right)}{1 - \exp\left(-\frac{zF}{RT}V_m\right)}$$

The current density, I, is obtained from the ion flux, J, by multiplication with z × F:

$$I = -\frac{uz^2F^2V_m}{d} \cdot \frac{C_1 - C_2 \exp\left(-\frac{zF}{RT}V_m\right)}{1 - \exp\left(-\frac{zF}{RT}V_m\right)}$$

The (ion) permeability, P, is defined as: D/d, with D the diffusion coefficient. Combining the definition with the Stokes–Einstein equation above gives the relation between permeability, P, and mobility, μ:

$$P = \frac{uRT}{d}$$

Replacing μ/d in the flux equation by P_i/RT, where P_i is the permeability of ion species i and indexing the flux, charge and concentrations gives:

$$I = -\frac{P_iz_i^2F^2V_m}{RT} \cdot \frac{C_{i1} - C_{i2} \exp\left(-\frac{z_iF}{RT}V_m\right)}{1 - \exp\left(-\frac{z_iF}{RT}V_m\right)}$$

The net current of all ion species combined equals 0 at equilibrium (i.e., at the current reversal potential) or:

$$\sum_i I_i = 0$$

and hence:

$$\sum_i \frac{P_i z_i^2 F^2 V_m}{RT} \cdot \frac{C_{i1} - C_{i2} \exp\left(-\frac{z_i F}{RT} V_m\right)}{1 - \exp\left(-\frac{z_i F}{RT} V_m\right)} = 0$$

If only the monovalent ions K, Na, and Cl are considered, z_i^2 is always 1. Noting that for anions it suffices to interchange C_1 and C_2 to keep the equation homogeneous with respect to the sign of z in the exponential because $\frac{C_1 - C_2 \exp(a)}{1 - \exp(a)} = \frac{C_2 - C_1 \exp(-a)}{1 - \exp(-a)}$:

$$P_K \left\{C_{K1} - C_{K2} \exp\left(-\frac{FV_m}{RT}\right)\right\} + P_{Na} \left\{C_{Na1} - C_{Na2} \exp\left(-\frac{FV_m}{RT}\right)\right\}$$
$$\ldots + P_{Cl} \left\{C_{Cl2} - C_{Cl1} \exp\left(-\frac{FV_m}{RT}\right)\right\} = 0$$

$$P_K C_{K1} + P_{Na} C_{Na1} + P_{Cl} C_{Cl2} = \{P_K C_{K2} + P_{Na} C_{Na2} + P_{Cl} C_{Cl1}\} \exp\left(-\frac{FV_m}{RT}\right)$$

or:

$$\frac{P_K C_{K1} + P_{Na} C_{Na1} + P_{Cl} C_{Cl2}}{P_K C_{K2} + P_{Na} C_{Na2} + P_{Cl} C_{Cl1}} = \exp\left(-\frac{FV_m}{RT}\right)$$

After taking the natural logarithm of both sides and inversion of the quotient to eliminate the negation in the exponent, the Goldman equation for monovalent ions, relating the intra- and extracellular ion concentrations to the reversal potential, $V_m(I = 0)$, is obtained:

$$V_m = V_1 - V_2 = \frac{RT}{F} \ln\left(\frac{P_K C_{K2} + P_{Na} C_{Na2} + P_{Cl} C_{Cl1}}{P_K C_{K1} + P_{Na} C_{Na1} + P_{Cl} C_{Cl2}}\right)$$

ION SELECTIVITY

Goldman's equation can be used to determine ion permeability ratios or ion selectivity of a given channel. Suppose we wish to determine the permeability ratio of Na over K of a cation channel. To do so, the cell containing the channels of interest is perfused both intracellularly and extracellularly with known concentrations of K^+ and Na^+ ions, and the reversal potential, V_{rev}, is measured using a voltage-clamp amplifier. Then according to the Goldman equation:

$$V_{rev} = \frac{RT}{F}\ln\left(\frac{P_K[K]_o + P_{Na}[Na]_o}{P_K[K]_i + P_{Na}[Na]_i}\right)$$

If the permeability ratio is $r = P_{Na}/P_K$ and hence $P_{Na} = r \times P_K$, it becomes, eliminating P_K:

$$V_{rev} = \frac{RT}{F}\ln\left(\frac{[K]_o + r[Na]_o}{[K]_i + r[Na]_i}\right)$$

from which, after exponentiation, r can be easily obtained. A special case is the situation in which only one cation species, say K^+, is perfused intracellularly, while the other, Na^+, is perfused extracellularly at the same concentration, also in the absence of other cations.

$$V_{rev} = \frac{RT}{F}\ln(r)$$

This gives:

$$r = \exp\left(\frac{FV_{rev}}{RT}\right)$$

ELECTRODES SENSITIVE TO H^+ AND OTHER IONS

Artificial membranes from glass or plastic may also develop membrane potentials, which have proved useful for ion activity measurement. The glass electrode used for the measurement of pH consists of a thin-walled glass sphere fitted with an Ag/AgCl electrode and filled with 0.1 N HCl. It was found early in the 20th century that a glass membrane, basically an insulator, is slightly permeable to H^+ ions and thus can serve as a pH-sensitive electrode. When the electrode is immersed in a solution of unknown pH, a Nernst potential of 58 mV per pH unit develops that can be used to measure the pH of the unknown solution.

The pH meter proper is in fact an electrometer (a voltmeter with a very high input impedance), with tick marks reading one pH unit per 58 mV. At pH = 1, the potential across the glass membrane is zero, since the hydrogen ion concentrations on both sides are equal. In this case, the potential of the reference electrode determines whether the potential measured will also be zero. Usually, the reference electrode consists of an Ag/AgCl electrode in a saturated KCl solution. Modern pH meters employ a combination electrode, in which the glass electrode and an AgCl reference electrode are combined into one probe (Fig. 5.11).

This principle of ion-selective membrane electrodes has been extended to other ions. First, types of glass were developed that have a higher permeability for Na^+ than for K^+ or the reverse. These can be used as Na-electrodes and K-electrodes, respectively, provided that the pH is buffered. This is necessary because the H^+ permeability is still higher than for

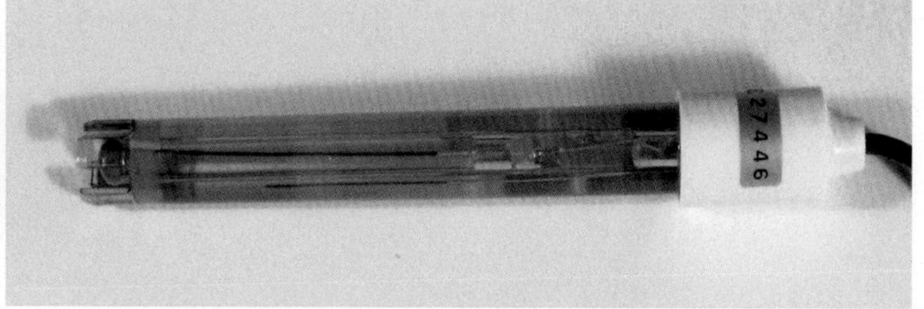

FIGURE 5.11 A pH electrode with built-in reference (AgCl) electrode.

Na^+ and K^+. Later, several oil-like substances were developed, where each shows a selective permeability for one ion species. These are known as liquid ion exchangers (LIX) and were developed for a number of ion species, such as hydrogen, potassium, chloride, calcium, and magnesium. To be used as ion-selective membranes, such substances are soaked up in thin plastic fabric membranes or put directly into the tip of a glass micropipette. The latter form allows intracellular ion activity measurements. Since the ion permeability, and hence the conductance, of a liquid ion exchanger is very low with respect to the concentrated salt solutions used to fill micropipettes, the tip resistance of an ion-selective electrode is very high: $10^9\ \Omega$ or more. Ion-selective microelectrodes may even have a resistance of $10^{12}\ \Omega$ or higher! Therefore, only electrometer-grade amplifiers employing MOSFETs can be used to measure the Nernst potentials of these electrodes.

THE GLASS MICROPIPETTE

Strictly speaking, a glass "microelectrode" is not an electrode at all: It is a salt bridge with a (sub-)microscopic tip diameter. Of course, the pipette fluid is in contact with an Ag or Ag/AgCl electrode mounted together in a fluid-filled holder fitted with a cable. The whole assembly behaves as an electrode.

Glass micropipettes are manufactured starting from thin-walled capillary tubes of 1–2 mm diameter by means of a pipette puller, a device made especially to pull pipettes with different properties to fulfill the needs for different types of recording. Intracellular recording and patch-clamp recording, for instance, demand different shapes and tip diameters.

In general, the form of a micropipette can be described as shaft, taper, and tip (Fig. 5.12, top). The angles of taper and tip vary between designs, but the tip angle is usually only a few degrees of arc. The most conspicuous property of a glass micropipette is its high resistance. Because of the conical shape, virtually all resistance resides in the tip. This can be computed with the following equation, which expresses the (partial) resistance R_{tip} of the pipette tip up to a distance of l from the tip.

$$R_{tip} = \frac{\rho}{\pi \cdot tg^2(\alpha)} \left(\frac{2tg(\alpha)}{d} - \frac{1}{l + \frac{d}{2tg(\alpha)}} \right)$$

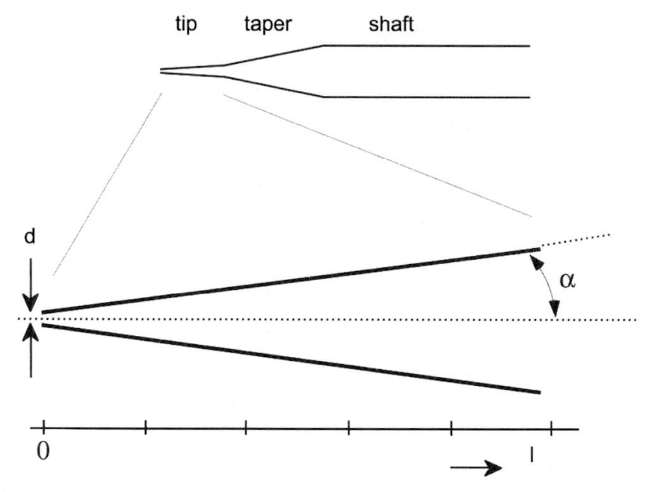

FIGURE 5.12 General outline of micropipette (top) and derivation (bottom) of tip resistance.

The resistivity of the filling fluid is designated as ρ (rho), the tip angle as α, and the tip diameter as d. The situation pertaining to the equation is illustrated in Fig. 5.12, bottom.

A graph of this equation is shown in Fig. 5.13. Fig. 5.14 shows a practical form of micropipette and holder.

Apart from the potential of the Ag/AgCl electrode and the tip resistance, both of which must be measured before use. The wall of a glass microelectrode brings along stray capacitance shunting the signal to a certain extent. The capacitance value depends on the wall thickness and the perimeter and since both thickness and perimeter are reduced toward the tip, the capacitance per unit length is approximately constant.

Finally, glass micropipettes have an annoying habit to develop a so-called tip potential. This is an added potential jump at the tip that occurs often after impalement of a cell. It is thought to arise by the partial clogging of the tip with large molecules such as proteins or

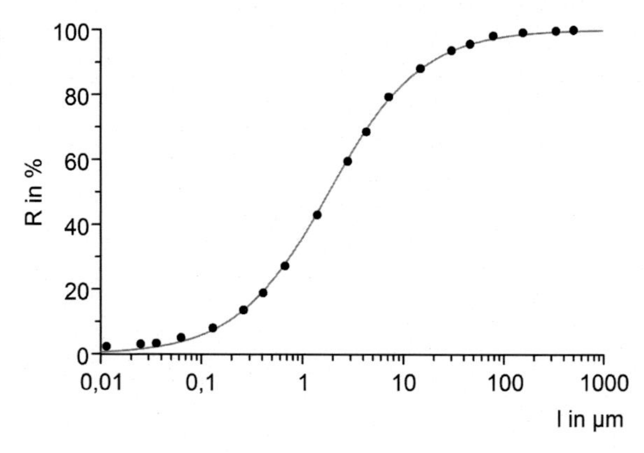

FIGURE 5.13 Graph of (relative) resistance (R) contribution versus length (l).

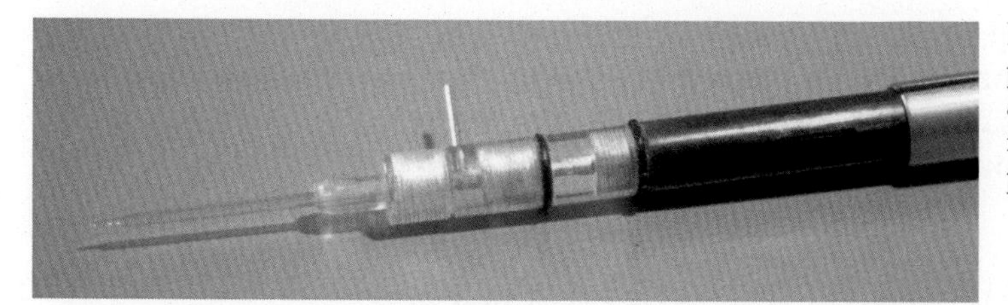

FIGURE 5.14 A glass micropipette in a holder. The thin side tube allows to apply pressure (tracer injection) or suction (membrane patch attachment).

cell fragments. Tip potentials are very capricious and can amount to 30 mV. They develop only at tip diameters below about 0.5 μm and so can be assessed after an experiment by breaking off the tip to a diameter well over this size.

PATCH PIPETTES

The procedure to make patch electrodes is very similar to pulling microelectrodes. Often a multistep electrode puller is used such that the taper of the pipette is short, which makes it easier than long-tapered pipettes to back-fill and to remove air bubbles. The pipettes are made on the day of the experiment. If they are kept too long, it is difficult, if not impossible, to make good seals with the cell membrane, possibly due to the slow hydration of the glass. The wall of the tip may be enlarged for improved sealing using a microforge after pulling. This step is not essential and is often omitted. As the tip of the pipette is relatively large (up to a few micrometers), ions and small molecules in the pipette solution interchange freely with the cell contents. It takes small fluorescent molecules like fluorescein or Fura only a few seconds after rupture of the cell membrane to spread from the soma of a cerebellar Purkinje cell to far into the dendritic arborization. Ions are likely to diffuse even faster into the cell. Therefore, the patch pipette is filled with an iso-osmotic solution (typically 150 mM KCL, 1 mM Mg^{2+}, a pH buffer such as HEPES, and a Ca^{2+} buffer such as EGTA). If filled with this solution or a solution of equal ionic strength, patch pipettes should have a resistance between 1 and 10 MΩ, depending on the size of cell to patch. The pipette resistance can be estimated by applying air pressure to the interior while keeping the pipette tip in 95% methanol. The pressure required to expel air bubbles from the tip increases with electrical resistance, 2 bar (200 kPa) corresponding approximately to 5 MΩ. The presence of divalent ions (usually Mg^{2+}) in the pipette solution is important in order to obtain and maintain a seal.

When preparing pipettes for excised patches, it is, in general, necessary to reduce capacitive currents (which are in the order of nA) to be able to measure the unitary currents of interest (which are in the order of pA). This is done by coating the glass pipettes with a resin such as Sylgard, excluding the last 100 μM close to the tip. The syrupy liquid is easily applied to the electrode using a fire-polished "Pasteur" pipette and then hardened by heating with a microforge. The same procedure also reduces noise in the recording due to random movements of charges in the glass (see Noise in Chapter 3).

Subsequently, the patch electrode is filled in two steps. First the tip is filled by suction, usually by fixing the pipette on a syringe and pulling the piston. Second, the pipette is backfilled using either a syringe needle or a plastic pipette tip that is pulled over a Bunsen burner to a diameter that can enter the glass patch pipette and reach down to the taper. Air bubbles are then removed by tapping lightly on the pipette with a fingernail. A more sophisticated way to remove air bubbles is to pass a bolt along the side of the pipette in a kind of "sawing" movement, the pitch of the bolt causing the pipette to vibrate.

THE SEMIPERMEABLE PATCH

One of the advantages of the whole-cell suction pipette technique is that the ion composition of medium of the cell interior is well defined, as ions and molecules in the cell equilibrate rapidly with the pipette contents. Although it is a benefit to be able to control ion gradients over the plasma membrane, the washout of the enzymatic machinery of the cell and its second messengers is not always a blessing. It may be especially a nuisance if one intends to study the regulation of ion channels by intracellular factors. Washout of larger molecules such as proteins may be delayed by using pipettes with small tips, degrading at the same time voltage-control. Even then, smaller molecules such as ATP and cAMP are still rather effectively removed from the cytosol.

To overcome this difficulty, pore-forming proteins may be included in the pipette solution (Fig. 5.15, left). After having made a seal with the plasma membrane (the "cell-attached" configuration), these proteins gradually insert themselves in the membrane allowing electrical contact and exchange of ions with the cell interior, while keeping larger molecules inside the cell. The molecular cut-off size depends on the choice of the pore-forming protein. The α-toxin from *Staphylococcus aureus* forms pores that let pass ions and molecules up to 1000 Da. Digitonin permits the exchange of structures the size of a mitochondrion. The most popular pore-forming substance among electrophysiologists is nystatin. It conducts

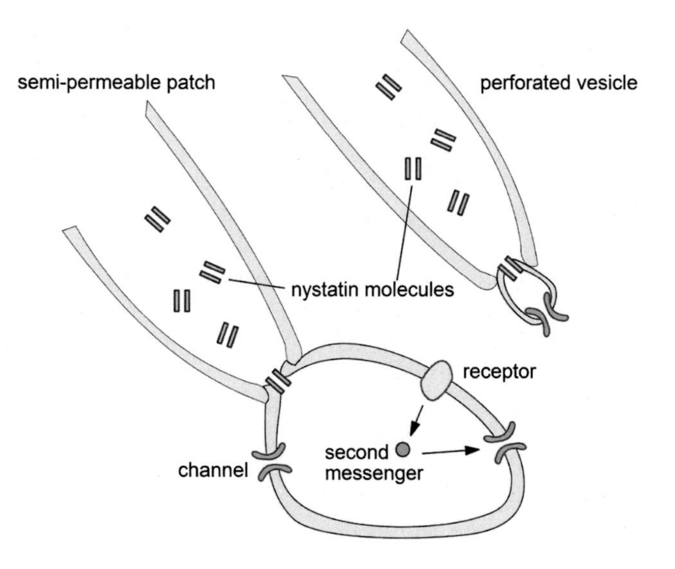

FIGURE 5.15 Two methods for making semipermeable patches.

monovalent ions but bars (at least to a large extent) the way to Ca^{2+}. Nucleotides cannot pass the nystatin pore either. Unfortunately, nystatin is not easy in its use. It is sensitive to light and almost insoluble in water. Moreover, it prevents formation of a seal. For this reason, the tip of the pipette should be filled with a solution devoid of nystatin and backfilled with a freshly sonicated solution containing (typically 100 μg/mL) nystatin. Then the seal is made (and microscope illumination switched off), after which nystatin slowly diffuses down the tip and integrates itself in the patch membrane. Its integration may be followed by monitoring the membrane capacitance that should increase during a period of about 20–30 min and then remain stable.

An interesting variation on the semipermeable patch technique is the perforated vesicle (Fig. 5.15, right). Rather than making an inside-out patch, a vesicle is excised from the cell using a nystatin back-filled pipette. This results in a configuration similar to the inside-out patch but with cytosolic constituents concentrated close to the plasma membrane.[5] This technique is not for the impatient or the weak of nerve.

GROUND ELECTRODES

Any ground electrode needs to have a low impedance, and so they are usually made relatively large. A large Ag/AgCl pellet is sometimes used to ground the preparation, but more often a metal wire, plate, or wire mesh serves as the ground connection (Fig. 5.16). The use of naked metals may seem odd at first sight, since we saw earlier that metal surfaces are polarized electrodes that generate relatively large and variable polarization voltages. However, sufficiently large metal electrodes have fair DC properties, and this can be understood by the following argument. As we discussed earlier, small metal surfaces have random potentials, centered on a certain mean or expected value. Since the deviations from the mean are random variables, the average value of a number of such potential values will be closer to the mean than the individual values.

Since a large electrode surface area can be considered to be the average of a large number of smaller areas, a large metal plate has a lower and more stable polarization potential than a

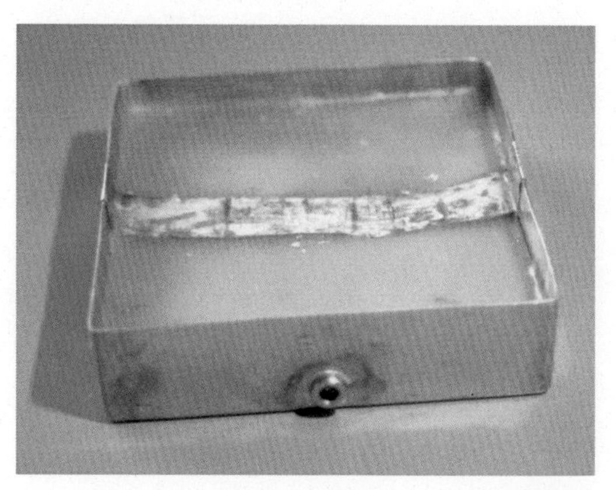

FIGURE 5.16 A silver strip ground electrode in an insect preparation tray.

smaller one. In addition, a silver surface will be chlorinated spontaneously during usage with chloride containing solutions, causing the DC properties to improve over time. This effect is also used in the "ground" electrode of a building, which consists usually of a long copper tube or pole, dug deeply into the ground (below the subsoil water level). For very critical DC recordings, a nonpolarized ground electrode is still recommended, however.

References

1. Landolt H, Landolt-Börnstein NA, Börnstein R. *Physikalisch-Chemische Tabellen*. Heidelberg: Springer-Verlag; 1912. Berlin Heidelberg.
2. Parsons R. *Handbook of electrochemical constants*. London: Butterworth; 1959.
3. *Handbook of chemistry and physics*. 82nd ed. Boca Raton: CRC-Press; 2001.
4. Berman JM, Awayda MS. Redox artifacts in electrophysiological recordings. *Am J Physiol Cell Physiol* 2013;**304**(7):C604–13.
5. Levitan ES, Kramer RH. Neuropeptide modulation of single calcium and potassium channels detected with a new patch clamp configuration. *Nature* 1990;**348**(6301):545–7.

CHAPTER

6

Volume Conduction: Electric Fields in Electrolyte Solutions

OUTLINE

Sources of electric current in water or in salt solutions give rise to stationary electric fields. The theory of electricity was developed in the 18th century with static forms of electricity. Charges in space, maintaining electrical potential differences, were kept from leaking away by dry air and by insulators like glass or ceramics. In a conducting medium, however, an electric current would flow and discharge any static source within a fraction of a second. Therefore, to generate an electric field in a fluid, we have to supply current continually (DC and/or AC). The mathematical descriptions of electric fields in water and in an insulating space are similar, but in a conducting medium, current sources replace the charges in a static field, and current densities replace field strengths.

Three main forms of stationary electric field can be distinguished and can be used or encountered in the electrophysiological practice:

1. Homogeneous field
2. Monopole field
3. Dipole field

These are illustrated in Fig. 6.1 and will be described briefly below.

Introduction to Electrophysiological Methods and Instrumentation, Second Edition
https://doi.org/10.1016/B978-0-12-814210-3.00006-2

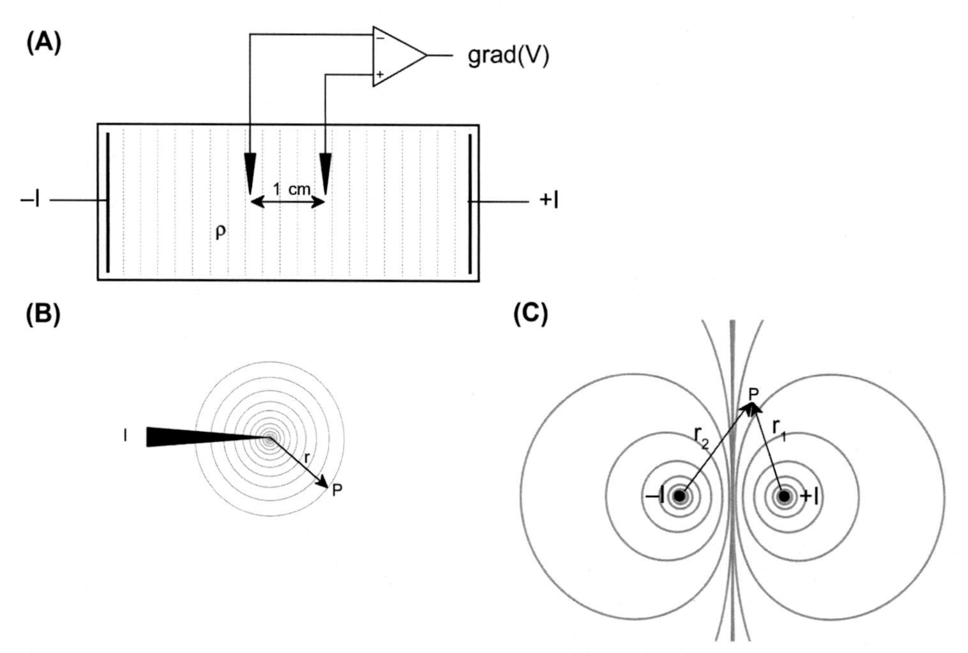

FIGURE 6.1 (A) Homogeneous, (B) monopole, and (C) dipole fields.

HOMOGENEOUS ELECTRIC FIELD

A homogeneous field arises when an electric current is fed through two parallel electrodes that cover the short sides of an elongated tray. This configuration, sometimes called an electrolytic trough, is shown in Fig. 6.1A. It can be used to illustrate the quantities used to describe electric fields in conducting fluids. Let us assume that we feed a direct current (of strength I) through the trough from right to left, by making the right electrode positive with respect to the left one. Because of the constant cross-section of the tank and because the electrodes fit the short sides entirely, current flows in straight lines, parallel to the long sides of the trough. Since the properties of the field are the same throughout the tank, it is called a homogeneous field.

The quantities that describe the electric field in the liquid are analogous to the quantities that describe current through an object, viz. an electrical resistance, but expressed per spatial unit. In analogy to resistance R, we have the resistivity ρ (rho), which is usually expressed in Ωcm. In analogy with current, we have the current density J, expressed in A/cm^2. Finally, the voltage gradient, $grad(V) = \partial V/\partial x$ expressed in V/cm, replaces the voltage. Resistivity, current density, and voltage gradient thus describe the properties of a unit cube of 1 cm side, which has $R = \rho$, $I = J$, and $U = -\partial V/\partial x$. Obviously, these quantities are related via an analogue of Ohm's Law:

$$U = IR \leftrightarrow -\frac{\partial V}{\partial x} = J\rho$$

To describe the potential, we need a reference value. In a practical circuit, one of the electrodes would be grounded, but theoretically, it is more attractive to take zero at the middle of the trough (note that otherwise the electrode polarization would introduce an error). Thus, the potential (U) can be determined as the voltage between a measuring electrode and a central reference electrode, l cm apart, and follows from:

$$U = -l\frac{\partial V}{\partial x} \quad \text{or} \quad U = J\rho l$$

Lines connecting points of equal potential, or equipotential lines, are parallel to the electrodes that generate the field. The homogeneous electric field can be used to test and calibrate one's measurement setup.

THE MONOPOLE FIELD

The situation at an isolated electrode, or monopole, is completely different (Fig. 6.1B). A monopole field arises, theoretically, when the reference, or ground electrode, is situated at infinity (and is infinitely large). In practice, to approximate a monopole field sufficiently, the ground electrode has to be far away only with respect to the distances involved in the measurements. For example, the field around a point-shaped electrode in a $10 \times 10 \times 2\ cm^3$ tray is sufficiently monopole-like over a radius of at least a few millimeters. The electrode itself must, theoretically again, be infinitely small, i.e., pointlike. If one uses the tip of a glass micropipette, the monopole situation is sufficiently approximated, even close to the tip.

According to a general principle first established by Gauss, the current injected, I, spreads out in all directions and the current leaving a closed surface around the source must be identical to the current injected. Now, if the medium is isotropic, spherical symmetry may be assumed. The current density, $\vec{J}$ at a distance r from the tip of the electrode is then given by:

$$\vec{J} = \frac{I}{4\pi r^2}\hat{r}$$

The arrow over the J indicates that $\vec{J}$ is a vector, i.e., it has both a size and a direction, r is the radius of a sphere with surface $4\pi r^2$, and $\hat{r}$ is a unit vector pointing in the same direction as $\vec{J}$ (i.e., from the center of the sphere inward or outward depending on the sign of the current). If ρ is the resistivity of the medium, then according to Ohm's law as we have seen above:

$$\vec{J} = \frac{\vec{E}}{\rho} = \frac{-1}{\rho}\frac{\partial V}{\partial r}$$

where $\vec{E}$ is the electrical field and $\partial V/\partial r$ is the potential gradient induced by the current. Note that V, the potential, is a scalar and not a vector. Combining the two equations gives:

$$\frac{\rho I}{4\pi r^2} = -\frac{\partial V}{\partial r}$$

In order to obtain an equation in V rather than its gradient, we need to integrate:

$$\int \frac{\rho I}{4\pi r^2} dr = -\int \frac{\partial V}{\partial r} dr$$

which simplifies to:

$$V(r) = -\frac{\rho I}{4\pi} \int \frac{1}{r^2} dr$$

and hence,

$$V(r) = \frac{\rho I}{4\pi r}$$

THE DIPOLE FIELD AND CURRENT SOURCE DENSITY ANALYSIS

A dipole can be considered as two monopoles, the first of which carries a current +I. The other one "absorbs" the same current and hence carries a current −I. The electrodes are also known as source and sink, respectively. To compute the voltage gradient, current density, or potential at any point, the contributions of both poles simply sum, taking the respective distances r_i into account.

Therefore, current density, voltage gradient, and potential at any point p follow from:

$$J_P = \frac{I}{4\cdot\pi\cdot r_1^2} + \frac{I}{4\cdot\pi\cdot r_2^2} \qquad -\text{grad}(V)_P = \rho\cdot J = \frac{\rho}{4\cdot\pi}\left(\frac{I}{r_1^2} + \frac{-I}{r_2^2}\right)$$

$$V(p) = \frac{\rho}{4\cdot\pi}\left(\frac{I}{|r_1|} + \frac{-I}{|r_2|}\right) \tag{6.1}$$

This leads to the pattern shown in Fig. 6.1C.

Since in the middle between the poles the contributions of the two poles cancel, the line through the center of the dipole, perpendicular to the line connecting the poles, has zero potential. This midline is also shown in Fig. 6.1C. Close to the electrodes, the equipotential lines are almost circular (monopole-like) and too closely spaced to be distinguished in the figure. Multipole fields can be considered as a combination of poles, by extension of the dipole equations. Dipole fields arise in a number of situations, of both natural and technical origin.

FIGURE 6.2 Iron filings showing magnetic field lines of a (hidden) bar magnet.

However, because of their complex form, dipole fields are less suited for testing and calibration purposes. Note that, properly speaking, current density and voltage gradient are vector quantities, having both a magnitude and a direction. For simplicity, the values given here are the maximum values that pertain to the mentioned directions. In directions perpendicular to the stated ones, both current density and voltage gradient are zero. To be more precise, they follow a cosine law: As an example, the current density at an angle of 45 degrees to the direction of the field lines, in the abovementioned homogeneous field, would be J cos(45°), or about 0.71 J.

The shape of the dipole field is best known from that of a bar magnet. In fact, the electric and magnetic dipole fields share a common mathematical form. In Fig. 6.2, the magnetic field lines of a hidden bar magnet are made visible by the layer of iron filings.

MULTIPOLE FIELDS AND CURRENT SOURCE DENSITY

The situation where there are many sources and sinks is a generalization of Eq. (6.1). These current sources, I_k, located at $\vec{r_k}$, contribute to the potential at $\vec{r}$ by summing:

$$V(\vec{r}) = \sum_k \frac{\rho I_k}{4\pi\left|\vec{r} - \vec{r_k}\right|} \tag{6.2}$$

It is easy to calculate the field potential distribution in space with Eq. (6.2) if the distribution of current sources and sinks is known. This has been done as an exercise in Fig. 6.3A and B for one dimension. However, the problem is usually the inverse: given the field potential

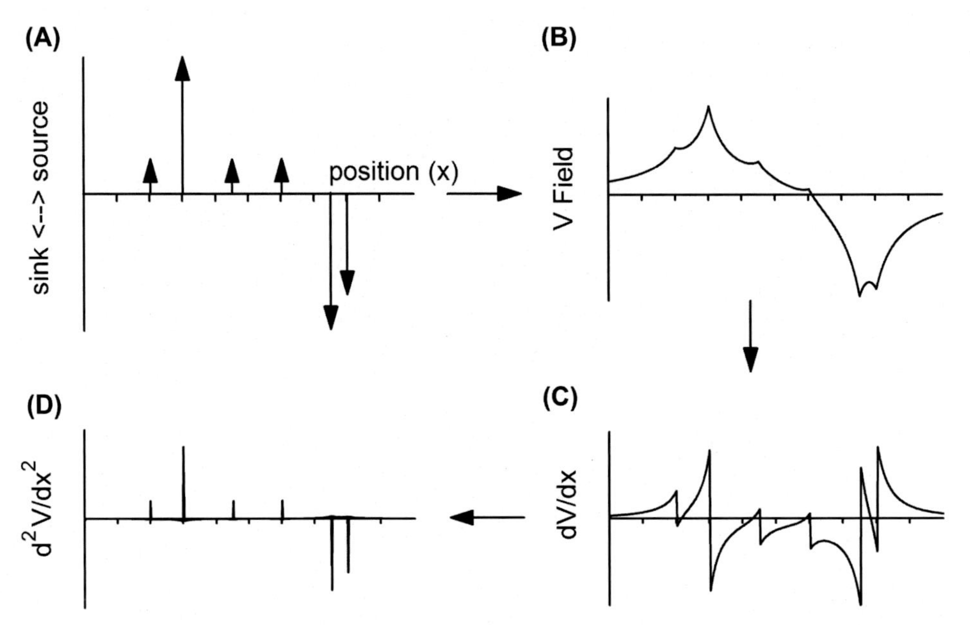

FIGURE 6.3 Simulation of field potentials and simplified current source density analysis in one dimension (x). Suppose six sources and sinks are distributed as in (A). The length of the *arrows* is proportional to the current. (B) The field potential (V) at each point along the x-axis can be calculated with Eq. (6.2). If the field potentials at an ensemble of points are known, the location of current sources and sinks can be estimated by taking the second derivative of V (D). The first derivative is shown in (C). Note that the second derivative of V is proportional to the original current source distribution in (A).

distribution where are the current sources and sinks located? That is because these indicate, for example, active brain regions and sources of EEG potentials (see Chapter 9).

The methods of analysis that solve this problem are collectively called current source density (CSD) analysis. Unfortunately, CSD math is complicated and not very straightforward.[1,2] We will therefore not attempt to develop it here but instead give an important result that can be obtained:

$$C(\vec{r}) = \frac{-1}{\rho}\left\{\frac{\partial^2 V}{\partial x^2} + \frac{\partial^2 V}{\partial y^2} + \frac{\partial^2 V}{\partial z^2}\right\}$$

This is the Poisson equation, which states that if one takes the second partial derivative of the field potential with respect to the spatial coordinates, one obtains the current source density distribution, $C(\vec{r})$ (see Fig. 6.3C and D). The dimension of C is A/m^3, hence current per volume, which is not the same as current per surface or current density (J). C(r→) can be thought of as the current emanating from a small volume. In the above equation, ρ is supposed to be uniform in all spatial directions. If ρ(x,y,z) depends on spatial coordinates, the equation becomes somewhat more intricate. However, CSD analysis is useful and often applied even in the absence of any knowledge of eventual resistivity variations in the tissue. In that case, $C(\vec{r})$ cannot be deduced. For this reason, usually only the second derivative is taken and the result, expressed as V/m^2.

This description of the spatial distribution of electric currents in ion solutions ends our discussion of electrochemical processes in this and the two preceding chapters. More detailed information can be found in the excellent book on this subject by Bard and Faulkner.[3]

References

1. Einevoll GT, Kayser C, Logothetis NK, Panzeri S. Modelling and analysis of local field potentials for studying the function of cortical circuits. *Nat Rev Neurosci* 2013;**14**(11):770–85.
2. Nicholson C. Theoretical analysis of field potentials in anisotropic ensembles of neuronal elements. *IEEE Trans Biomed Eng* 1973;**20**(4):278–88.
3. Bard AJ, Faulkner LR. *Electrochemical methods: fundamentals and applications*. New York: Wiley; 2001.

CHAPTER

7

The Analysis Toolkit

OUTLINE

INTRODUCTION

In general, electrophysiological signals have to be processed and analyzed. However, there are many types of electrophysiological signals and recording techniques, so the methods to process and analyze them will be equally diverse.

Introduction to Electrophysiological Methods and Instrumentation, Second Edition
https://doi.org/10.1016/B978-0-12-814210-3.00007-4

For analysis, signals can be subdivided into the following broad categories: intracellular voltages, extracellular voltages, and transmembrane currents. One of the signals most frequently used and processed is the action potential, or spike.

Analysis might deal with the shape of the voltage or current, i.e., the amplitude and time course of a sensory or postsynaptic potential, an action potential, or the opening and closing of an ion channel. Such signals are called analog, in contrast with digital (finite-precision numeric) signals. Alternatively, for many purposes, each action potential may be considered as a point process, of which size and shape are taken for granted. Here, only aspects of the distribution in time are the all-important quantities. One might be interested in the average spike frequency, in stimulus-related occurrences of spikes, in simultaneous occurrences of spikes, and so on.

As an example, a photoreceptor cell, stimulated with light flashes, responds by a depolarization, which may be picked up by an electrode in (or near) the cell. The nerve fiber conducts a varying spike rhythm, which may be recorded too. Both signals need specific processing methods. This is illustrated in Fig. 7.1, where the receptor potential is averaged, and the spike series is processed into a so-called instantaneous frequency plot.

We will start with the most straightforward analysis techniques, applicable when both input and output of the structure under study can be considered as continuous functions of time, like the light intensity and the receptor potential in the above example. Specialized methods to analyze an action potential series (colloquially called a spike train) and methods

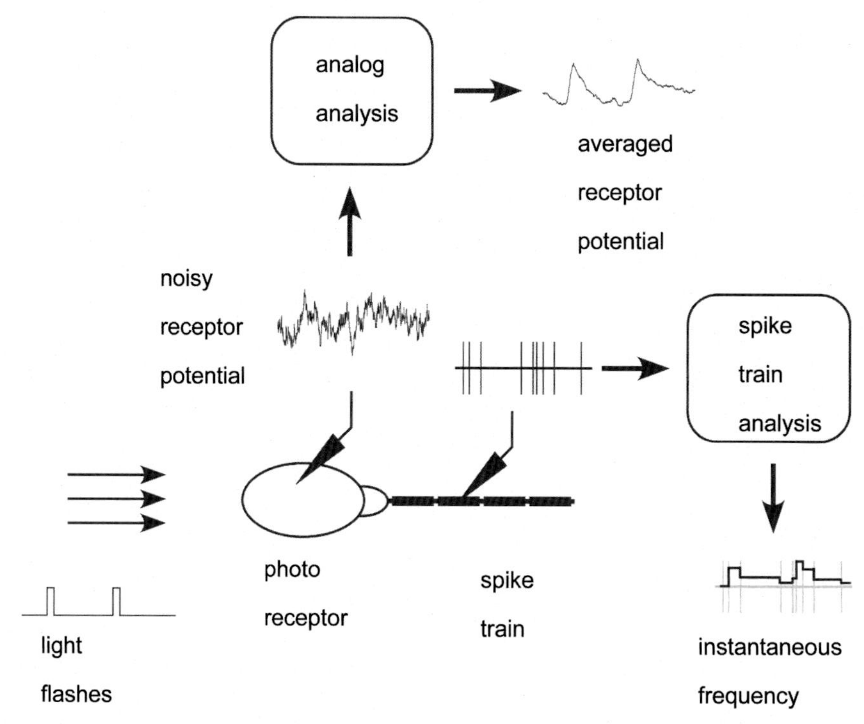

FIGURE 7.1 Signal analysis summarized for a photoreceptor preparation.

to analyze the opening and closing statistics used to characterize ion channel types will be treated later on.

SYSTEMS ANALYSIS

Definition: what we call a "system" is some part of the real world that is defined by the input(s) and output(s) studied. Systems, in general, may have multiple inputs and outputs. For simplicity, however, we will deal here only with systems having one input and one output.

The standard procedure of systems analysis is to feed an input signal into the system, record the output signal, and try to find a relationship between the two. The found relation is called a model of that system. In everyday language, hearing the notion of a model, we think of a scale model, reduced in size, such as a ship model, or sized up, like the models of atoms and molecules. In systems analysis, the model may be a physical one, such as an electronic filter that mimics some property of a neuron, or, more generally applicable, a mathematical model. The concomitant mathematical theory is called systems theory; we will discuss some of its basics later on.

The methods of systems analysis are by no means limited to electrophysiology, or even to physical science: The principles are equally useful for technical systems, in ecology or economics. To imagine the latter case, an input might be a change in interest rate, and the observed output, the development of investments.

The principle of systems analysis is illustrated in Fig. 7.2, left side.

The arrow from system to model reflects that the model is based on the found properties of the system.

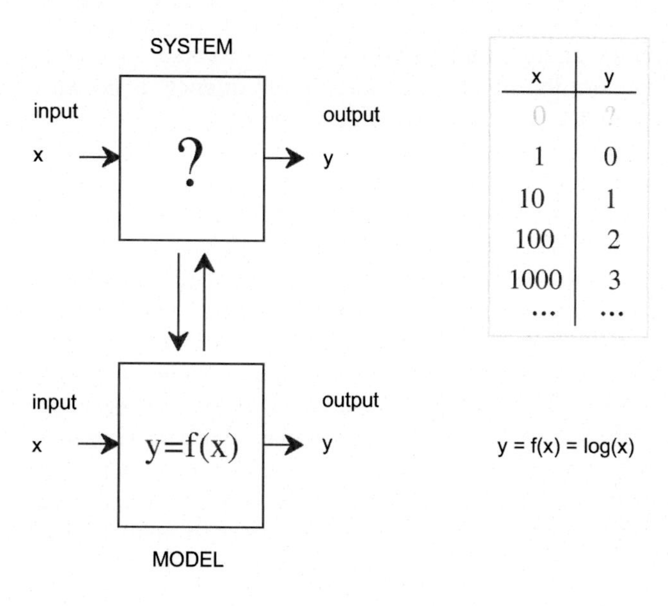

FIGURE 7.2 Systems analysis principle (left) and an example result (right).

Once a model has been formulated, it can be treated like the real system: feed it with an input signal and record the output. In most, if not all cases, however, the model will be neither sufficiently precise nor complete, which necessitates going back to study the real-world system. Therefore, the second arrow points back from model to system.

As an example of the systems-analytical practice, imagine the case of the photoreceptor cell again. The cell has been stimulated with light flashes of different intensities, while measuring the amplitude of the cell's response. The outcome could be as depicted in the short table of Fig. 7.2 (right): At a stimulus intensity of 10 lux, the potential is 1 mV; at 100 lux, it is 2 mV; and so on. This simple example is chosen to illustrate two main points:

1. The relation between the output and input quantities is derived tentatively from the results of an experiment, unless there is a fundamental reason to pick a specific mathematical function. Here, a logarithmic relation is an obvious choice, but its success is never warranted.
2. Once a model has been formulated (here $y = \log_{10}(x)$), it leads to predictions of results that would be obtained under circumstances different from the ones used in the experiment. If the model equation leads to unlikely results, the model must be replaced, or refined, by new measurements. In our example, the log relationship would lead to an impossible situation at zero illuminance, where the model predicts log(0). Returning to the real system, the actual response in the dark (stimulus zero) might give us the necessary correction to the model equation.

This ends our simple example. Because we compare the amplitudes of the input and output signals, this form of analysis is said to be performed in the amplitude domain. This is useful by itself, but systems theory is centered on time functions. The reason for this will become clear soon.

A crucial notion is that, in all but a few trivial cases, the input and output signals change in time, and the details of the time behavior reflect the most important aspects of functioning. Therefore, systems theory is centered on the analysis of time functions, which is said to occur in the time domain. As we will see later on, the inverse of time, viz. frequency, is equally important, and so much of the theory will deal with the frequency domain.

If we analyze a system, we may make some change, such as a raise in temperature, a light flash, an electrical pulse, and so on. All of them are functions of time, and so the input signal is given the symbol f(t). The output is also a function of time and is called g(t) (the g being simply the letter following f in the alphabet). This is illustrated in Fig. 7.3.

Systems analysis is the technique to get information about the action of a system, expressed as a relation between output and input. This system property will also be a function of time and is called the **transfer function** h(t). The transfer function reflects an important property of any system, viz. the way it responds to changes of the input, hence of some circumstance or quantity in the environment. Therefore, systems analysis is sometimes called dynamic systems analysis. If we would choose a simple, static component like a resistor as a system, the response to any input would follow from Ohm's law and hence would be rather uninteresting. However, most real-world systems, be it physical, chemical, or biological, show rather complex dynamical behavior, and so it will need some effort to analyze and describe.

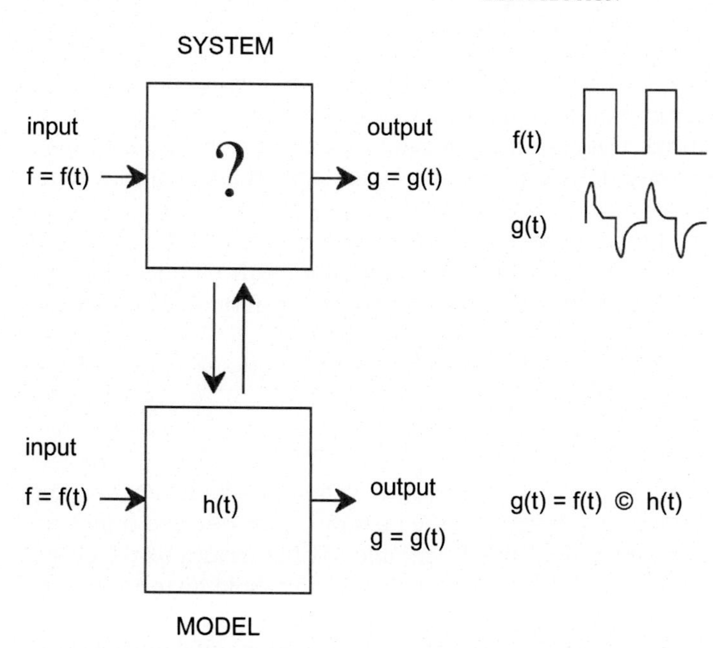

FIGURE 7.3 Systems analysis principle for time functions (left); the system function h(t) (lower left) and example input and output functions (upper right).

Thus, g(t) = some relationship between f(t) and the system behavior h(t). This relation will prove to be a "convolution," a mathematical term that needs further explanation. This is where the symbol © in Fig. 7.3 stands for.

CONVOLUTION

To explain the process of convolution, we will use the well-known low-pass RC filter.

If a very short pulse is fed into this circuit at $t = 0$, the output jumps to some value, taken here to be unity for simplicity. Immediately after the pulse, the capacitance starts to discharge, causing the output to decrease in an exponential way: $g(t) = 1 - e^{-t/\tau}$.

The pulse is "forgotten" slowly, and the longer one waits, the less the output reflects the original pulse strength (Fig. 7.4A). Principally, the pulse never dies out entirely, but in practice, after some time, we consider the output to be zero again.

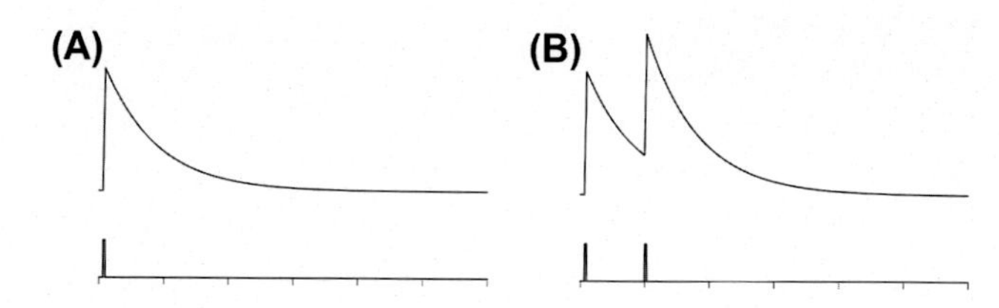

FIGURE 7.4 (A) An input pulse and the response of an RC filter. (B) The response to two successive input pulses.

If we administer a second pulse, the same process repeats, but in this case, the output voltage is the sum of the remnants of both pulses, taking their respective times of occurrence into account (see Fig. 7.4B). The summation of input amplitudes can be generalized to apply to any input signal, continuous in time. The output is the sum of all input signals, taking their times of arrival into account. For continuous time functions, the sum changes into a time-integral.

Apparently, we have two kinds of time functions intertwined, or "convolved": The input signal is a time function, and the "memory" function of the RC filter is also a time function.

To compute the output, we need to reckon with both time functions at once. This leads to the so-called convolution integral:

$$g(t) = \int_{-\infty}^{t} f(t) \cdot h(t - \tau) d\tau$$

This is a mathematical way to express the way in which two time functions are "sliding along one another." Since there are two time functions, the integral comprises two time variables, t and τ, respectively. The regular or "real time" variable t is the instant at which we want to know the output g(t), but this single value depends on the whole history, i.e., the time-integral from $-\infty$ to t. This is the meaning of the integration variable τ.

To get the flavor of convolutions, imagine two pieces of cardboard, each with a rectangular hole in it, stacked on top of each other, and positioned between the observer and a lamp (Fig. 7.5, left). In this situation, the total amount of light that will pass the cardboard barrier is determined by the amount of overlap between the two holes. If we slide the first piece of cardboard along the second one horizontally (x-direction), the amount of light grows slowly until the overlap is maximal and then diminishes again.

The amount of light l is determined by the convolution of the two hole functions. If the two rectangles are identical, the convolution is a triangle (top right figure). If one of the holes is wider than the other, there is a region of "slack," in which the light output remains constant. In that case, the convolution is trapezium shaped (bottom right figure).

The process of convolution is used so much in systems analysis that it is given a separate symbol. Originally, this was the asterisk (the symbol *), but since the advent of the computer,

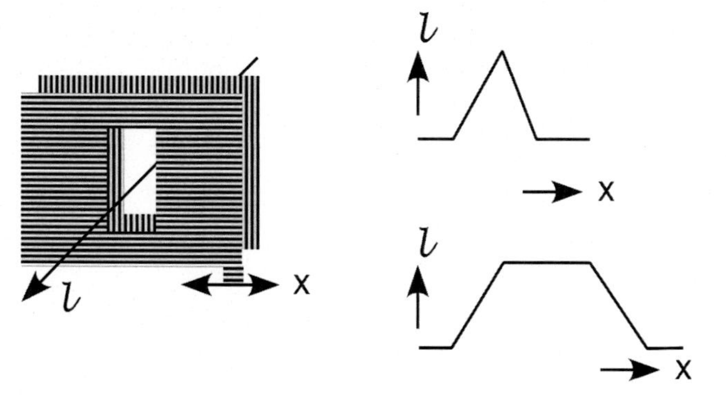

FIGURE 7.5 Convolution explained with a light ray and sliding cardboard masks.

the asterisk is often used as multiplication sign (to prevent confusion with the letter x). Therefore, we have chosen the © symbol here.

In the case of our low-pass filter, the convolution integral that has to be determined is between the input signal f(t) and the filter property h(t). It will be notated as follows:

$$g(t) = f(t) \copyright h(t)$$

Unfortunately, in all but the simplest cases, derivation of convolution integrals is tedious, even for people with sufficient mathematical skills. And things get still worse: to derive the transfer function of a system, the measured output signal and the input signal must be deconvolved:

$$h(t) = f(t) \copyright^{-1} h(t)$$

Here, the symbol $\copyright^{-1}$ stands for a deconvolution integral, which is the inverse operation of the convolution we saw before.

An attractive solution is to transform the functions into the frequency domain. In that case, a convolution is replaced by a simple multiplication, a deconvolution by a division.

How does one transform an entire time function into the frequency domain?

Transforming a single number from time to frequency domain is simple of course: inversion. If an alternating current has a period of 20 ms, its frequency is 1/0.02, or 50 Hz. The inverse transform, from frequency to time domain, is identical: inversion.

Here, however, we have to transform complete functions or signals between the two domains, and so we need other mathematical tools. These tools, or mathematical procedures, are known as integral transforms. The two most frequently used transforms are due both to 18th-century mathematicians: Laplace and Fourier. These will be explained below. Note that the convention used to distinguish time domain functions from their corresponding frequency domain counterparts is to notate time functions with lowercase letters (f, g, and h) and frequency functions with capitals (F, G, and H, respectively).

For both transform types, an inverse form exists that serves to transform functions backward, i.e., from frequency domain to time domain. These inverse transforms are almost identical to their respective forward counterparts. Although the Laplace and Fourier transforms resemble one another, their respective uses differ in practice. Laplace transforms are used mostly in analytical (i.e., algebraic) work, whereas the Fourier transform is used mostly in numerical analysis (i.e., calculations). This is shown in the table below.

	Laplace Transform	**Fourier Transform**
Mathematical form:	$F(s) = \int_0^\infty f(t)e^{-st}\,dt$ $s = \sigma + j\omega$	$F(j\omega) = \int_{-\infty}^{\infty} f(t)e^{-j\omega t}\,d(t)$
Signals:	Transient and periodical	Periodical
Time axis:	$0 < t < \infty$	$-\infty < t < \infty$

(Continued)

	Laplace Transform	Fourier Transform
Practical application:	Analytically (determining of transfer functions)	Numerically (computation of frequency characteristics)
Resources:	Table of Laplace transforms	Computer with FFT (Fast Fourier Transform) algorithm

We will discuss both transforms in more detail below.

THE LAPLACE TRANSFORM

The procedure for the application of Laplace transforms in order to derive the output of a system from the input and transfer functions is as follows:

f(t)	h(t)	g(t)
⇓	⇓	⇑
L	**L**	$\mathbf{L}^{-1}$
⇓	⇓	⇑
F(s)	H(s)	G(s)

Of course, usually the system transfer function is the unknown and must be derived from the input and output functions:

f(t)	h(t)	g(t)
⇓	⇑	⇓
L	$\mathbf{L}^{-1}$	**L**
⇓	⇑	⇓
F(s)	H(s)	G(s)

The inverse Laplace transform, shown between braces, is often omitted because we are satisfied to have the frequency domain transfer function, H(s). This is called the frequency characteristics and renders the system properties in an intelligible way.

At this stage, one may wonder whether the situation is improved at all: To replace the problematic convolutions and deconvolutions by more simple operations, we have to perform a number of forward and/or inverse Laplace transforms in addition. These transforms are likewise complex mathematical procedures. At a first glance, this action does not seem to help at all. The improvement, however, stems from two facts:

1. Many Laplace transforms have been derived already and are tabulated in books. This means that the task of deriving a specific Laplace transform reduces to finding it in the tables.

2. Since, in the frequency domain, functions can be combined using multiplication and division rather than (de)convolutions, complex functions not in the tables can be split up often into component functions that are tabulated.

In a way, this compares to an artist's palette on which the variety of colors needed can be derived from a small number of primary colors.

Below is a short list of Laplace transforms, necessary for the example illustrated subsequently.

Note that the variable of a Laplace-transformed time function is called s. This variable is a complex variable (see Appendix for complex numbers). The meaning of complex numbers, in general, and complex frequency is treated in Appendix in section Complex Numbers and Complex Frequency. For the moment, it suffices to consider s as the equivalent of frequency (in Hz). The variable t stands for time (in seconds), whereas a and c are constants.

Description	**Time Domain Function**	**Frequency Domain Function**
Ramp	$f(t) = ct$	$F(s) = \frac{c}{s^2}$
Step	$f(t) = c$	$F(s) = \frac{c}{s}$
Impulse	$f(t) = \delta$	$F(s) = 1$
Low-pass RC filter	$h(t) = e^{-at}$	$H(s) = \frac{1}{s + a}$
Output, see text	$g(t) = 1 - e^{-at}$	$G(s) = \frac{1}{s(s + a)}$

Example: a step input to a low-pass RC filter.

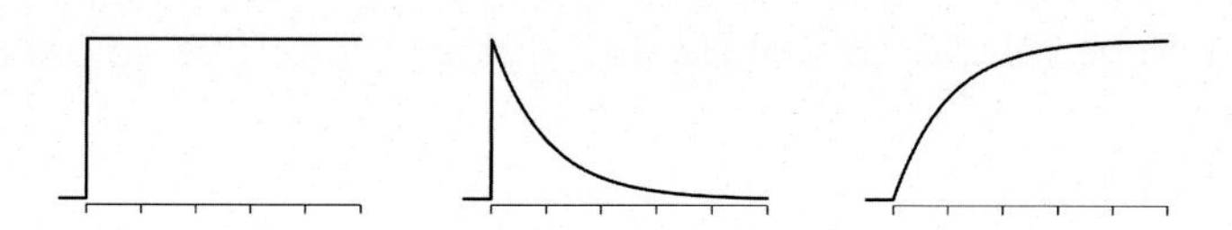

$f(t) = 1$	©	$h(t) = e^{-at}$	(=)	$g(t) = 1 - e^{-at}$
⇓		⇓		⇑
L		**L**		$\mathbf{L}^{-1}$
⇓		⇓		⇑
$F(s) = \frac{1}{s}$	(x)	$H(s) = \frac{1}{s + a}$	(=)	$G(s) = \frac{1}{s(s + a)}$

Thus, the well-known output of a low-pass filter to an input step is derived here by means of a multiplication in the frequency domain of the step input and the filter transfer function.

Another interesting property follows from the transform table above, namely that multiplying by s is equivalent to differentiation, and division by s is equivalent to integration of a function. This is illustrated by the frequency domain function of a ramp (a linearly rising voltage). The step function, $f(t) = \text{constant}$, is just the derivative of the ramp, and $1/s$ is $1/s^2$ multiplied by s. The next stage, comparing the step function with the impulse (or delta) function, may need a further explanation. A step is a sudden change of voltage (or any other quantity). At a certain moment, the quantity is supposed to jump (instantaneous, or infinitely fast) to a different value. In other words, the derivative is an infinitely high change in an infinitely short time. This derivative function, the impulse or delta function, is a very useful mathematical notion, notwithstanding the fact that is not realizable physically. In practice, however, the function can be approximated by a finite (but very short and very high) rectangular pulse.

The beauty of the impulse as an input signal lies in the frequency domain: here it is unity: $F(s) = 1$. A spectrum that has the same strength at all frequencies is called a "white" spectrum.

Sending an impulse into any system is equivalent to stimulation with all frequencies at once. The result is that the output signal G(s) is identical to the system response H(s). In the time domain, if the input signal is an impulse, then $h(t) = g(t)$. In this case, the response g(t) is called the impulse response. The attractiveness of an (approximated) impulse lies not only in the identity of output and system transfer function but also in the "whiteness" of the spectrum: instead of stimulating patiently with a number of sine stimuli of different frequencies in succession and collecting the response amplitudes and phases, stimulating a system with an impulse yields a full-blown frequency characteristic in one fell swoop.

The impulse has also disadvantages, however. The main problem is that an infinitely short, infinitely strong signal is best approached by a very short and hence very strong signal. The large amplitude may cause several problems. These will be dealt with later, together with the alternative, white noise. For the moment, it suffices to remember that the mathematical impulse can be approximated by a physical pulse that is shorter than the timescale one wants to investigate. The spectrum of such a pulse is white up to the highest frequency of interest. As a rule of thumb, the timescales should differ by about one order of magnitude: If we want a system response to be valid up to 100 Hz (i.e., $[10\ \text{ms}]^{-1}$), it is OK to use a pulse of about 1–2 ms duration.

THE FOURIER TRANSFORM

In recording electrical signals, one has the input and output signals, not as explicit functions, but as changing voltages in time. Digitized with a computer, these signals form arrays of numbers in memory. This is where the Fourier transform comes in.

Basically, the Fourier transform (FT) is an integral transform much like the one by Laplace, and so it can be used for essentially the same kind of operation: transforming time data into the frequency domain and vice versa. However, as indicated in the table given earlier, the practice is different. The frequency variable of the Fourier transform is $j\omega$, and so it is applicable only to periodical signals. In Appendix, Fig. A.6, this consists of all signals lying on the vertical axis ($j\omega$ axis).

Performing an FT can be compared to matching the input signal to a number of sinusoids with different frequencies and determining the amplitude and phase of the signal content at each of these component frequencies. This will be explained later.

Usually, the Fourier transform is performed numerically, using a computer. Since computers work with a finite number of limited-resolution numbers, the integral transformation describing the FT is converted into its discrete version, DFT (discrete Fourier transform). DFT will be treated later on.

The FT has its inverse function, or inverse Fourier transform (IFT), that converts a frequency domain signal into its time domain counterpart. Since performing an FT neither adds nor reduces the information content of the signal, the operation is fully reversible, and the inverse transform of the inverse transform is again a forward transform: I(IFT) = FT. Thus, one can collect FT pairs: pairs of signals, one in the time domain and one in the frequency domain that are related to one another by an (I)FT.

Fig. 7.6 depicts the most frequently used pairs. Note that the figure may be read left-to-right and right-to-left. This means that, if the left column represents time functions, the right column yields the corresponding frequency functions.

The converse is also true: If the left column is interpreted as frequency functions, the right column gives the corresponding time functions.

The reader will have noticed that any representation in the frequency domain should consist of both an amplitude and a phase graph. Here, in the right column, we attempt to depict both amplitude and phase in a single graph. This is possible because in the examples given here, the phase is either 0 or pi radians (counterphase; 180 degrees). So in the right column spectra, energy at 0 phase is drawn above the horizontal (frequency) axis, whereas counterphase spectral energy is below the axis. Note also that it is a custom to draw spectral energy in narrow bands as vertical lines starting at the f-axis (hence spectral lines), whereas continuous spectra are drawn as curves. Another important fact is that the left and right graphs are not drawn to scale and have arbitrary time and frequency calibrations. Speaking about the right column, we have to explain yet that the graphs are drawn with the origin (0 Hz) in the middle. Mathematically, the Fourier transform yields negative frequencies and positive ones. In discussing real-time electrical signals, this may seem odd because it is unclear what could be meant with "a frequency of minus 100 cycles per second." In this situation, negative and positive frequencies can be interpreted as rotating clockwise and anticlockwise, respectively. Using an arbitrary reference, we can speak of a shaft rotating at 100 rpm when it rotates clockwise, and −100 rpm when it rotates anticlockwise. Similarly, a car driving in reverse gear could be assigned a speed of −5 km/h. In optical applications of the Fourier transform (see Chapter 11), positive and negative frequencies represent simply light rays to the right and left of the optical axis, respectively, and so need to be distinguished. In the analysis of signals in time, however, the energy content at negative frequencies may be "folded back" onto the positive frequency axis, which means adding up their energies. The Fourier transform of a 1 W electrical signal of 50 Hz consists of 0.5 W at 50 Hz and 0.5 W at −50 Hz.

We will inspect each row (Fourier pair) separately.

A depicts an impulse at t = 0. The spectrum is white, as we have seen from the Laplace transform. Since the function value of the Laplace-transformed impulse is unity at all frequencies, an impulse is the neutral element in the frequency domain, irrespective of the choice of transform; it has a "white" spectrum. Reading **A** backward reveals that a signal that is always the same, i.e., a direct current, has a spectrum that shows only one

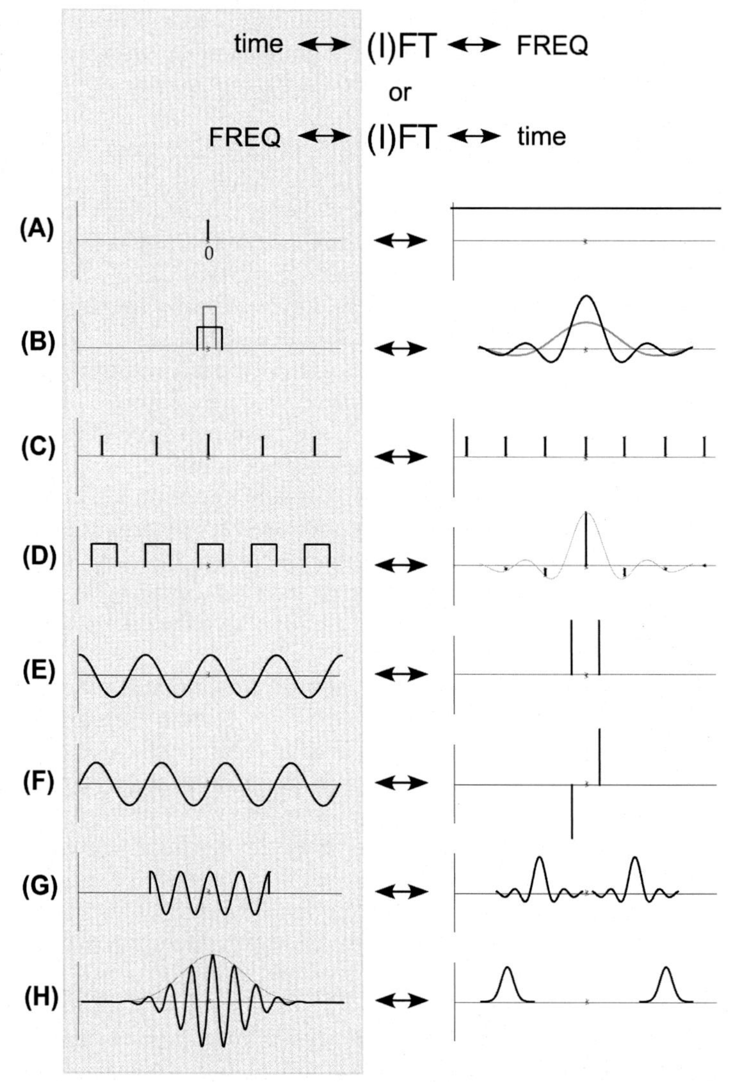

FIGURE 7.6 Fourier transform pairs. Left column: time function. Right column: frequency function, or the other way round.

line, situated at 0. In other words, all energy is at a frequency of 0, which agrees with our notion of a direct current. As stated earlier, an impulse cannot be made physically and must be approached by a finite pulse.

B depicts a finite pulse (symmetrically around $t = 0$) and its frequency spectrum. A long pulse has a relatively narrow spectrum (black lines), whereas a shorter pulse has a wider spectrum (gray lines). In the limiting case, an infinitely short pulse, this reduces to the situation at **A**, an infinitely wide (i.e., white) spectrum. The rule of thumb given earlier for the use of a finite pulse to represent an impulse can be derived from the

relations given here. The form of the spectrum (amplitude as a function of frequency) has a sinc-shape (sinc being an abbreviation of the sin(x)/x function).
C shows the situation in which a pulse is repeated periodically (an impulse train). As can be seen, the frequency domain representative is also an impulse train. All spectral lines are equally high, so it can be considered to be a "white" spectrum that only exists at certain frequencies.
D is a combination of B and C: a repeated, finite pulse. The spectrum is a combination of the line spectrum caused by impulses, now with the sinc function as an envelope (gray curve). Most energy is at zero, so such a pulse train has a strong DC content.
E shows a cosine function, i.e., an example of the well-known sinusoid, or "pure tone" signal function. As is expected, the spectrum consists of a single line at the frequency, f, that corresponds to the inverse of the period (2π radians). This is in keeping with the idea of a sinusoidal (harmonic) motion at a single frequency. However, mathematically, the energy is divided evenly between the frequencies $+f$ and $-f$ hertz.
F shows the sine function. Intuitively, one would expect the spectrum to be the same as the one at **e**, since the sine and cosine functions differ only in their relative phase. Indeed, the spectrum consists again of two lines at $+f$ and $-f$, now with opposite phases. This reflects the difference between so-called even and odd functions, which will be explained below.
G depicts a tone burst, i.e., a sinusoidal signal that exists for a limited time. Note that, mathematically speaking, the sine and cosine signals shown at **E** and **F** are supposed to exist indefinitely. The tone burst can be considered as a combination of **B** and **E:** a cosine limited to a finite, rectangular pulse shape, or window, as it is called. Indeed, the frequency domain shapes of both time functions can be seen: a sinc-shaped spectrum, centered on the sine frequencies $+f$ and $-f$ hertz. Note that the left and right graphs are not drawn to scale. This is necessary to show the essential shapes in both time and frequency domain. Note also the implication that a short tone pulse strictly does not possess a single frequency but has a frequency band (or range of frequencies) instead. In hearing research, for instance, one tries to measure the performance of our hearing system as to tone frequency discrimination. It must be taken into account that a tone burst has a less sharply defined frequency *physically*, i.e., irrespective of the skills of our hearing system. In addition, the spectrum is not monotonous: the sinc function that forms the envelope of the spectrum has side lobes. The spectrum has several peaks, some of them with inverted phase. This led to the development of other shapes of tone bursts. Switching a tone smoothly on and off, by increasing and decreasing the amplitude gradually (rather than by a switch), improves the shape of the spectrum in certain ways.
H shows a Gaussian-shaped tone pulse. Mathematically, an elegant way of smoothing in the time domain is the Gauss function, the bell-shaped curve known from statistics. The Fourier transform of a Gauss curve is again a Gauss curve. For the spectrum of a Gaussian-shaped, or "windowed," tone pulse, this means that there are no side lobes (at the expense of a somewhat wider main lobe). In signal analysis, more window shapes are used. The art of "windowing," i.e., shaping the time course of a pulse to improve the spectrum, is dealt with later on.

ODD AND EVEN FUNCTIONS

The difference between the sine and the cosine functions (**E** and **F**, above) illustrates the difference between odd and even functions. As can be seen in **E**, the cosine function is symmetrical with respect to the origin (time zero). This can be formulated as $\cos(t) = \cos(-t)$. The cosine is called an even function because of this property. The sine function belongs to the other category called odd functions that show antisymmetry around time zero. Thus, $\sin(t) = -\sin(-t)$.

Many more examples could be given. However, most other cases can be derived from the ones given here, using the following, simple rules.

- Time and frequency are each other's inverse. Shorter times correspond to wider spectra and vice versa. The Fourier pairs may be interpreted backward. Examples: a sinc-shaped pulse yields a rectangular spectrum (**B** read backward); two impulses at opposite sides of time zero yield a sinusoidal spectrum (**E** or F read backward).
- Addition is linear. The sum of two signals yields a spectrum that is the sum of the constituents (taking the phases into account).
- Time differences leave the amplitude spectrum untouched but change the phases. An impulse at any moment other than zero has the same white amplitude spectrum but phases that increase with frequency. For example, an impulse at $t = 1$ ms has a phase of 360 degrees at 1 kHz.

These rules can be combined. As an example, take two successive pulses, identical in amplitude and duration, but one positive and one negative one. In this case, the DC components cancel, whereas other parts of the spectrum do not, in general.

LINEARITY

The foregoing principles and methods of systems analysis rely on a special property of the system in question: the system must be linear. It means that the response to a sum of input signals must be the sum of the individual responses. If we call the input signals A and B and the outputs α and β, respectively, the linearity principle can be depicted as in Fig. 7.7.

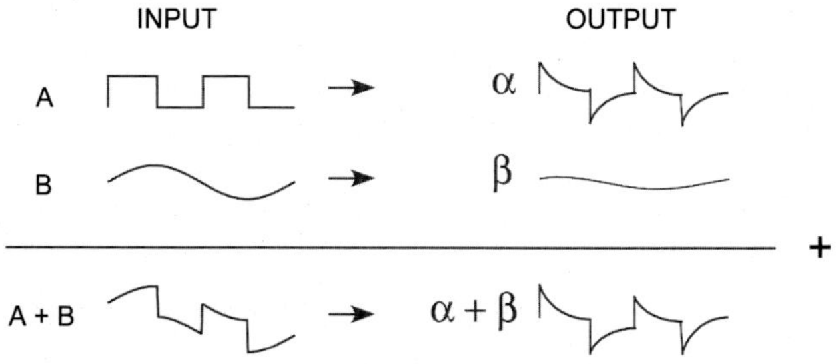

FIGURE 7.7 The response of a linear system to the sum of two input signals.

In this example, A is a square wave signal, and B, a low-frequency sine. From the output signals depicted, one can deduce what kind of system we are dealing with: a high-pass filter.

The demand for linearity seems a rather strong limitation of the applicability of system analytical methods. However, if the linearity principle does not hold, mathematical manipulations such as the Fourier transform and convolution to predict the system response would not work, since they implicitly assume that this response can be described by the summation of sine waves or past inputs, respectively. One may wonder whether the systems we want to analyze are linear and what will be the consequences if they are not. Fortunately, many systems are linear, at least over a fairly large range of amplitudes. Passive electronic components such as resistors and capacitors and circuits made thereof behave in a linear fashion. To the contrary, many active circuit elements, such as diodes and transistors, and instruments, such as amplifiers, do not behave linearly, or do so only in a limited amplitude range. The same holds for ion channels, synapses, and other neurobiological processes.

In Chapter 3, we have met the saturation of an amplifier, leading to distortion of the (sinusoidal) signal. Note that saturation, or distortion, is a useful measure of (and a warning for) nonlinearity. Distortion implies the generation of harmonics, i.e., of frequencies not present in the input signal. If a sinusoid is distorted by a saturating amplifier, part of the signal power will appear in the harmonics. The stronger the distortion, the more power is in the harmonics.

Especially biological systems may show very complex, nonlinear forms of behavior. The electrical responses of nerve cells, in particular the time course of action potential series, are often strongly nonlinear. How then do we know whether the methods of linear systems analysis can be applied?

There are several ways to cope with nonlinearity:

1. *Approximate linearity*. Fortunately, the linearity demands are not very strict, and many systems are *approximately* linear. If one feeds a sinusoidal signal in, say, an amplifier, the output signal may look fairly sinusoidal when viewed on an oscilloscope screen. If one compares output and input signals simultaneously on the screen, however, any small distortion will show up. The question to be answered is: How much distortion can be accepted by the methods of systems analysis? A pragmatic answer, used as a guideline in systems analysis, is: 10% distortion. Since distortion implies the generation of other frequencies, one can express the degree of distortion by the relative contribution of the harmonics to the total signal power. If a pure sinusoid is distorted 10%, that fraction of the signal power is converted into higher frequencies (harmonics). It means that still 90% of the signal power is at the fundamental, i.e., at the input frequency. Note that, even at less than 10% distortion, a signal may look quite distorted when viewed on an oscilloscope screen, so if a signal looks fairly sinusoidal on-screen, one can be sure that linearity is warranted to a fair degree.
2. *Small-signal analysis*. Even in explicitly nonlinear systems, approximately linear responses can be obtained if signal amplitudes are kept low. As an example, the voltage change of an animal photoreceptor cell is approximately logarithmically dependent on the light intensity. However, to determine the frequency response of such a sense organ, one simply stimulates with a light intensity that is modulated only partially.

3. *Linearization*. If one knows the mathematical form of the nonlinearity involved, the output signal may be transformed with the inverse of that function before performing the analysis. If the output depends, say, on the square of the input signal, one simply takes the square root of the output signal before applying systems-analytical methods. In the above example, one could linearize the sensory cell's response by taking the antilog of the output. If a signal is linearized mathematically, the signal amplitude does not need to be kept low.
4. *Nonlinear systems analysis*. There are methods developed explicitly to treat (strongly) nonlinear systems. These will be treated later on.

ANALOG-TO-DIGITAL AND DIGITAL-TO-ANALOG CONVERSIONS

So far, we have treated signals that are truly continuous waveforms in time. Analogue apparatus, such as amplifiers, filters, pen recorders, and the like, keep the continuity in time. Of course, we cannot measure these signals with infinite precision, but the thresholds of detection and precision are gradual, depending on noise levels, interference, etc. In contrast, most signal processing nowadays takes place within computers, instruments that treat only finite series of finite-precision numbers. This means that the first step in signal processing (after amplification and, usually, display on an oscilloscope screen) consists of converting the continuous signal into a discrete and digital one (i.e., existing only at a finite number of instances and expressed as numbers). This is done by an analog-to-digital converter or ADC for short. The result is a series of numbers, kept in computer memory (random access memory, RAM) and eventually written to disk or to any suitable storage medium. A sample is thus defined as a single numerical measurement from the continuous world. Thanks to the generous amounts of RAM in present-day computers and the fast processing speeds, a signal can be digitized in a sufficiently large number of sufficiently precise numbers. Nevertheless, it must be kept in mind that digitization limits the resolution both in time and in amplitude, and the choice of digitization parameters is crucial to the validity of the digital data.

The process of sampling a continuous, real-world quantity can be considered as a mathematical operation on that quantity. To find the essential properties of sampled signals, a theory called sampling theory was developed in the period 1930–50, by celebrities like Shannon and Nyquist. From these theories, we fortunately need to remember only a few basics. A first, basic rule is that we cannot make inferences on processes we did not sample. A sampled signal has two important quantities: the sampling interval (notated with a lowercase t) and the total sampling time, notated with a capital T (see Fig. 7.8).

For a series of k samples, spaced t seconds apart, the total sampling time is $T = kt$ seconds. A slight confusion might arise because sometimes the entire series of samples is also called "a sample" (which it is, of course; a small piece of a real-world signal). To avoid the confusion, a series of consecutive samples (sampling points) is better called a sweep or trace (see oscilloscopes in Chapter 4).

The very fact of sampling a real-world quantity poses limits on the information content of the resulting digital signal. First, processes that take more than T seconds have been sampled only partially so that insufficient information about them is available. In the frequency

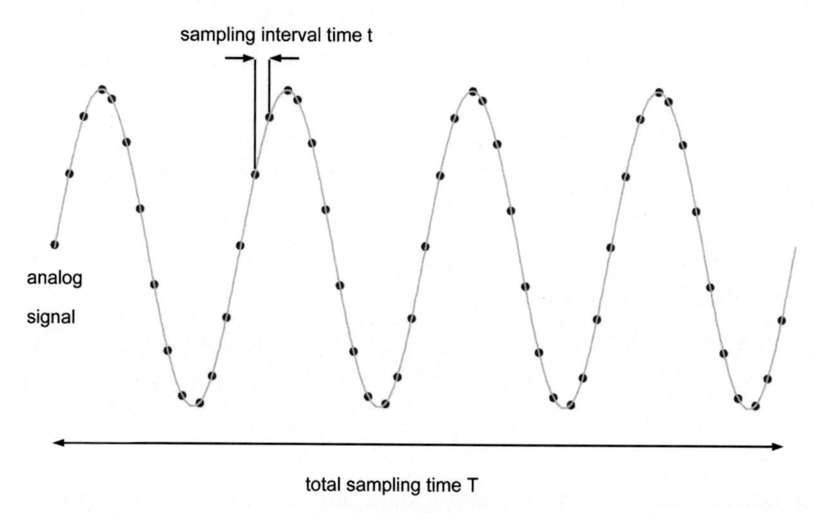

FIGURE 7.8 Digitizing an analog waveform. The analog signal, here a sine, is depicted in gray, together with the digitization times (*red dots*). The total time T might be 1 s; the sampling interval t, 1 ms (note: not drawn to scale for clarity).

domain, this implies that we have no information about frequencies lower than 1/T Hz. Second, the sampling interval limits the time resolution: between two samples, essentially nothing of the original signal is known. These limitations are both obvious and important and can be explained best with a numerical example.

Imagine a 1-s sound sample having 1000 samples, spaced 1 ms apart. Since we have only 1 s, we do not have information about frequencies lower than 1 Hz. In addition, the sampling interval of 1 ms limits the time resolution of our sample: the "spaces" between our time samples are empty, i.e., unknown. This limits the upper boundary of the frequency spectrum fundamentally to 500 Hz. Remember that one needs at least two samples per period to characterize a frequency. This is the important Nyquist criterion. If a signal is sampled too infrequently, i.e., with two samples per period (of the highest frequency present) or less, insufficient information on the sampled waveform is obtained. Worse still, a phenomenon called aliasing occurs. Both the Nyquist frequency and the pitfalls of aliasing were treated in Chapter 4. A voice signal sampled as in Fig. 7.8 would be barely useful, since it would contain only frequencies up to 500 Hz. This implies that the higher frequencies that help to recognize and understand the voice signal are absent.

Note that the Nyquist frequency is the absolute, fundamental limit of digitization.

It is better to choose a sampling frequency somewhat higher than the Nyquist value. Digital audio, for example, uses sampling frequencies of 44.1 or 48 k samples/s to cover the audio spectrum, i.e., frequencies up to 20 kHz.

For the electrophysiological practice, the sampling frequency has to be chosen carefully, dependent on the type of signal one observes. Since fast AD converters are affordable and computer storage is cheap today, sampling frequency is usually chosen higher than the required minimum value.

Thus, for the examples mentioned in Chapter 4 on amplifiers, useful sampling rates would be as in the table below.

for nerve membrane potentials:	about 10 k samples/s,
for electrocardiograms:	about 100 samples/s,
for electroencephalograms:	about 100 samples/s,
for nerve or muscle spikes:	about 10 k samples/s,
for plant action potentials:	about 10 samples/s,
for the analysis of spike shape:	about 200 k samples/s,
for single-channel recording:	about 100 k samples/s.

Notes:

1. Oversampling too much only yields larger data files, without giving more information. Despite the availability of cheap storage media, collecting useless amounts of digital data should be avoided.
2. To avoid aliasing, one must filter the analog signal in order to avoid frequencies higher than half the Nyquist sample frequency. Remember that some AD converter cards have anti-alias filters built in, but most of them have not.

SIGNAL WINDOWING

In digital signal processing, the data consists of a finite sample of discrete values, taken from a practically continuous physical world. As discussed earlier, this limits the amount of information that can be deduced from the sampled signal both in time and in amplitude. Essentially nothing can be said about the physical reality before the first and beyond the last sample point. Worse still, the abrupt beginning and end of a sample introduce artifacts when performing transformations such as the Fourier transform. This is illustrated in Fig. 7.6, where Fourier pair **E** shows an infinitely long sinusoid and its Fourier transform which consists of two spectral "lines" at + and −, the frequency of that signal. A short sample of such a signal is shown in trace **G**, and indeed, it has a spectrum very different from the former one: Instead of lines at certain precise frequencies, the spectrum consists of two continuous functions, spanning a range of frequencies (the sidebands) and having side lobes where the phase is reversed. Apparently, the properties of a short sample of a continuous function may deviate substantially from the properties of the original. A first conclusion of this observation is that a long sample is better than a short one. Indeed, if the sample length is about 10 s, the width of the sidebands is reduced to about 0.1 Hz so that the spectrum resembles the one of Fig. 7.6, **E**, better. No matter how long the sample, it will always be bounded by a beginning and an end.

Windowing functions are invented to alleviate this problem. The trick consists of reducing the amplitude of the signal sample values in the vicinity of the bounds. This is performed by multiplying the sample values with a function that tends to zero at the boundaries. A useful

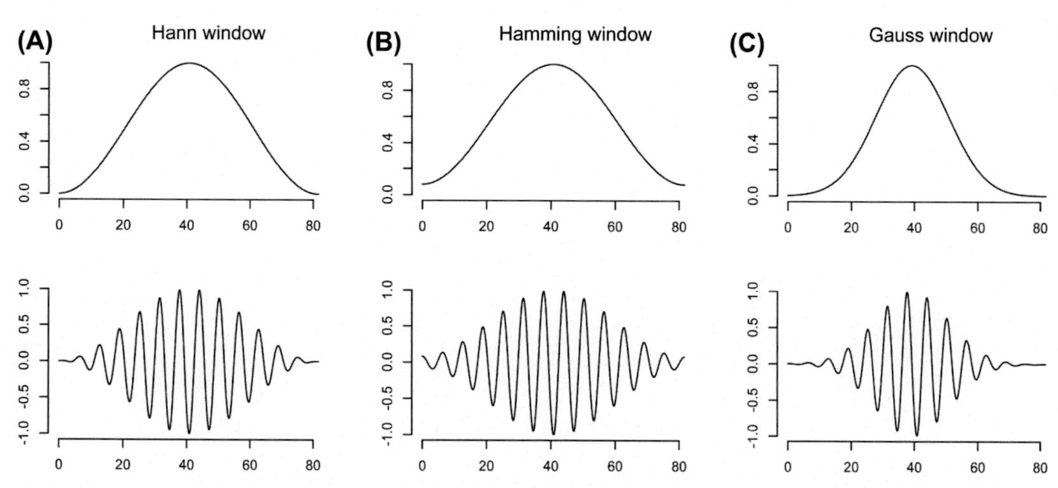

FIGURE 7.9 A window function (upper graph) and its resultant signal sample shape (bottom graph) for three popular window functions: (A) Hanning, (B) Hamming, and (C) Gaussian.

function is a shifted cosine, called the Hanning window after its inventor, the German J. von Hann. This is depicted in Fig. 7.9A.

The other functions shown are the Hamming (Fig. 7.9B, after R.W. Hamming) and Gaussian window (Fig. 7.9C). Note that the Gaussian envelope is necessarily truncated so that the amplitude at the ends of the sample is not exactly zero. Still more windowing functions exist, all having the same purpose: to reduce the signal amplitude at both ends of the sweep (window) to zero or to a small value. All these window functions yield better representations in the frequency domain, as is illustrated in Fig. 7.10. Finally, applying no such window function at all is in fact a window by itself, i.e., a rectangular window, because the signal remains unattenuated during the entire sample time. The effect of using different window functions show up dramatically in the frequency domain. This is shown in Fig. 7.10.

In principle, a pure sine signal, such as in Fig. 7.10A, should show up in the frequency domain as a single spectral component, or "spectral line." In other words, all energy exists at a single frequency, provided the sine wave exists at all times. However, since any sample spans a finite time, the spectral lines have a nonzero width, equal to the inverse of the sweep duration. Things get worse when the sweep contains a signal truncated in an arbitrary way, and indeed, this is almost always the case when sampling real-world signals. An example is illustrated in Fig. 7.10B: both ends of the sample do not connect, since the sweep starts at a large positive value and ends at a large negative one. The effects in the frequency domain are dramatic. The spectral line (or lines, at frequencies of both +16 and −16 units) are widened substantially, thus reducing the precision with which the signal spectrum can be analyzed. In addition, the wider lines may occlude any nearby frequency components. The Hanning window (Fig. 7.10C) reduces the problem to a large extent. Note, however, that the amplitude spectrum is rendered with a logarithmic y-axis, reading decibels. At very low amplitude values (more than 40 dB below the peak), the Hanning-windowed signal spectrum is wider rather than narrower than the rectangularly windowed one. Although such small artifacts are

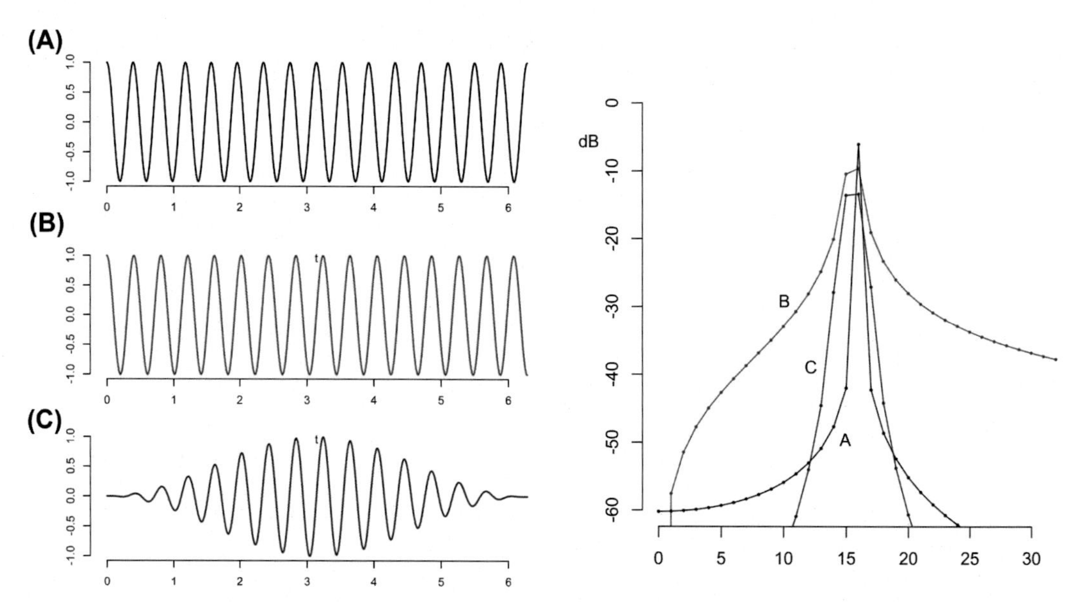

FIGURE 7.10 Signal samples (256 points) with a rectangular versus a Hanning window. Left: shape of signal sample; right: frequency spectrum. (A) Sinusoid sample with exactly 16 cycles in the rectangular window. (B) A sinusoid with about 15.5 cycles in the rectangular window. (C) The same as (B) but multiplied with the Hanning window.

usually not a great concern, at times better results may be wished for. In the analysis of loud sounds, for instance, −50 dB might very well be within the audible range. Investigators of sound and hearing must choose their window functions carefully.

DIGITAL SIGNAL PROCESSING

Signal Averaging

Signal averaging, also called signal recovery, is the most straightforward method of digital signal processing. It can be applied in all those cases in which a repetitive signal is administered to a human, an animal, or a preparation. Brief electrical pulses, tone bursts, or light flashes are usually given, say, once every few seconds. If the response of a neuron, a sensory cell, or a brain nucleus is buried in noise, signal averaging may improve the signal (more precisely the signal-to-noise (S/N) ratio) to a large extent. In many cases, a recognizable response only emerges after substantial averaging, such as in the case of event-related potentials (ERP) in the brain. A good example is the weak response of a photoreceptor cell to a dim light flash, as shown in the introduction to this chapter. The response of the cell (say, a slight depolarization) is obscured by the noise in the recording. This may include membrane noise stemming from the cell and noise from the electrodes and preamplifier.

Fig. 7.11 shows a weak pulse buried in noise and the improvement in S/N ratio obtained by the averaging of different numbers of sweeps.

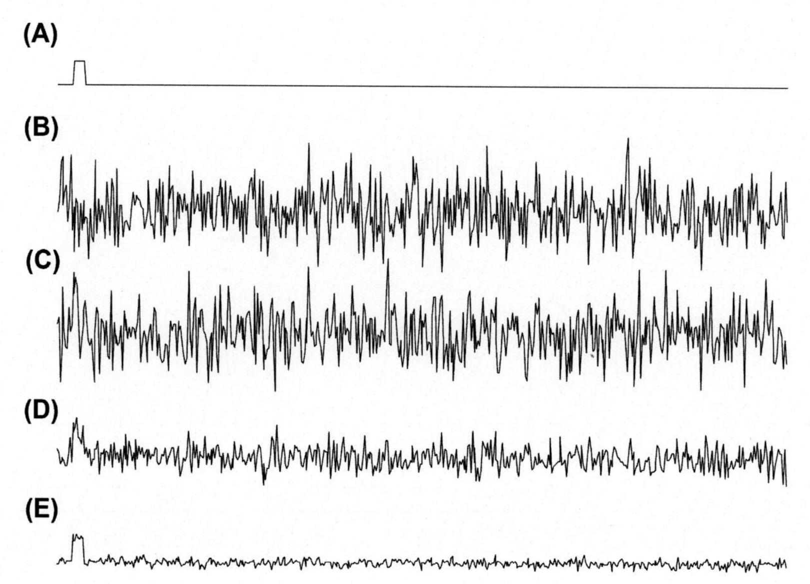

FIGURE 7.11 A weak pulse buried in noise and the improvement in S/N ratio obtained by the averaging of different numbers of sweeps: (A) signal, (B) noise, (C) signal + noise, (D) average of 5 sweeps, and (E) average of 50 sweeps.

This form of signal averaging is used widely for the analysis of physiological and other biological signals. However, its use depends entirely on the availability of a clean trigger signal that marks the time when the averaging should start. In neurophysiology, this is often easy to come by, since we have a stimulus signal or some other event, time-related to the signal investigated. However, in situations where there is no distinct starting command, other techniques are needed. One such possibility lies in correlating one or more signals. This is the next subject.

AUTOCORRELATION

Consider a noisy signal (such as in Fig 7.11) that has been sampled fast enough to see its details. If the signal were a purely random signal, then the value of each sample would be completely different from the preceding or the following sample. What we see instead is that each sample is followed by one of a similar value: adjacent samples are correlated. A measure of the degree of correlation between samples in a signal x(t) that are separated by a distance t is given by multiplying the signal with a time-shifted version of itself:

$$A(\tau) = M[x(t) \cdot x(t - \tau)] \tag{7.1}$$

where τ is the time shift and M[ ] reads "the mean of." Because x(t) is being correlated with itself, A(t) is called the autocorrelation function (ACF). It is clear that A(t) takes on positive values if x(t) and $x(t - \tau)$ are on the average very similar, whereas if x(t) and $x(t - \tau)$ vary

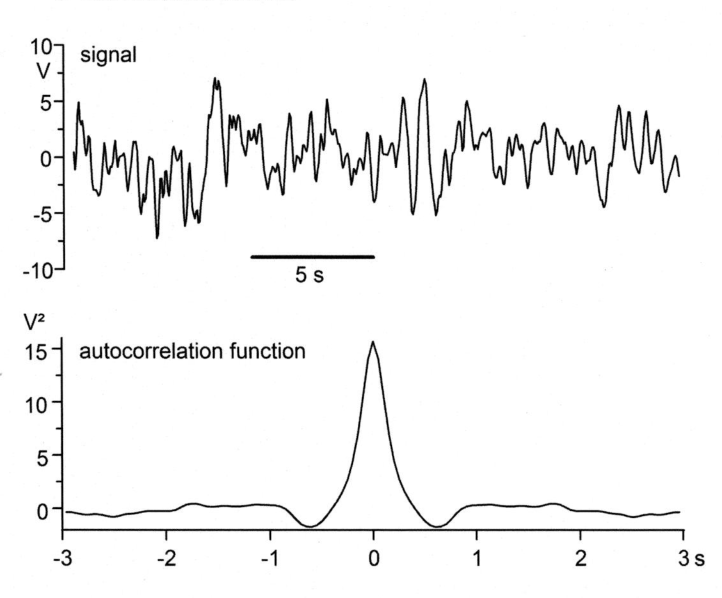

FIGURE 7.12 A noisy signal and its autocorrelation function.

independently of one another, it will be 0. A(t) for the signal in Fig. 7.12 is shown in the same figure. As might be expected, it is a function that is symmetrical around $t = 0$.

Going from a sampled signal to a continuous signal, Eq. (7.1) becomes:

$$A(\tau) = \int x(t) \cdot x(t - \tau) dt$$

Things become different when the noisy signal contains a periodic component, such as a sine, square, or pulse. For periodic signals, the ACF does not die out at larger time shifts. Since a sine wave repeats itself every period, the ACF shows peaks at those values. In addition, it will be clear intuitively that the autocorrelation of a sine wave yields negative values after half a period because there one multiplies a positive with a negative value. The result seems simple: The ACF of a sine wave is again a sine wave. Note, however, that the horizontal axis of the ACF reads time shift rather than time. Other waveforms are not preserved in their ACFs: for example, the ACF of a square has a triangular shape. Apparently, some information as to the signal shape is lost. Nevertheless, a lot of information on the signal can be inferred from the shape and scaling of ACFs, which is illustrated in Fig. 7.13.

In the case of pure, band-limited noise signals, the ACF shows the bandwidth (Fig. 7.13A) and the steepness of the bandwidth limitation (Fig. 7.13B). The ACF of periodic signals does not die out at higher time shifts and retains some information on the waveform (Fig. 7.13C). Finally, most real-world signals will contain both a periodic component and noise, which can be segregated by looking at the ACF (Fig. 7.13D). From the latter example, one can determine amplitude and bandwidth of the noise component from the peak at low time shifts, as well as amplitude and frequency of the sinusoidal component from the ACF at higher time shifts.

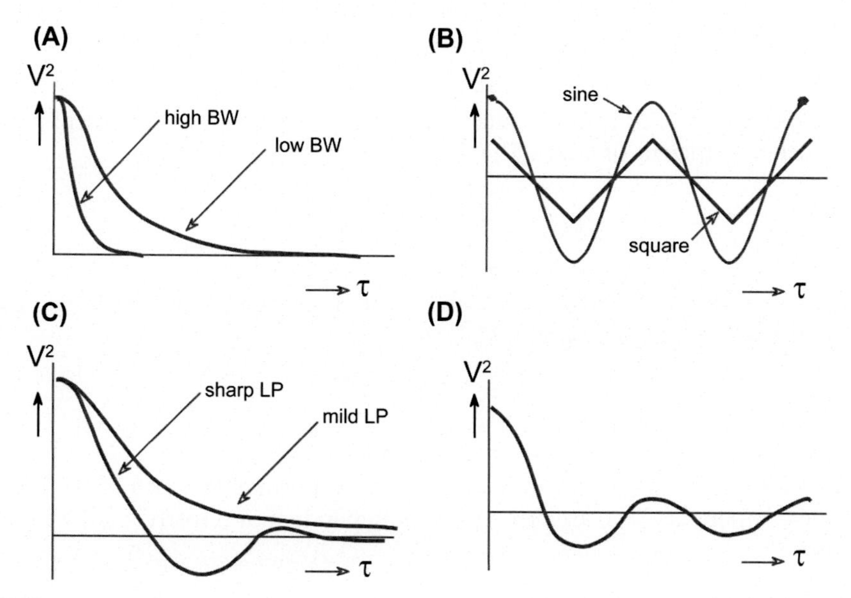

FIGURE 7.13 Autocorrelation functions of noisy and periodic signals. (A) Noise signals with different bandwidths, (B) mild versus steeper BW limitation (first-order and higher-order low-pass, respectively), (C) periodic signals, and (D) a combined sine + noise signal.

The autocorrelation procedure implies averaging and is a powerful way of detecting *any* periodic signal buried in noise. Note that, contrary to the signal averaging method described above, no clean stimulus or trigger signal is necessary. Therefore, autocorrelation may also be used to find unknown spontaneous periodic signals in signals from neurons, sense organs, or brain. In astronomy, the existence of pulsars (pulsating stars) was found by the saving grace of autocorrelation: the pulsations are far too weak to be detected directly.

CROSS-CORRELATION

In addition to time-shifting a single signal, we may also wish to correlate x(t) with a different signal y(t). We then obtain the cross-correlation function (CCF):

$$C(\tau) = \int x(t) \cdot y(t - \tau) dt \tag{7.2}$$

Cross-correlation functions can be used, for instance, to establish whether two neurones are functionally coupled and if so, to determine the time lag between activity in one neurone and the other. Eq. (7.2) is also called convolution integral. As we will see later, it is a powerful tool in data analysis, signal detection, and filter design.

A special case of cross-correlation is the Fourier transform. If we replace y by a sine wave function of frequency f, we get

$$A(f) = \int x(t) \cdot \sin(2\pi ft)dt$$

A(f) represents the amplitude of the sine wave having a frequency f as present in the signal x, i.e., the Fourier coefficient of sin(f). The cosine series is found similarly:

$$B(f) = \int x(t) \cdot \cos(2\pi ft)dt$$

Using complex notation, the two last equations can be combined in a single, compact equation (see Appendix for complex numbers):

$$X(\omega) = \int x(t) \cdot e^{-j\omega t}dt$$

with $\omega = 2\pi f$ and j is the imaginary unit, $\sqrt{(-1)}$. An important property of the Fourier integral is that the components $X(\omega)$ are orthogonal (independent or uncorrelated) with respect to each other or:

$$\int e^{-jst} \cdot e^{-j\omega t}dt = 0$$

This property indicates that its inverse function exists. In fact, reconstitution of the signal x from its spectral components $X(\omega)$ is simply a summation of all sines and cosines in the spectrum:

$$x(t) = \int X(\omega) \cdot e^{j\omega t} \cdot d\omega$$

The convolution theorem gives a straightforward relation between convolution of two functions in the time domain and (complex) multiplication in the frequency domain. It states that if g(t) is the convolution (©) of x(t) and y(t), then the Fourier transform of g(t), $G(\omega)$, is the product of the Fourier transforms of x(t) and y(t), or:

$$\text{if } g(t) = x(t) \copyright y(t) \quad \text{then} \quad G(\omega) = X(\omega) \cdot Y(\omega) \tag{7.3}$$

Thanks to this theorem, it is relatively easy to carry out the inverse of convolution. This is called deconvolution and it is often used to compensate for signal deterioration due to known sources, for example, in improving the image of a star seen through an imperfect lens. If in Eq. (7.3) we know g(t) and y(t) and wish to know x(t), then:

$$x(t) = F^{-1}\{G(\omega)/Y(\omega)\} \tag{7.4}$$

where F^{-1} stands for the inverse Fourier transform.

If the convolution integral of Eq. (7.3) is rewritten with slightly different symbols, then it can be regarded in a general way as the response y of a (linear) system to a stimulus x. The transfer function h(t) embodies the properties of the system under study:

$$y(t) = \int h(t - \tau) \cdot x(t) dt$$

The advantages and disadvantages of the impulse as a signal to probe the transfer function were described earlier. Since the theory holds only if one may assume linearity, the impulse should remain relatively small. It carries little energy in that case, and the output might be difficult to extract from the ever present background noise. A way to circumvent this problem is to use white noise as the input signal. Although it may seem strange, in the frequency domain, white noise is very much like an impulse: it has a white spectrum, i.e., all frequencies have equal amplitudes. Only the phases differ: in an impulse, all phases are 0 at $t = 0$, whereas the phases of the components of white noise are distributed randomly.

As with the ACF, the shape and scaling of a CCF yields information on the process to be analyzed. This is illustrated further in Fig. 7.14, where the cross-correlation of noise and deterministic signals is shown. With cross-correlation, the common component in two signals can be revealed, even in the presence of much noise. Obvious applications in neurophysiology are the detection of event-related potentials, which are usually both buried in brain activity that is unrelated to the stimulus.

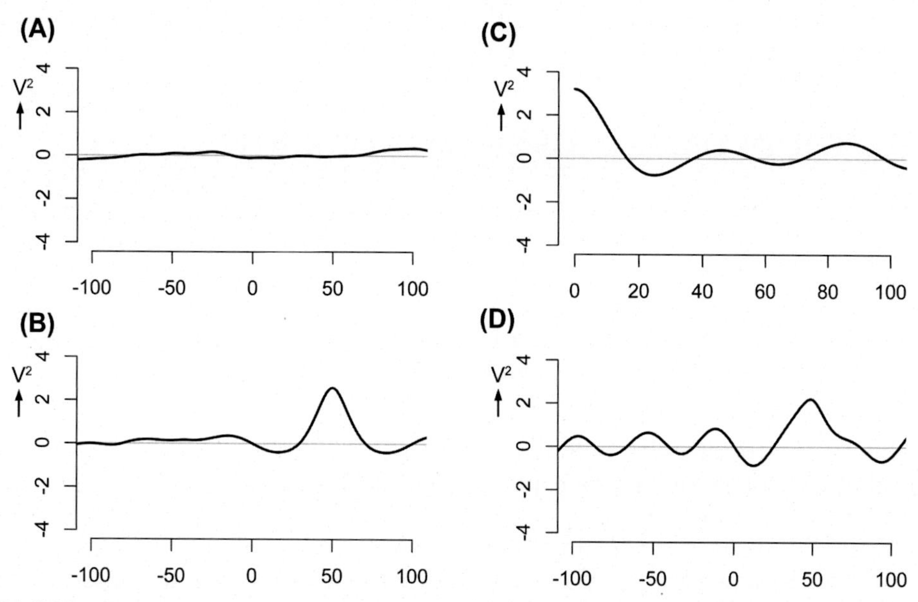

FIGURE 7.14 Correlation functions of different signal situations. (A) Cross-correlation function of two uncorrelated, band-limited noises. (B) Cross-correlation function of noise and a delayed version of that noise. (C) Autocorrelation function of a sine with noise (noisy sine). (D) Cross-correlation function of a noisy sine with a delayed version of that signal.

THE DISCRETE FOURIER TRANSFORM

Having dealt with the do's and don'ts of signal digitization, we return to the digital, or discrete, form of the Fourier transform (DFT). Algorithms to implement the FT exist for the most popular programming languages such as Basic and C. In addition, math packages for personal computers, such as R, Matlab®, and Sysquake, have built-in routines for performing the FT and IFT.

Let us take our former example again: a 1-s record having 1000 samples at 1 ms intervals. The DFT would contain 500 spectral lines, 1 Hz apart. This is in keeping with the limitations stated by the Nyquist criterion. If we do not have any information on the signal where this record was taken from, we have no information on frequencies lower than 1 Hz, so the "space" between the spectral lines is empty (unknown). Only if we would extend our sample to span 2 s, we would have spectral lines every 0.5 Hz. And at the other end, we have spectral information up to 500 Hz. Taking samples every 0.5 ms would extend our spectrum to 1 kHz and so on.

In principle, performing a DFT of sample size n involves n^2 computations (involving sine and cosine functions, so many time-consuming floating point operations). For the early computers, in the 1950s and 1960s, this was a tedious task. In 1965, however, Cooley and Tukey published a far more efficient algorithm, known since as the fast Fourier transform, or FFT for short. The FFT avoids any superfluous computations, but to obtain the maximum efficiency, the sample size must be a power of two. Thus, an FFT might comprise of 256, 512, or 1024 samples. The last value is called "1k." Our modern computers allow FFTs of 8k or more to be performed routinely. The size chosen will depend usually on the nature of the input signal, on the time it takes to perform the FFT, and the desired precision of the result (the frequency domain version of the signal).

THE DETECTION OF SIGNALS OF KNOWN SHAPE

Eq. (7.4) can help the detection of signals of known shape but unknown size. An example of such a detection problem presents itself when we wish to analyze a record containing postsynaptic currents. If only one type of neurotransmitter is released, then the decay times are often similar between individual currents. They are not identical, however, due to differences in the distances of the sites of transmitter release and the soma. The voltage-clamp record shown in Fig. 7.15 has been recorded from cerebellar granule cells in culture.

To detect an event, we do the following:

1. Take the Fourier transform of both the stretch of data and the template. The template is an idealized prototype of a postsynaptic current without noise and of unit amplitude.
2. Divide the two spectra.
3. Take the inverse Fourier transform of the resulting spectrum. The lower trace in Fig. 7.15 shows the deconvoluted data. Sharp peaks indicate where the onset of synaptic currents occurred.
4. Finally, using a threshold to separate the peaks from the background noise, we get the event times we were looking for. Note that the amplitudes of the deconvoluted spikes

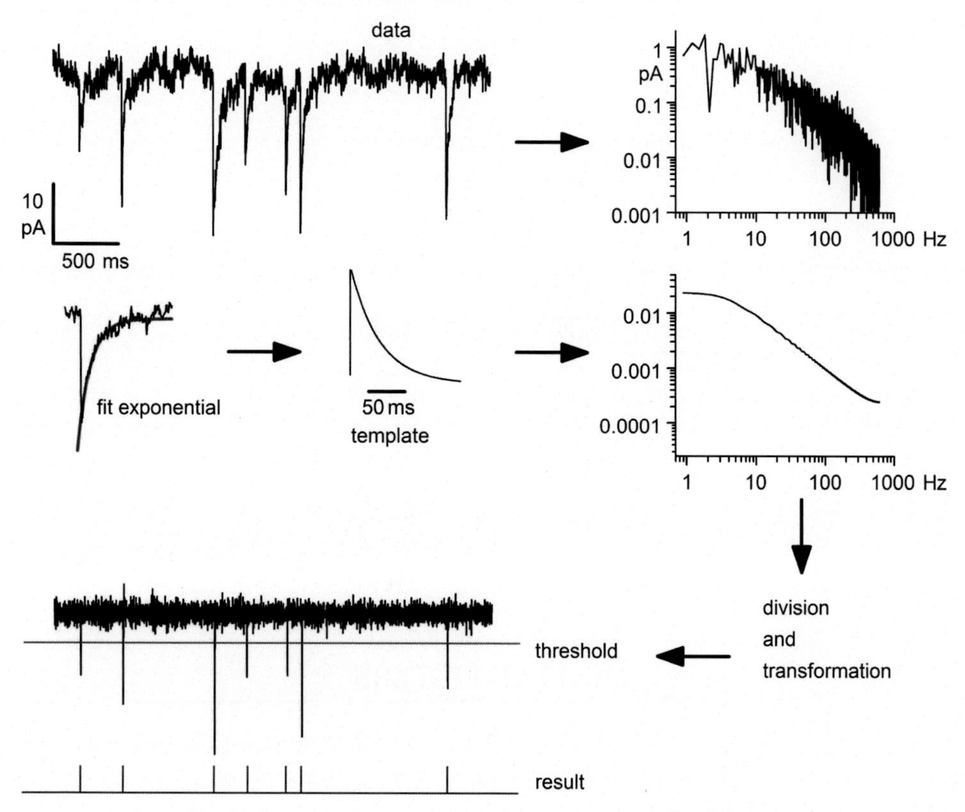

FIGURE 7.15 Detection of signals of known shape (explanation in text).

are not always proportional to the original synaptic currents. This is due to an imperfect match between the template and the synaptic current in question that has for instance a less typical decay time.

The template may be created in two ways. A few of the larger synaptic currents may be fitted with an exponential function and the average decay time may then be used to create the template (the method used in this example). The second method consists of fitting the Fourier spectrum of the data with the appropriate function. Note that in Fig. 7.15, the data spectrum and the template spectrum are similar. This is because the relaxation time constant of the synaptic currents dominates the data spectrum. On top of that, the spectrum contains white and pink (1/f) noise (instrument and thermal noise). To get the Fourier transform of the exponential, we integrate from 0 to ∞:

$$E(\omega) = \int_0^{\infty} e^{-kt} \cdot e^{-j\omega t} dt$$

yielding:

$$E(\omega) = 1/(k + j\omega)$$

which after separation of real and imaginary parts yields:

$$E(\omega) = k/(k^2 + \omega^2) - j\omega/(k^2 + \omega^2) \tag{7.5}$$

If we now create the power spectrum, $E^2(\omega)$, we have a function that is easy to fit:

$$E^2(\omega) = 1/(k^2 + \omega^2) \tag{7.6}$$

This function, a spectrum resembling the frequency characteristic of a low-pass filter, is called a Lorentzian. By adding white (w) and pink (p) noise, the final function to fit to the power spectrum will be:

$$F(\omega) = a/(k^2 + \omega^2) + w + p/\omega^2$$

with $1/k$ being the relaxation time constant of the "average synaptic current."

DIGITAL FILTERS

The above procedure is an example of a digital filter. Since data acquisition and analysis is done almost always with computers, it is worthwhile to examine the principles of digital filtering on a more general basis.[1]

The most elementary form of a digital filter that is useful to reduce wide-band noise is to average a small number of consecutive samples and use the series of averages as a filtered signal. This is illustrated in Fig. 7.16. Of the noisy signal, every five samples are averaged. This reduces HF noise in the sample somewhat. However, we end up with far less time resolution (samples per second) than the original signal had. Samples 1–5 yield the first data point; samples 6–10, the second; and so on. There is a better way, however, that keeps the time resolution almost unchanged. In averaging consecutive samples, it is not forbidden to have some overlap. Thus, one may average samples 1–5, samples 2–6, samples 3–7, and so on. This procedure is known as a running average or moving average. In this case, the output signal has almost as many samples as the input (at the start and/or at the end of the sweep one loses a few points). As you can imagine, this behaves as a low-pass filter, since any abrupt change of the signal will be "smeared out" over a small number of output samples. The well-known "blurring" of digital photos is just a two-dimensional version of a running average. This serves to reduce image noise (or grain), at the expense of sharpness.

[1]Although digital filtering is more powerful and flexible than the old analog filters, one should not underestimate the importance of a simple RC filter in the preamplifier. If an electrophysiological preamplifier is saturated by a strong polarization voltage, or by a 50 (60) Hz hum or any other form of unwanted signal, the signal may get distorted beyond recognition, and no digital operation afterward can help.

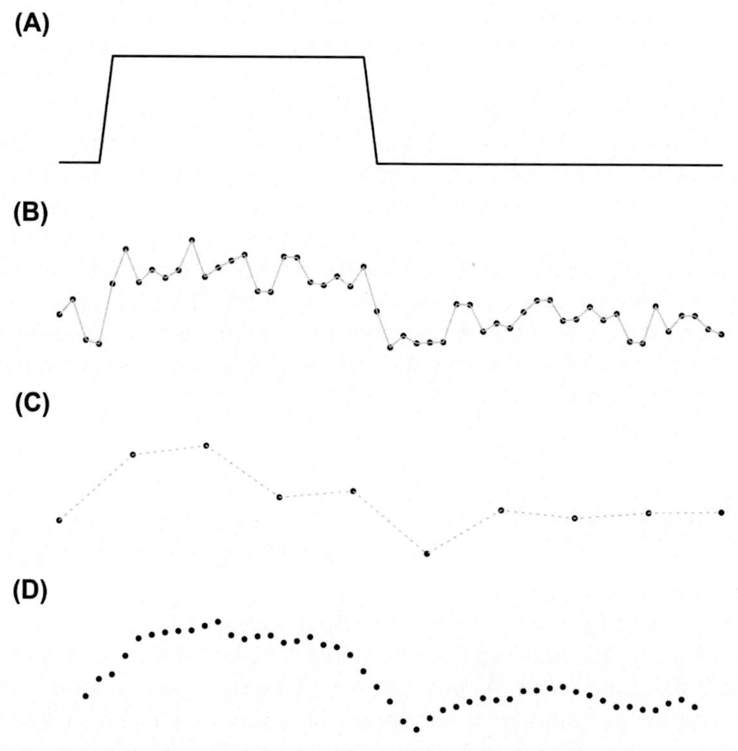

FIGURE 7.16 Effect of simple digital filters: (A) a real-world pulse, (B) a noisy, sampled version thereof, (C) a series of average of 5 samples each, and (D) a running average more than 5 samples.

This type of filter is known more generally as a "finite impulse response" or FIR filter. As you can see in Fig. 7.17, the response to an impulse dies out after as many samples as the averaging interval includes.

In this example, all five samples are averaged with equal contributions. This is by no means the only possibility: One might let the center samples prevail, by assigning the different samples different weights. For instance, sticking with the same example, we can average $0.25\ y_1 + 0.5\ y_2 + y_3 + 0.5 y_4 + 0.25\ y_5$ (and divide by 2.5). The series of weights

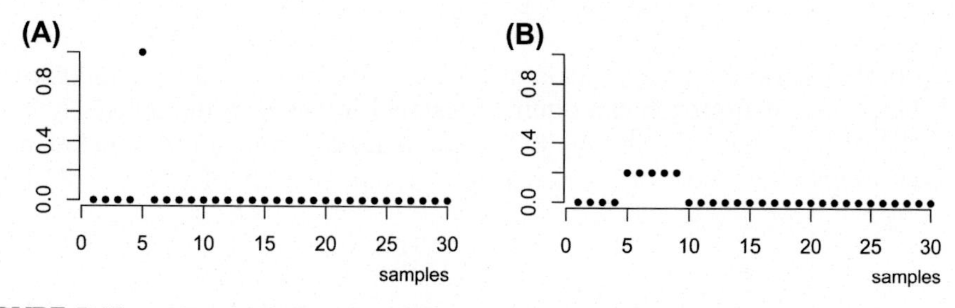

FIGURE 7.17 A digital impulse (A) and the response of a 5-sample running average filter to it (B).

(0.25 0.5 1 0.5 0.25) is again called a window (see section Signal Windowing above). The window with all values equal is a rectangular window. Different window shapes may be used to get specific results.

Analog filters, such as an RC filter, have an unlimited (infinite) impulse response. For example, a capacitor, charged at t = 0, will lose its charge asymptotically, its charge formally never reaching exactly 0. Although digital computations are limited to discrete steps, it is nonetheless possible to simulate such asymptotical, real-world behavior with digital filtering algorithms. These are called (you might have guessed) infinite impulse response (IIR) filters. They are also known as recursive filters because the filter output is used over and over again. The simplest version is a simulation of an RC low-pass filter. The procedure is simple indeed: at each sample point, take 20% of the new sample and add it to 80% of the existing value. This is the new, filtered, value. Repeat the process over the whole digitized signal (i.e., the sweep).

In an equation, each output value Y(t) at time t depends on the previous output value Y(t−1) and the input value X(t):

$$Y(t) = 0.2 \times X(t) + 0.8 \times Y(t-1), \text{ or, more general: } Y(t) = p \times X(t) + (1-p) \times Y(t-1)$$

The factors may be varied over a wide range, provided the two factors add up to 100%, to keep the signal amplitude unchanged (hence p and $1 - p$). Using larger values would introduce a gain, which should not pose a problem, by the way.

Taking $p = 20\%$ again, the impulse response would consist of the numbers 1, 0.8, 0.64 (i.e., 0.8×0.8), 0.512, 0.4096, and so on. As you can see, this series approaches 0 in an exponential way. However, despite the "infinite" in the name, the response will reach 0 in a finite time, by rounding errors, i.e., when the smallest quantization level has been reached. The time constant is approximately equal to the inverse of the factor p.

This simple example may serve to show the principle of digital filtering. However, the art of digital filtering has grown into a specialized branch of digital technique. So far more sophisticated and specialized filtering methods can be applied with success to electrophysiological data. The following section is an example of such a method.

FOURIER FILTERS AND NONCAUSAL FILTERS

Electrophysiological data is almost always recorded on hard disk by a computer for offline analysis. This creates the possibility to acquire the data with maximal bandwidth and filter it later depending on the type of analysis chosen. We start with a set of data in digitized form. In general, there are two approaches to the problem of digital filtering: (1) the frequency domain approach and (2) the time domain approach. However, the distinction between the two is not sharp, as one approach can often be restated in terms of the other.

After the discovery of the FFT, filtering in the frequency domain has become the most popular one. According to the theory of Fourier, a signal can be thought of as a sum of sine waves of different frequencies, amplitudes, and phase shifts. It is then possible simply to remove unwanted frequencies. This process can then be resumed into three steps:

1. Calculation of the sine wave spectrum (Fourier transform).
2. Removing or otherwise manipulating of the sine wave parameters, such as their amplitudes.
3. Reconstitution of the filtered signal by an inverse Fourier transform.

A simple example as shown in Fig. 7.18 may illustrate the procedure. A pulse-shaped signal was recorded against a background of 50 Hz mains interference (upper left). The sine wave amplitude spectrum was then taken (step 1, upper right). The large peak at 50 Hz and its harmonics at 100, 150, 200, and 250 Hz were then removed (step 2, lower right), and the signal was reconstituted by Fourier transformation (FT) (step 3, lower left).

Filtering in the time domain is based on the convolution integral of Eq. (7.2):

$$y(t) = \int_0^\infty x(t)\cdot h(t-\tau)d\tau \quad \text{or in shorthand:} \quad y(t) = x(t) © h(t)$$

where x contains the input data, h is the impulse response of the filter, and y is the filtered signal. In the case of a simple RC filter, the function h is:

$$h(t) = k\,e^{-k\cdot t}$$

where the RC time of the filter equals 1/k.

According to the convolution theorem (13), the Fourier transform of our RC filter, $h(t) = k\,e^{-k\cdot t}$, should give us the exponential time characteristics as shown in Chapter 2.

$$H(\omega) = k\int e^{-kt}\cdot e^{-j\omega t}dt$$

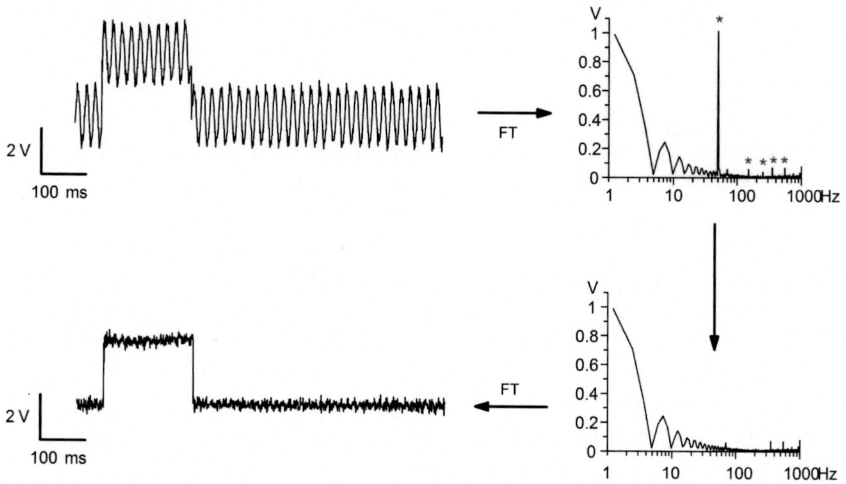

FIGURE 7.18 Frequency domain filtering. 50 Hz hum and its harmonics (*red asterisks*) are removed from the spectrum. The signal without 50 Hz is reconstituted to give a "clean" signal.

When integrating from 0 to ∞ and after separation of real and imaginary parts, this yields the same result as in Eq. (7.5), scaled by a factor of k:

$$H(\omega) = \frac{k^2}{k^2 + \omega^2} - \frac{jk\omega}{k^2 + \omega^2}$$

As we have seen in Eq. (7.6), the real part, or $k^2/(k^2 + \omega^2)$, is the so-called Lorentzian function. The amplitude spectrum of the RC filter can now be found by multiplication of $H(\omega)$ with its complex conjugate and taking the square root (for an explanation of the complex conjugate, see Appendix):

$$|H(\omega)| = \sqrt{H(\omega) \cdot H^*(\omega)} = \frac{k}{\sqrt{k^2 + \omega^2}}$$

We see that for low frequencies, $(\omega \ll k)$ $H(\omega) = 1$, and for high frequencies, $(\omega \gg k)$, the equation reduces to k/ω. Hence the slope of the filter after the cut-off point is 6 dB/octave (see Chapter 2).

This little exercise shows that filter algorithms that work in the frequency domain can be designed starting from impulse responses in the time domain and vice versa. Now let us take a closer look at the time domain RC filter. When we rewrite h(t) in discrete form it becomes h[i]:

$$h[i] = ke^{-ki}\Delta t$$

and the convolution becomes:

$$y[j] = \frac{k}{T}\sum_i x[j - i] \cdot e^{-ki}\Delta t$$

where x is an array that contains the data to be filtered and y, the filtered data. The array h contains the filter coefficients. T is the time span of the summation over n, and Δt is the sample interval between the data points. The process of filtering is shown in Fig. 7.19. The contents of array h and the input array x are shown on the top. The input (impulse) signal is 0 everywhere except at index 5. The filtering starts (step 1) with the x and h arrays aligned at index 0. The sum of products, $x[i] \cdot h[i]$, which is 0 in this example, is stored in y[4]. Then the array h is shifted one position to the right and the sum is stored in y[5], etc., until the end of the x array is reached. Upon exit, y contains the filtered impulse signal, which, in our case, is a scaled copy of the filter function itself. This illustrates why the function h is called the impulse response.

Note that the filter in Fig. 7.19 uses only points that are located in the past to calculate the present output. Such filters are called causal filters. These filters can also be constructed with electronic elements in real life. However, since digital filters are applied on data that is stored on disk, it is possible to use future data to calculate the present filter output. This is called noncausal. It is especially useful to resolve problems concerning the beginning of the output array. In the last example, the elements 0, 1, 2, and 3 of the output array were ignored.

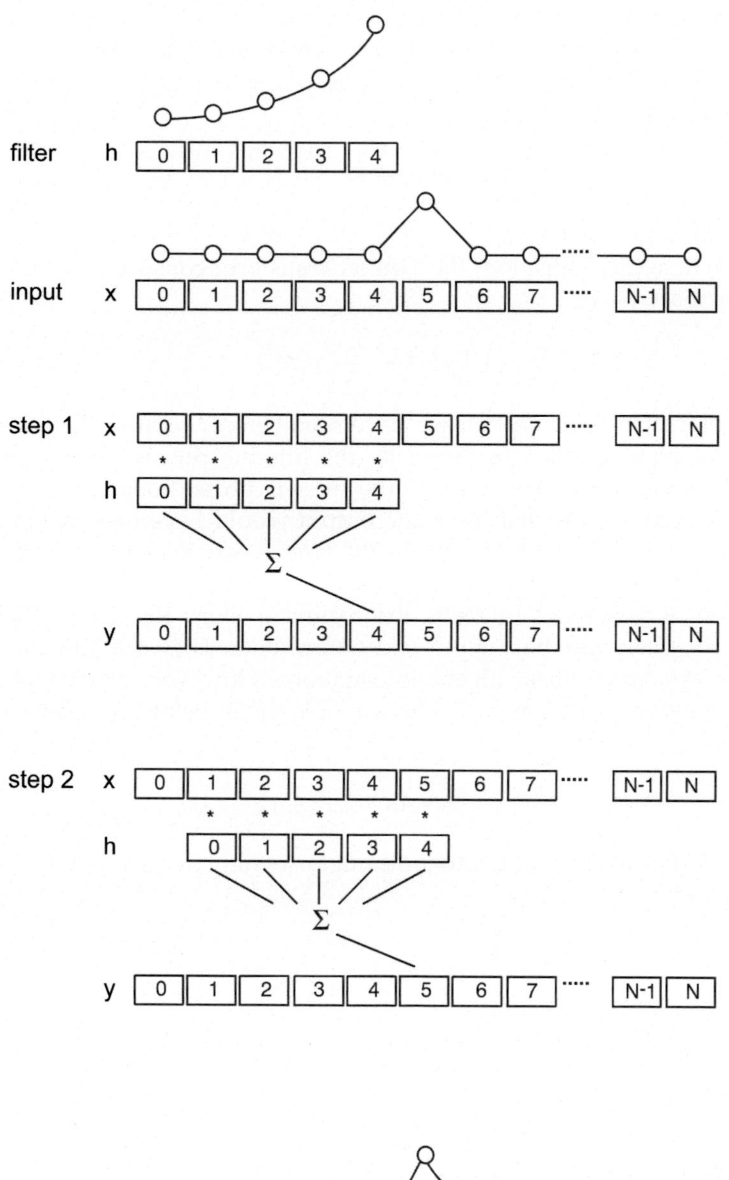

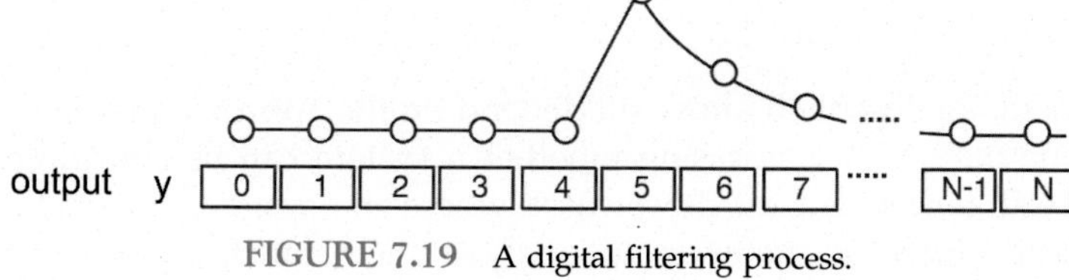

FIGURE 7.19 A digital filtering process.

Now that it is known what will come, it is easy to predict that the contents of the first four elements should be 0. An example of a noncausal filter is a simple extension of the causal RC filter discussed above. Suppose that the impulse response h is again an exponential, but now symmetrical around the $t = 0$ axis:

$$h(t) = ke^{-k}|t|$$

where $|t|$ is the absolute value of t.

As this is a symmetrical function, the Fourier transform consists only of a cosine series (there is no imaginary part):

$$H(\omega) = k^2/(k^2 + \omega^2)$$

This is the Lorentzian function that we have seen before. Because there is no imaginary part, there are no phase shifts introduced by the filtering process. It is a phaseless filter. For high frequencies ($\omega \gg k$), the slope of the filter is proportional to $1/\omega^2$, which corresponds to 12 dB/octave. Note that the same output would have been obtained if the data would have been passed through the causal RC filter twice, once in forward direction and once backward.

A special case of noncausal filters is the recursive filter. Recursive filters, which we met already as IIR filters, use both input and output data. They are difficult to design and imply solving a system of linear algebraic equations. However, once their characteristics are established they are very compact. The example given below is, again, a low-pass RC filter:

$$y[n] = (1 - a) \cdot x[n] + a \cdot y[n - 1]$$

Note that in contrast to the causal filter, this filter uses only two array cells at n and $n-1$. Its transfer function is[1,2]:

$$H(\omega) = \frac{1 - a}{1 - a \cdot e^{-j\omega\Delta t}}$$

from which it is not too difficult to find the relation between cut-off frequency (3 dB point) and the coefficient a.

NONLINEAR SYSTEMS ANALYSIS

The analysis methods described above all depend on the approximate linearity of the systems involved. The methods of assessing whether a system can be considered to be linear have been discussed earlier. For all those cases where the system under study is strongly nonlinear, we need different methods to analyze their function. In electrophysiology, nonlinear systems abound. Action potential generation of course is a very nonlinear process. The fact that some ion channels open or close when the membrane potential is changed means that membrane resistances are very variable and voltage dependent. This is quite different from the simple, constant resistors we assumed when treating filters, voltage dividers, and so on.

To analyze neuronal function, both analogue potentials and action potential series can be evaluated. In the latter case, some measure of spike activity, such as frequency or instantaneous frequency, must be derived to apply systems analysis. Spike trains behave in a nonlinear way very often. Even if a peripheral sense organ shows a spike frequency that is approximately linearly dependent on the stimulus, later stages of processing in the brain will generally react in a nonlinear way. Many brain cells for instance are silent until some condition is fulfilled. Therefore, we need a general method to analyze and describe nonlinear systems. These methods exist but are inevitably more complicated than their linear counterparts.

The Formal Method: Wiener Kernel Analysis

We saw earlier that the transfer characteristics of a linear system can be derived from the impulse response. If a system is essentially linear, the impulse response can predict the response to any arbitrary signal. For instance, if we would stimulate with two pulses in succession, the response will be simply the sum of two impulse responses, taking the time between them into account. However, this rule is not warranted at all in a substantially nonlinear system. As an example, we will analyze a simple nonlinear system that behaves linearly up to a certain hard boundary. In fact, *any* physical or physiological system will show this behavior when stimulated with strong inputs. Amplifiers, for instance, are bounded by their power supplies, often + and −12 to 15 V. Neuronal potentials are limited usually by the equilibrium potentials of sodium (about +60 mV) and potassium (−90 mV).

The output of such a system when confronted with one of its boundaries is illustrated in Fig. 7.20.

In the linear version, the response to a second impulse is shown to be added to the remains of the earlier impulse. It is easy to understand that the shape of the response is changed as a function of both the size and the relative timing if the input impulses. The latter is shown in Fig. 7.20B. It can be understood intuitively that, to describe the nonlinearity, one has to take both the amplitudes and the relative timing of the two impulses into account. In principle, however, an input signal is not limited to two impulses, and we should analyze the responses to three impulses with their amplitudes and relative timing, then four impulses, and so on.

This seemingly tedious task can be described formally by a series of functionals called Wiener kernels, named after Norbert Wiener and developed by Wiener after Volterra.

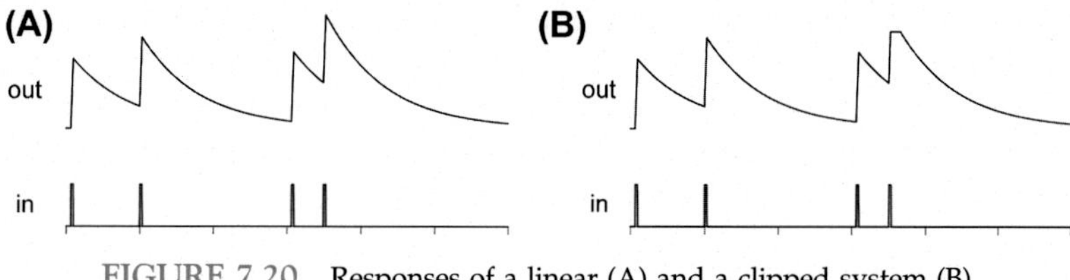

FIGURE 7.20 Responses of a linear (A) and a clipped system (B).

With this method, the full response of any system may be described by a series of kernels k_0, k_1, k_2, k_3, and so on. "Kernel" is mathematical jargon for a functional (a function of a function). For a complete description of this method, the reader is referred to the book by Marmarelis & Marmarelis.[2] A simplified explanation is given below.

The zero-order kernel, k_0, is time independent, constituting simply the DC value of the output in the absence of an input signal. The next two kernels depend on one and two times, respectively (the times of the input impulses shown above), and can be expressed as follows:

$$G_1 = \int_0^{\infty} h_1(\tau) \times (t - \tau)d\tau$$

$$G_2 = \int_0^{\infty}\int_0^{\infty} h_2(\tau_1, \tau_2) \times (t - \tau_1) \times (t - \tau_2)d\tau_1\tau_2 - P\int_0^{\infty} h_2(\tau_1, \tau_1)d\tau_1$$

Here, h_1 is the impulse response, hence the linear part of the transfer function, whereas h_2 reflects the nonlinear interactions between two input impulses. P is the power density (power/Hz) of the input signal. Without going into further detail, the reader can guess what the next term will look like: $G_3 = \int\int\int h_3$ (form with three taus) ... etc.

At first, one will be disappointed to learn that any nonlinearity can be described by an infinite series. Fortunately, the practice proves to be less hopeless. Early investigators computed the first three to four kernels and found out that most nonlinear systems, at least in physiology, can be described by the first two kernels: next to h_1, which describes the linear part of the transfer function, the nonlinearities show up mostly in h_2. Higher-order kernels are said to be almost "empty," i.e., they contain a very small fraction of the energy of the output signal, hence can be neglected for all but the most demanding situations.

Next, we will discuss what Wiener kernels look like. The zero-order kernel, h_0, is time independent and so constitutes a single number (viz. the DC value of the output). The first-order kernel, h_1, is a function of time (i.e., one time variable, τ) and so can be depicted as a line in a 2D graph of $h_1(\tau)$ versus τ. The second-order kernel, h_2, is a function of two time variables τ_1 and τ_2, and so can be depicted as a surface in a 3D graph of $h_2(\tau_1,\tau_2)$ against τ_1 and τ_2. This surface depicts the nonlinear interactions between two impulses. An example from neurophysiology, taken from a paper by one of the authors,[3] may elucidate the significance of the second-order Wiener kernel. Fig. 7.21A shows the response of secondary neurons in the brain of a catfish, processing electrosensory information. These neurons are often silent, responding only to some features of the sensory input (alternating electric fields in the water surrounding the fish).

The peaks in this "nonlinear landscape" show that the strongest nonlinearity occurs at certain combinations of τ_1 and τ_2. At first sight, this way of reporting the nonlinear behavior of a neuron may seem strange and hard to interpret. A convincing application of this technique lies in modeling, i.e., in predicting the response of the neuron in question from the found transfer function, in this case with the nonlinearity included. This is shown in Fig. 7.21B. The measured response of the neuron is simulated better if both h_1 and h_2 are used, in other words, when the approximated nonlinearity is taken into account.

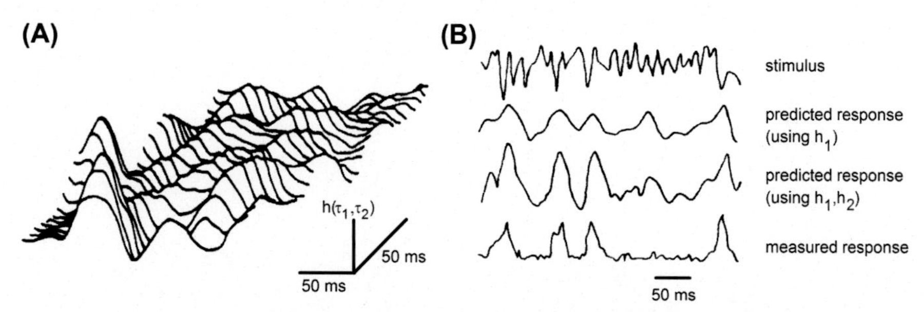

FIGURE 7.21 (A) Second-order Wiener kernel of a secondary electroreceptive neuron. Horizontal axis: time in ms, vertical axis: $h_2(\tau_1,\tau_2)$ in (arbitrary) units of spike frequency. (B) Prediction of responses of a secondary electroreceptive neuron to a pseudorandom electric stimulus (upper trace) using either h_1 or both h_1 and h_2. Bottom trace: actual response (averaged). Response peaks about 50 spikes/s. *From de Weille JR. Electrosensory information processing by lateral-line lobe neurons of catfish investigated by means of white noise cross-correlation. Comp Biochem Physiol. 1983;74A(3):377–680.*

The Informal Method: Output Shape Analysis

Independent of this formal mathematical approach, nonlinearity can be described qualitatively, viz. by a sort of taxonomic determination of the type of nonlinearity involved. This means that the type of distortion is first determined by graphical inspection of the data, preferably followed by assigning a quantitative value to the degree of nonlinearity.

This can be applied not only to the electronic instruments used in recording but also to neurobiological subjects. Even then, examples are often taken from electronics. In an interesting, but often poorly understood paper, MacKay[4] tried to explain the so-called "power law" (a.k.a. Stevens's law) found in perception research with the hypothetical nonlinear behavior of neurons.

The informal identification method of nonlinear systems (processes) works by comparing the type of output nonlinearity of the system in question with a palette of shapes of known nonlinear processes. Often, the shape of the output signal contains strong cues to the nonlinearity involved. Fig. 7.22 shows graphics pairs of an I/O curve (upper graph) and the concomitant signal shape (lower graph) in time.

In the upper left corner, the figure shows the linear I/O curve, together with a sinusoidal (i.e., undistorted) signal shape. The various forms of nonlinear processes lead to different forms of distortion, which can be recognized from a physiological signal.

Note that a truly logarithmic curve cannot exist: It has a singularity (infinite output) when the argument (input) is zero. In physiology, however, approximate logarithmic processes do occur often, albeit in a limited amplitude domain. The other nonlinearities shown have no singularity but may nevertheless be limited to certain amplitude domains. The square root curve for instance is limited mathematically to nonnegative values.

Apart from being more complex, nonlinear systems analysis is also more exciting, which we hope you will agree upon after reading the following section.

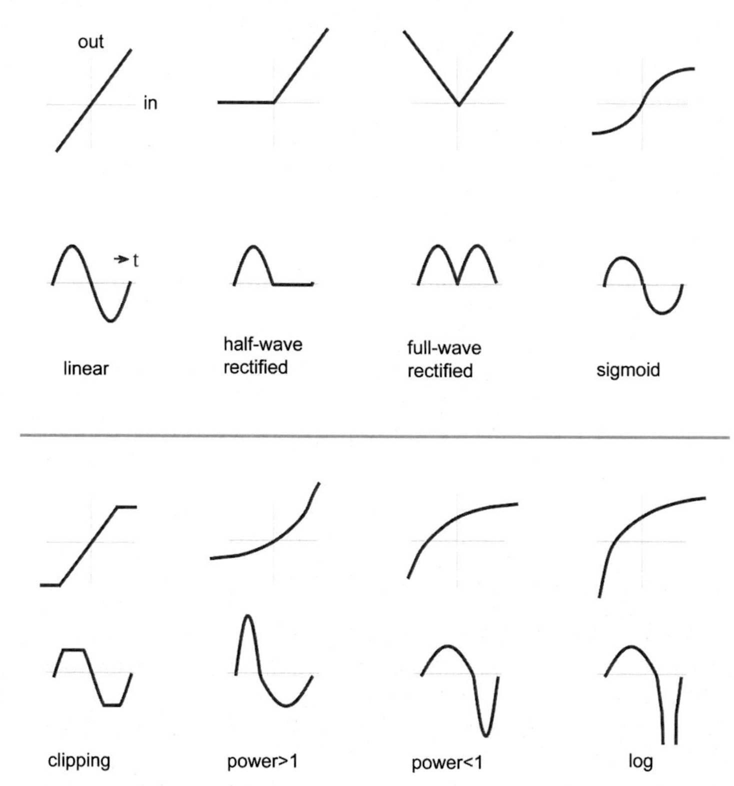

FIGURE 7.22 Palette of nonlinear operations. For each case, the upper graph shows the I/O characteristic, and the bottom graph, the resultant distortion of a sinusoidal signal. The linear case, top left, is given for comparison.

The Importance of Nonlinearity

In view of the various complications when dealing with nonlinear systems, it will come as no surprise that systems analysis is not very popular, at least in biology. In many cases, keeping stimuli small, and exploiting the other tricks mentioned before, will simplify the analysis by keeping the responses approximately linear. However, limiting ourselves to linear systems would sell our analytical capabilities short. In fact, linear systems can be summarized largely as high-pass filters, low-pass filters, and gain, or any combination thereof.

Nonlinear systems can be analyzed to a greater extent than linear ones. The principle is illustrated in Fig. 7.23. The upper row, Fig. 7.23A, shows how a sinusoidal signal is altered (i.e., filtered) by a cascade of a low-pass and a high-pass filter. The (qualitative) graph shows the output being both attenuated and phase-shifted. In the next row (Fig. 7.23B), the order of the two filters is reversed, which yields the same output. Therefore, a physiologist investigating a cascade of linear processes cannot determine, at least by input/output analysis, what the order of the components is. This is regrettable, since we want to "make the black box transparent."

The nonlinear cascade shown in Fig. 7.23C and D shows a high-pass filter and a clipping circuit. This yields completely different results depending on the order. If the clipping is

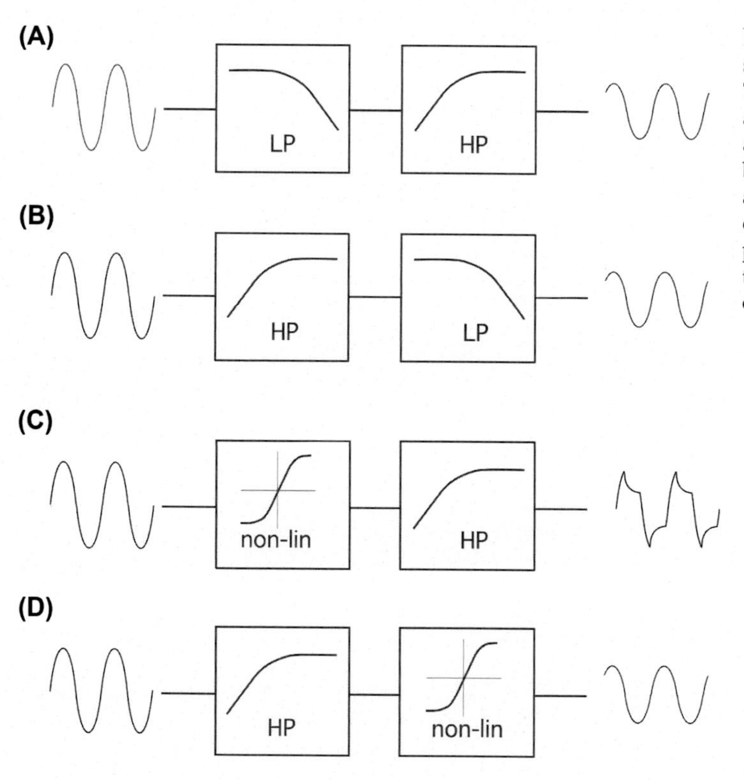

FIGURE 7.23 The order of subsystems: linear versus nonlinear. The order of linear subsystems (A and B), in this case a low-pass (LP) and a high-pass (HP) filter, cannot be determined by input/output analysis, since they yield identical outputs. If one or more nonlinear processes (C and D) are involved, the output can depend on the order of the subsystems.

performed first (Fig. 7.23C), the sine wave is distorted (flat tops). This distorted signal is then high-pass filtered, which leads to the odd signal shape shown. If the filter is the first segment, however, the sine may be attenuated so much that no clipping occurs afterward (Fig. 7.23D).

For physiologists, this enhances systems analysis into a tool that may reveal the order of subsystems without the need for an electrode or other physical contact in the middle, i.e., between the subsystems.

References

1. Hamming RW. *Digital filters. Englewood Cliffs*. New Jersey, USA: Prentice-Hall; 1983.
2. Marmarelis PZ, marmarelis VZ. *Analysis of physiological systems. The white noise approach*. New York: Plenum Press; 1978.
3. de Weille JR. Electrosensory information processing by lateral-line lobe neurons of catfish investigated by means of white noise cross-correlation. *Comp Biochem Physiol* 1983;**74A**(3):377–680.
4. MacKay DM. Psychophysics of perceived intensity: a theoretical basis for Fechner's and Stevens' laws. *Science* 1963;**139**:1213–7.

CHAPTER

8

Recording of Electrophysiological Signals

OUTLINE

In Chapters 4–6, we have provided some elements of how and with what to record electric activity. In this chapter, the events that may actually be recorded are discussed. To do this, we start with the intracellular recording, since it is the least complicated to interpret. Intracellular activity is at the base of the small signals that can be picked up with extracellular electrodes. The interpretation of these signals is less straightforward and requires understanding of volume

Introduction to Electrophysiological Methods and Instrumentation, Second Edition
https://doi.org/10.1016/B978-0-12-814210-3.00008-6

conduction that was presented in Chapter 6. Variations in membrane capacitance can be measured with voltage-clamp equipment, thus giving a possibility to follow secretion from a cell. Alternate methods to do this are discussed. Some electrophysiological techniques require special attention such as recording ion transport over artificial lipid double layers or recording from brain slices. The chapter ends with approaches to automate voltage-clamp and an evaluation of high-throughput alternatives to electrophysiology.

THE INTRACELLULAR RECORDING

The intracellular records in Fig. 8.1 were obtained from motoneurons in the whole-cell patch-clamp mode. It can be seen from this figure that: (1) sodium currents are switched on more rapidly than the potassium currents, (2) notwithstanding a maintained depolarizing potential, the sodium current inactivates (turns off) after it has been turned on, and (3) sodium currents are switched on at lower membrane potentials (~ -40 mV in this example) than the potassium currents shown here (~ 0 mV). These properties give the action potential its shape. The inflow of sodium ions at the beginning of the action potential causes the plasma membrane to depolarize, thus causing an increase of the inward sodium current in a positive feedback loop. This comes to an end as the sodium current inactivates and the potassium current repolarizes the membrane potential to its resting level.

In 1952, Hodgkin and Huxley mathematically described the time and voltage dependencies of the sodium and potassium currents in the squid giant axon.[2] Fig. 8.2, which is adapted from their article, shows the contributions of sodium (I_{Na}) and potassium (I_K) currents to the action potential (V). Their model will be discussed in more detail further on in the next chapter.

THE EXTRACELLULAR RECORDING

Consider an electrode that is placed in the vicinity of an electric dipole in a conductive medium as in Fig. 8.3A. As we have seen in Chapter 7 in the section about volume conduction, electric field lines extend from the positive pole to the negative pole and hence a voltage gradient exists between the poles. The differential amplifier measures no voltage difference between probe electrode and distant ground electrode as long as both electrodes are far from the dipole. But that changes when the dipole moves to the right passing under the electrode. First, the positive pole causes a positive potential difference at the input of the amplifier, and second, when the negative pole passes by, a negative potential is sensed by the amplifier. This produces a biphasic voltage fluctuation at the output of the amplifier as shown in Fig. 8.3B.

A very similar situation occurs when recording extracellularly from a firing neuron. Even if the potential for the medium around a cell is defined to be zero reference, membrane currents create small potential differences that can be picked up by electrodes. In Fig. 8.3B, an action potential progresses along an axon. As sodium channels open up, Na^+ moves into the neuron thus creating a current sink. The sink is replenished from the close by surrounding fluid, which forms a current source. Hence, as in A, field lines between source and sink create a voltage gradient that can be sensed by the extracellular electrode. When the action potential passes, the electrode first senses the positive source, then the negative sink, and finally the

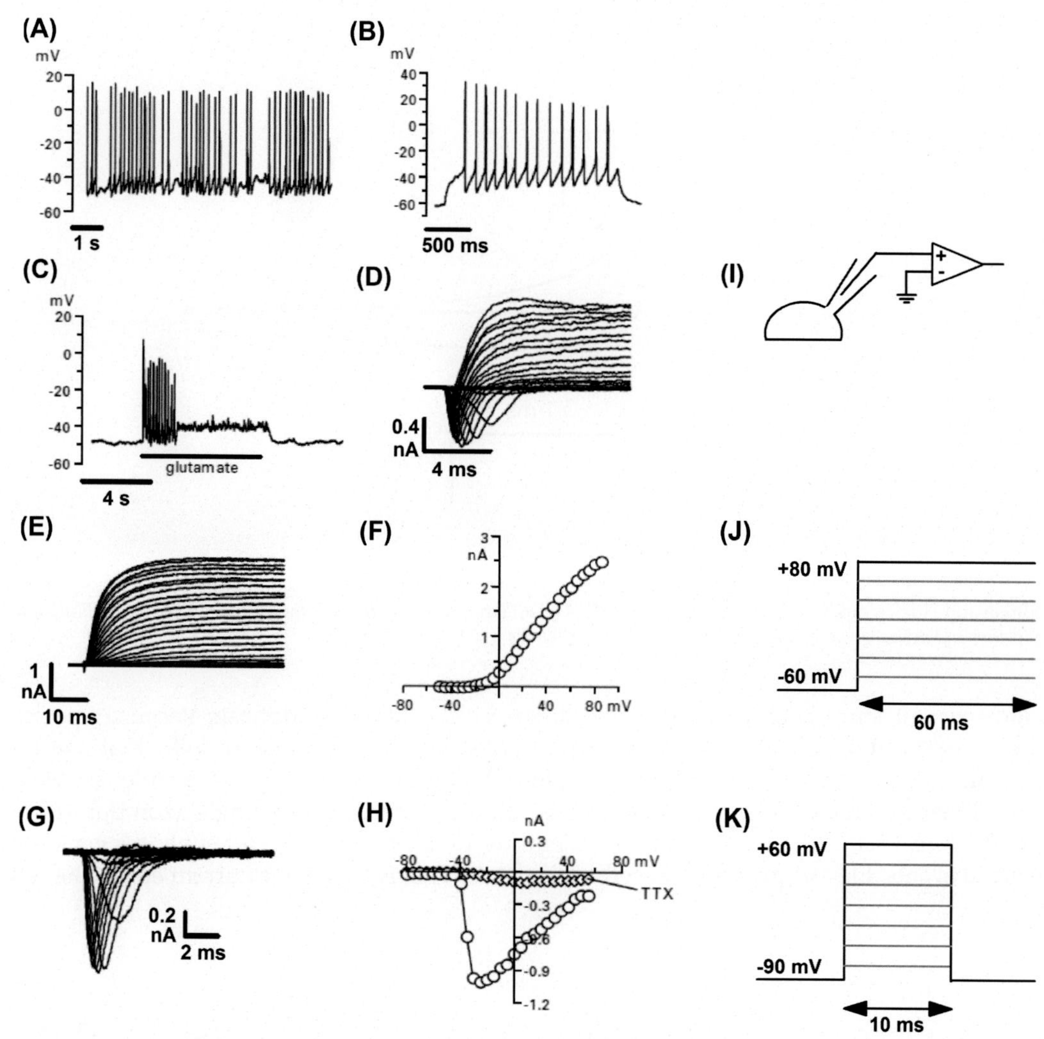

FIGURE 8.1 Membrane potentials (A–C) and currents (D–H) were recorded from isolated rat spinal cord motoneurons in vitro using the whole-cell patch-clamp technique as schematized in (I). The neuron in (A) is spontaneously active. An inward current, injected into a different neuron (B), raises the potential just above the threshold for action potential firing. Note the high threshold and sustained electrical activity, which is typical of motoneurons. The neuron in (C) was subjected to extracellular perfusion with 100 μM glutamate, depolarizing it. In (D–H), neurons were voltage-clamped and subjected to steps from the holding potential of −60 (D–F) or −90 mV (G–H) to various depolarizing potentials according to the protocols shown in (J) and (K). The records in (D) were obtained in K+- and Na+-rich solution. Na+ was replaced by choline+ in (E) and (F), thus suppressing inward Na+ currents. The steady-state K+ current shown in (E) at the end of the voltage impulse as a function of membrane potential is shown in (F). Hence, the K+ current increases with depolarization. In (G) and (H), intracellular and extracellular K+ was replaced by Cs+ and tetraethylammonium (TEA), thus suppressing outward K+ currents. The transient current that remains (G) is largely carried by Na^+, because it is almost completely suppressed by 100 nM of the specific inhibitor of voltage-dependent Na+ channels, tetrodotoxin (TTX) (H). (H) depicts the peak Na+ current as a function of membrane potential. The small current that remains might be carried by Ca2+. *Adapted from Rakotoarivelo C, Petite D, Lambard S, et al. Receptors to steroid hormones and aromatase are expressed by cultured motoneurons but not by glial cells derived from rat embryo spinal cord.* Neuroendocrinology *2004;**80**(5):284–97.*

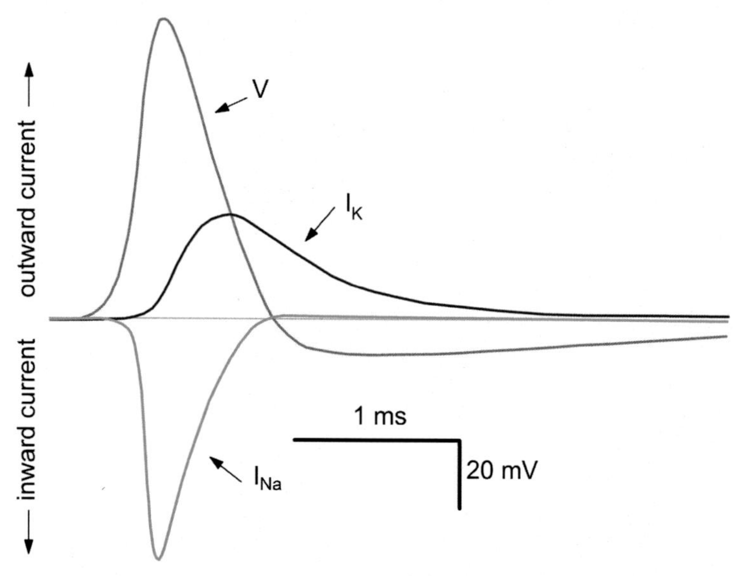

FIGURE 8.2 The Na^+ and K^+ currents during the course of the squid giant axon action potential according to the Hodgkin and Huxley model.[2] *Hodgkin AL, Huxley AF. A quantitative description of membrane current and its application to conduction and excitation in nerve. J Physiol 1952;117(4):500–44.*

trailing current source; the recorded extracellular spike is therefore triphasic. Because a potassium current activates at the end of the action potential, the extracellular spike that one records in practice has an accentuated trailing bump. Geometry plays a role as well. Extracellular electrodes are normally not being used to record from a single axon but rather from nervous tissue where a multitude of neurons, dendrites, axons, and synapses generate electrical events. Electric processes in cells are neutral. That is to say, if a current enters the cell

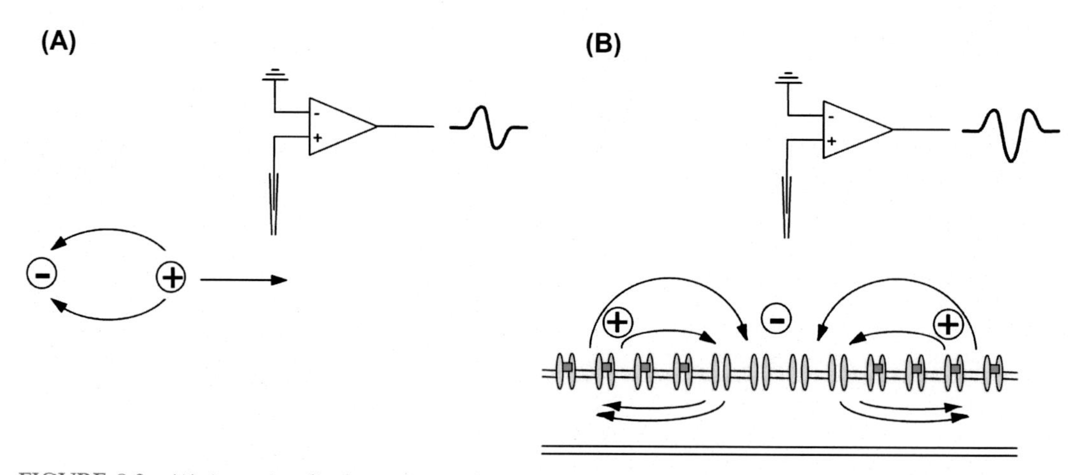

FIGURE 8.3 (A) A moving dipole passing under an electrode. (B) An action potential traveling down an axon sensed by an extracellular electrode.

in one place, the same amount of current must leave it at some other place. Therefore, a current source in one place is necessarily accompanied by a sink in some other place. For this reason, the potential measured by an extracellular electrode can be thought of as the summation of the fields of many dipoles located at different distances from the electrode. The potential thus measured is called the field potential, sometimes erroneously called the *local* field potential if it is recorded with a small electrode (e.g., microelectrode).

THE ELECTROCARDIOGRAM

The recording method best known in public is surface recording from the human skin. This started among others with the pioneering work of Einthoven, who recorded the human electrocardiogram (ECG) using large metal plate electrodes connected to the arms and legs of the person. The only recording device at that time was the string galvanometer, the tiny movement of the wire being recorded on photographic paper with a built-in microscope.

Today the ECG, recorded with a variety of recording methods, is an indispensable diagnostic method in the heart clinic. The basic set-up for a routine ECG is still the so-called "Einthoven triangle," using recording electrodes on the two arms and one leg of the subject (Fig. 8.4, left). The metal electrodes have been replaced by Ag/AgCl electrode foam pads such as those in Fig. 8.10.

This type of recording yields the familiar ECG shape with its sharp QRS peak and the more gradual P and T components (Fig. 8.4, right). Despite the large distance between the heart and the surface electrodes, the ECG is a form of single-unit recording. However, the shape of the ECG is largely determined not by the shape of the cardiac action potential but by the way the action potential, initiated in the sinoatrial node and propagated through all heart chambers, moves through space. Each phase of the ECG is an indicator of one of the electrical processes that together cause the heartbeat. The P wave is generated by the progressive depolarization of the atrium fibers. After a short delay imposed by the atrioventricular node, depolarization of the His bundle travels downward to the bottom of the ventricles causing the Q wave, after which the Purkinje fibers in both ventricular walls transmit the signal to the contractile

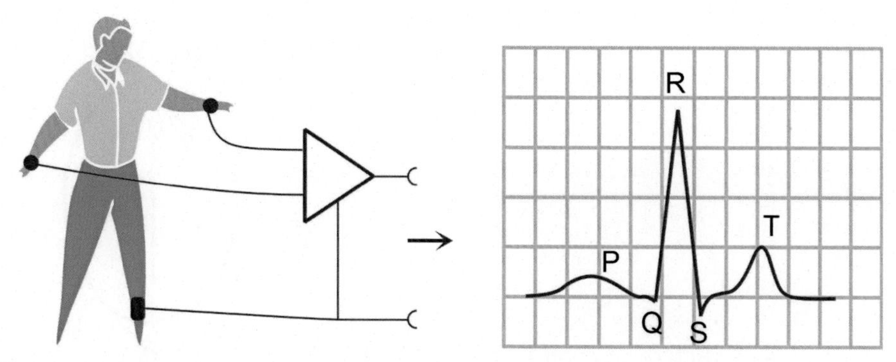

FIGURE 8.4 ECG recording with the Einthoven triangle (left) and the resulting ECG waveform (right). The amplitude of such an ECG is about 1 mV.

muscle cells causing the rapid R upstroke. The S wave coincides with the beginning of the ventricular contraction at the apex. The contraction continues until the muscle cells repolarize giving rise to the T wave. Like the extracellular recording discussed above, the electrodes attached to the limbs sense a moving dipole. The trajectory of the electric dipole during a cycle in two-dimensional space is shown on the bottom right of Fig. 8.5.

For a more detailed picture of heart functioning, so-called precordial electrodes are used, i.e., a row of electrodes on the chest and so in front and on the side of the heart (Fig. 8.6).

The electrical demands for ECG recording can be derived from the electrodes and the amplitude of the signal obtained: ECG electrodes are usually fairly large made of stainless steel, plastic foam pads containing carbon or AgCl contacts and connected to the skin with a salty, well conducting electrode gel. The relatively large surface area, together with the salt, warrants a low resistance (more correctly, impedance) and so does not put severe demands on the input impedance of the ECG amplifier: about 1 MΩ suffices. The voltage gain must be fairly high (say 1000 x), and filters must be built in to block hum and reduce noise. The necessary bandwidth is about 1–30 Hz.

Recording from human beings implies an electrical contact between the patient and the electrical apparatus. Since the latter is usually powered by the local mains, special medical instruments are needed, i.e., extra insulated and protected amplifiers, recorders, etc. The problems of medical recording and the specifications of the suitable equipment are discussed in Appendix.

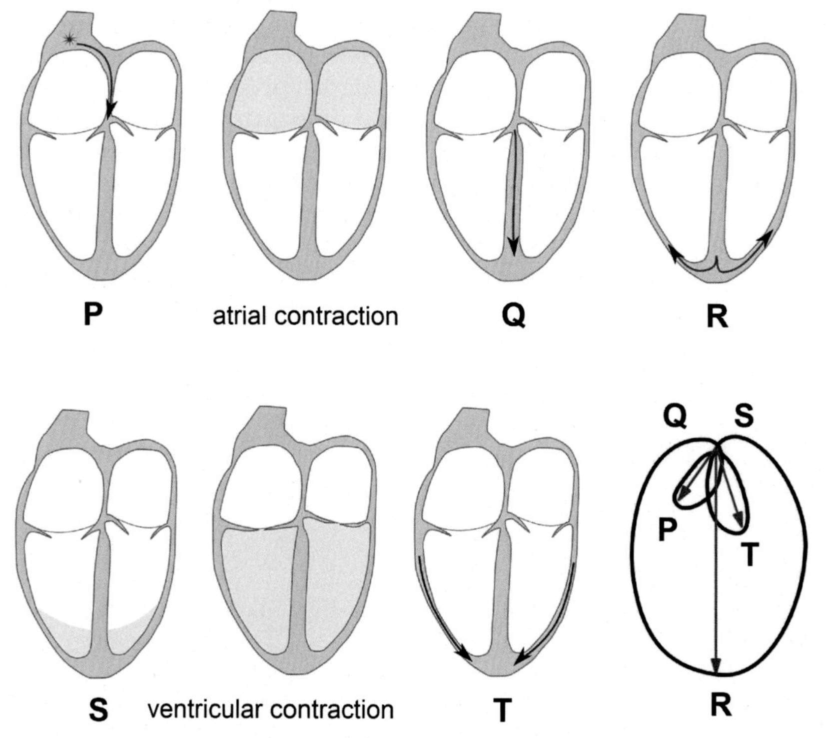

FIGURE 8.5 Different phases of the electric activity in a cardiac cycle. The figure at the bottom right shows the direction of the electrical dipole during the cycle.

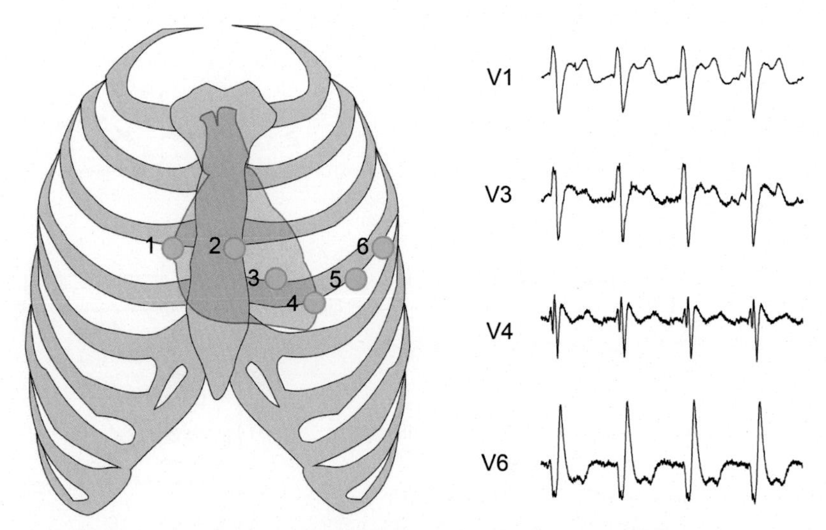

FIGURE 8.6 Placement of precordial electrodes in green on the surface of the chest with respect to the heart shown in red (left) and the electrical cardiac signals recorded at several positions (right).

The currents that make up the action potential of cardiac cells

Upon depolarization of the cardiac ventricular cell (from -90 mV to about -70 mV), a rapidly activating and slightly more slowly inactivating Na^+ current are elicited that further depolarize the cell to positive potentials. This is very similar to what happens when a neuronal action potential is elicited (Fig. 8.7, left). Outward rectifying K^+ channels (Kto) are transiently opening at the Na^+ peak, contributing to the inactivation of the Na current. The Na^+-induced depolarization opens slowly inactivating L-type Ca^{2+} channels, causing Ca^{2+} influx and muscle contraction. This is followed by Inactivation of the L-type Ca^{2+} current and slow activation of the delayed rectifier current (IKr) in conjunction with the inwardly rectifying IK1 current. These two potassium currents together repolarize the membrane. During the refractory period (RP) of a duration of about 200 ms, no new action potentials can be elicited by definition. Other currents play a role in the cardiac action potential, such as the Kh (hyperpolarization-activated current) also known as Kf (funny K^+ current), which is nonselective and mostly carries Na^+ at the resting potential. Since it is not a K current, If is a better name. This current slowly depolarizes the membrane potential eventually surpassing the firing threshold. Most of these ion currents are also present in the plasma membrane of atrial pacemaker cells albeit in different proportions. The IKf current is especially important for the pacemaker rhythm and explains why the heart can beat autonomously without any external input. Both the Ca^{2+} current and the IKf current are stimulated by adrenaline via cyclic AMP-activated protein kinase A phosphorylation, thus increasing heart rate and contraction force.

Several drugs modulate cardiac ion currents, for example:

Ivabradine is a drug that is used to treat high sinusoidal rhythm. It blocks IKf, thus reducing heart rate. Verapamil and Diltiazem inhibit L-type Ca^{2+} channels and produce a

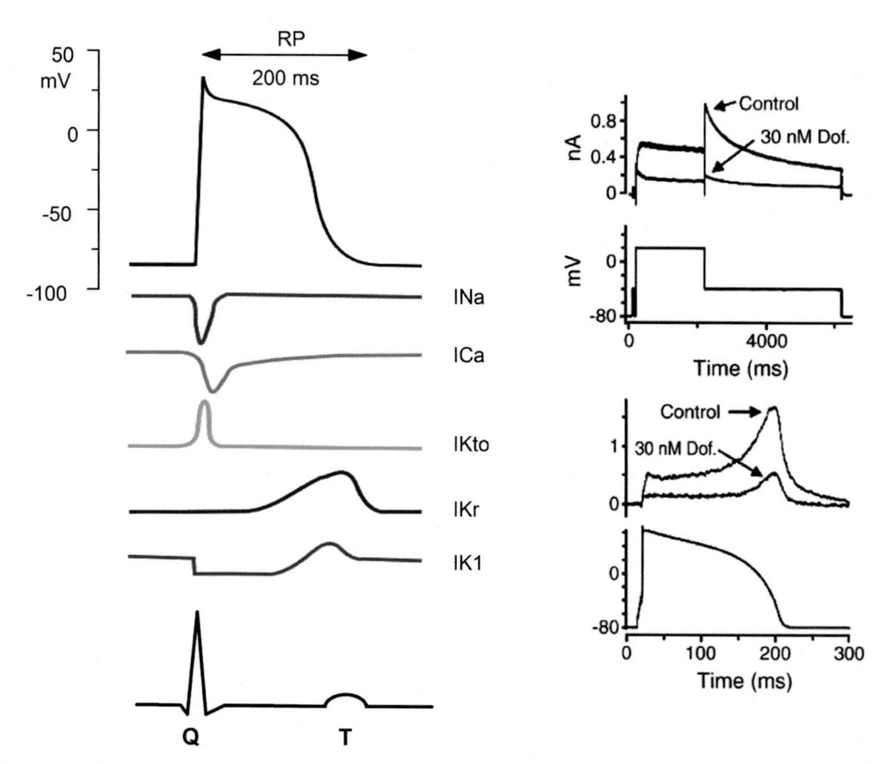

FIGURE 8.7 Left panel: Some of the most important ion currents that play a role in the action potential of a ventricular muscle fiber. RP = refractory period. The ventricular action potential coincides with the period between Q (INa activation) and T (IKr activation) waves. The panel to the right shows the activation of a hERG (IKr) current under voltage-clamp obtained from HEK 293 stably expressing hERG. The command voltage was given the form of a cardiac action potential in the lower figure. Note the difference of time scale in the two figures. The IKr current is shown here to be blocked by Dofetilide (Dof.). *Right panel adapted from Milnes JT, Witchel HJ, Leaney JL, Leishman DJ, Hancox JC. Investigating dynamic protocol-dependence of hERG potassium channel inhibition at 37 degrees C: Cisapride versus dofetilide. J Pharmacol Toxicol Methods 2010;**61**(2):178–91 courtesy Elsevier press.*

diminution in heart rate and force. Atropine blocks muscarinic acetylcholine receptors, thus increasing heart rate. The inhibition of the IKr current by the antiarrhythmic agent Dofetilide is shown in the right panel of Fig. 8.7, whereas in the case of Dofetilide, the effect on cardiac rhythm is what was aimed for; unfortunately, many drugs treating a disease that has nothing to do with cardiac function can also affect IKr. Drugs that block IKr, in particular hERG channels that make up a fast component of IKr, are potentially life-threatening because they prolong the repolarization period (QT interval). QT prolongation may lead to either early after-depolarizations and/or multifocal contractions (*torsade de pointes*). In Figs. 8.8 and 8.9, prolongation of the QT interval by the antipsychotic drug Haloperidol is illustrated.

Ventricular pressure is much reduced during a *torsade de pointes* causing the blood circulation to come to a stop. In the past, drugs that dangerously augmented the QT interval had to be withdrawn from the market such as the antihistaminic Astemizole, the anti-incontinence drug Terodiline, the gastroprokinetic drug Cisapride, and many others.[6] As

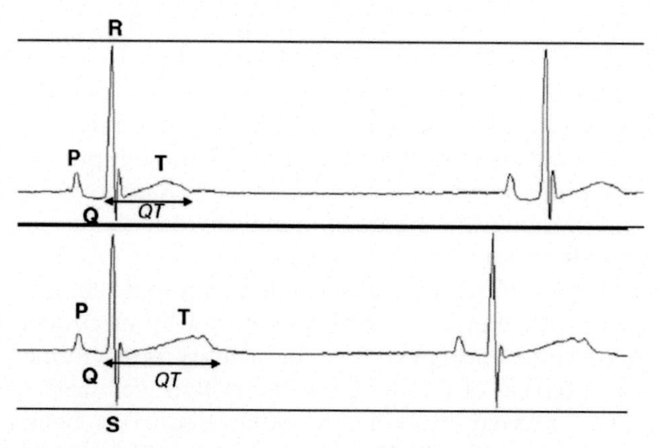

FIGURE 8.8 Upper panel: Electrocardiogram from a dog at predose baseline. Lower panel: Electrocardiogram approximately 5 h postadministration of 3 mg/kg of haloperidol.[4] *Chaves AA, Zingaro GJ, Yordy MA, et al. A highly sensitive canine telemetry model for detection of QT interval prolongation: studies with moxifloxacin, haloperidol and MK-499. J Pharmacol Toxicol Methods. 2007;56(2):103–114. Courtesy Elsevier press.*

of 2002, worldwide measures have been taken to prevent drugs having QT prolongation as a side effect to enter the market. Since then, in vitro electrophysiology measurements on hERG-expressing cells and ECG recording from nonrodent animals have become compulsory elements in preclinical development of new drugs. The primordial role of electrophysiological methods in this context cannot be underestimated. How the need of high-throughput clinical screening has changed electrophysiology methods will be explained in further detail in the section on automation.

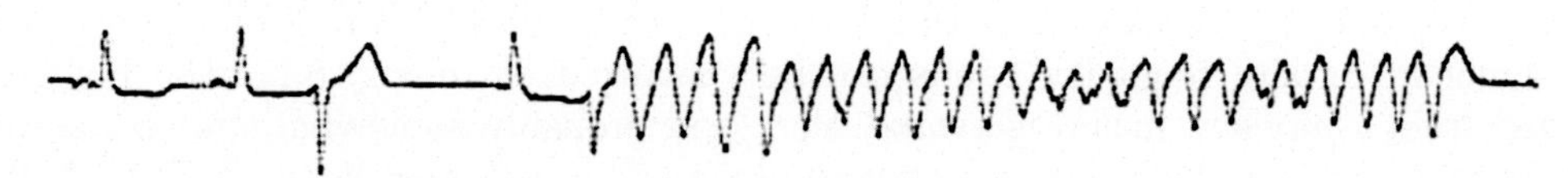

FIGURE 8.9 *Torsade de pointes.*[5] *Gowda RM, Khan IA, Wilbur SL, Vasavada BC, Sacchi TJ. Torsade de pointes: the clinical considerations. Int J Cardiol. 2004;96(1):1-6. Courtesy Elsevier press.*

THE ELECTROENCEPHALOGRAM

The beating heart generates only a relatively weak electrical field resulting in about 1 mV picked up by electrodes on the skin. The situation becomes worse when trying to pick up electrical signals generated by the brain's neurons at the surface of the skull. As we have seen above, extracellular recording of a single neuron is possible as long as the measuring electrode is close to the cell. The signal strength drops of rapidly with increasing distance and disappears in the background noise when it exceeds a few hundred micrometers. For this reason, it is impossible to record activity of single cells in the brain with electrodes placed on the skull. Notwithstanding these difficulties, it has been shown by the neurologist H. Berger in the 1920s that weak electrical signals can be recorded (see Fig. 8.12A). These signals arise from organized brain structures that fire their neurons in a synchronous fashion. The neocortex of mammals is such a well-organized brain tissue, which contains several layers of pyramidal cells that project their axons radially and receive input from horizontal fibers (see the exquisite drawing of the Spanish neuroscientist S. Ramon y Cajal, Fig. 8.11A). The combined electrical field of many thousands of identically oriented dipoles generates a field large enough to traverse the skull.

As can be seen from Fig. 8.11B, activity of a single pyramidal cell, recorded with a microelectrode, correlates well with the gross activity recorded by electroencephalography. The autocorrelation functions show moreover that the activity is rhythmic. The theta rhythm (4–7 Hz) in that figure is typical of the light anesthesia that was used to sedate the animal and normally appears in relaxed individuals. Other frequency bands are delta (below 4 Hz) occurring during sleep, alpha (8–13 Hz) associated with reflection, and beta (above 13 Hz) which shows during stress and alertness, in general. Alpha waves are also elicited by closing the eyes.

Electroencephalogram (EEG) electrodes are usually somewhat smaller than their ECG counterparts (often small metal discs with or without silver chloride coating, Fig. 8.10) but are still large enough to have a fairly low impedance (a few kiloohms at 50 Hz). Gain demands for an EEG amplifier are, however, very high because the signal is usually only a few tens of microvolts in amplitude. Limiting the bandwidth with higher-order filters is also important.

Both time domain and frequency domain are important for the processing of EEG signals. Like cardiograms, EEGs are usually analyzed visually by an expert. Some types of EEG show characteristic patterns called "spindles" in the time domain, where the amplitude is waxing and waning gradually. In addition, the amplitude spectrum, obtained by Fourier transforming the encephalic signal, may give additional cues. The concentration of the energy of an EEG signal in certain frequency bands can be visualized as in Fig. 8.12. The EEG was recorded from a monkey that was awake, alert, and probably somewhat stressed, as witnessed by the dominant beta activity. After administration of an anesthetic, the EEG rhythm was mostly theta.

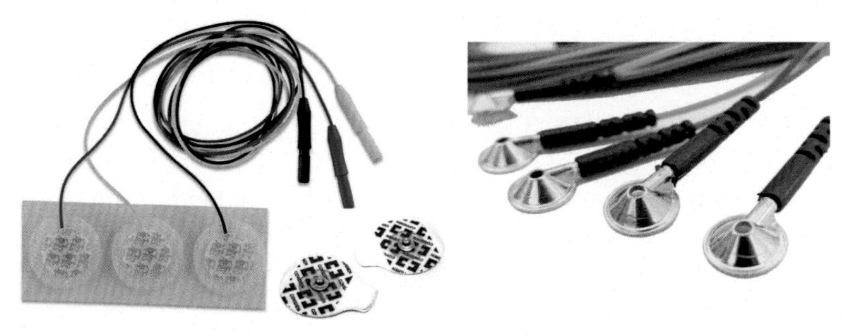

FIGURE 8.10 Typical foam pad ECG electrodes (left). The active flip sides of the ECG electrodes are pregelled EEG electrodes (right). The hole in the electrodes is to back-fill with conductive gel once they are glued onto the skull (with double-sided adhesive seals).

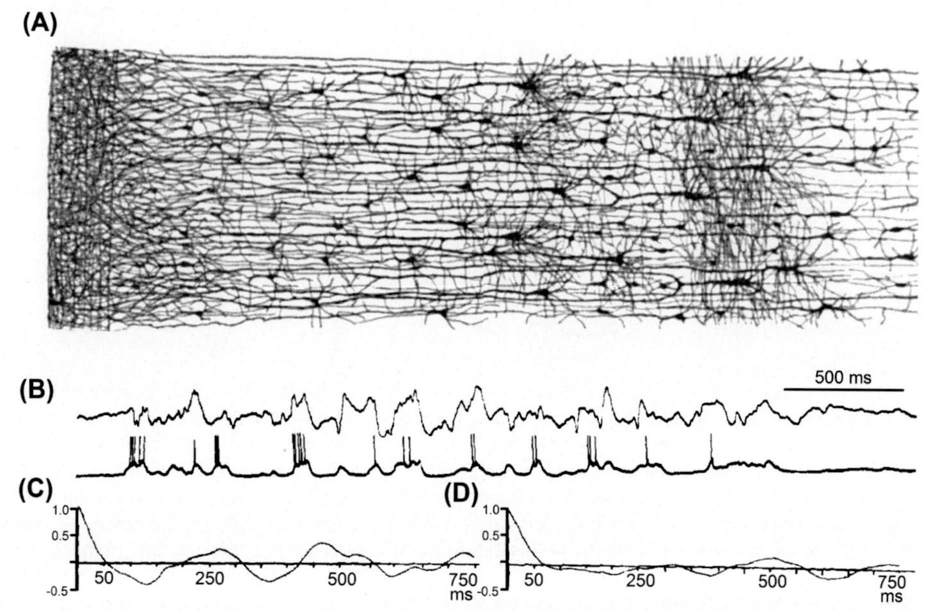

FIGURE 8.11 (A) Drawing of the Golgi-stained motor cortex of a child by Santiago Ramon y Cajal.[7] The skull (not drawn) is situated on the left of the dense layer of horizontal fibers and pyramidal dendrites (molecular layer). (B) Simultaneous recording of intracellular spikes from a pyramidal neuron in the motor cortex and EEG activity of a cat. Above: EEG; below: intracellular recording. (C) Autocorrelation of an EEG record. (D) Autocorrelation of intracellular slow potential changes (spikes are left out) of the same frequency as in (B). *Adapted from Creutzfeldt OD, Watanabe S, Lux HD. Relations between EEG phenomena and potentials of single cortical cells. I. Evoked responses after thalamic and erpicortical stimulation. Electroencephalogr Clin Neurophysiol 1966;**20**(1):1–18, permission Elsevier.*

In order to be able to compare results between experiments or medical consults, it is important that electrode placement is standardized. A classical way to do this is the so-called 10–20 system (Fig. 8.13). In the bipolar montage, the difference of signals coming from two neighboring electrodes is amplified. This gives rise to a multitude of possible wiring

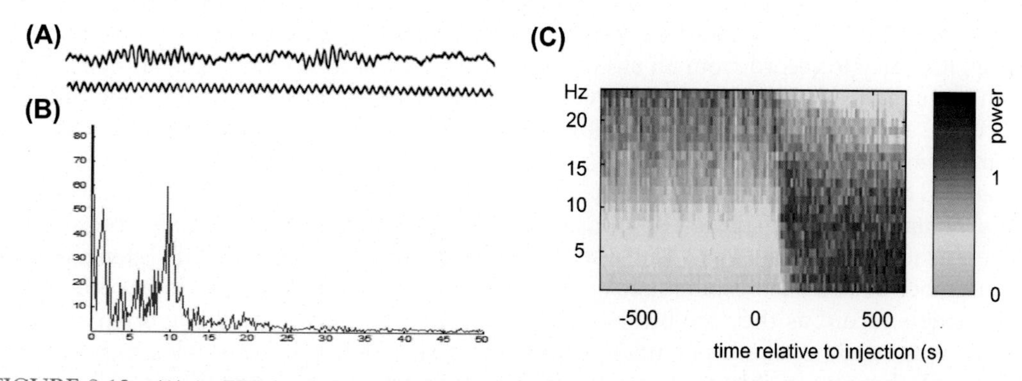

FIGURE 8.12 (A) An EEG (upper trace) by Berger of his 15-year-old son, together with a 10 Hz reference signal (lower trace).[9] *Berger H. Über das Eletrenkephalogramm des Menschen. Arch Psych Nerverkr 1929;87:527-70.* (B) Amplitude spectrum (0–50 Hz) computed from the graph, showing peaks at about 1 and 10 Hz. (C) The evolution of different frequency components in the EEG of a monkey before and after injection of an anesthetic. Note the shift from beta to theta rhythm.[10] *de Cheveigne A, Parra LC. Joint decorrelation, a versatile tool for multichannel data analysis. Neuroimage. 2014;98:487–505. Courtesy Elsevier.*

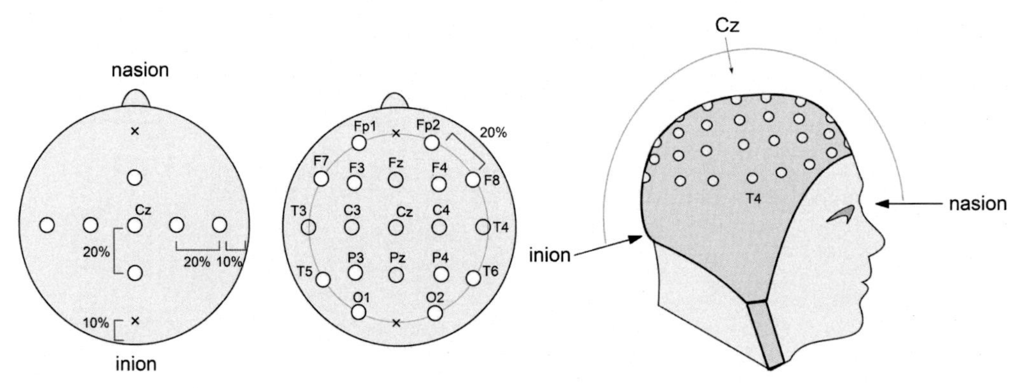

FIGURE 8.13 A standard placement of EEG electrodes according to the 10–20 method begins by locating the nasion (bridge of the nose) and the inion (the occipital protuberance of the skull) (left figure). The central point, Cz, is halfway between the former two points on the midline. Then, on the projection of the skull on the left, points are chosen such that the distance between nasion and the first point is 10% of the distance between nasion and inion. The other four points are at 20% intervals. The distance between the fifth point and the inion is 10%. A similar rule applies to the horizontal row of points: 10% between the left ear and the first point, then 20% for the others. An imaginary circle that passes through the four most extreme points determines the placement of intermediary points (such as Fp1 and F7), each at a distance of 20% of the whole circle (middle figure). The remaining points (e.g., F3 and F4) are determined by interpolation. All these points, except for those labeled with a cross, receive an electrode. The letters stand for Frontal (F), Temporal (T), Parietal (P), Occipital (O), and central (C). The electrodes may be integrated into a rubber cap as on the right.

schemes as each electrode has up to 4 neighbors. This method gives good results if the source of a signal is focal and peripheral. It performs badly if highly synchronized activity is to be recorded: The difference signal will be small, since both electrodes receive the almost identical electrical input.

In the referential montage, the inverting input of all differential amplifiers is common and may come from electrodes on one or both of the ears, a point far from the brain or the average of all electrodes. Taking two ears or the average of all electrodes as the reference helps to cancel the ECG component in the EEG signals.

In the old, analogous, days the montages were hardwired and the configuration had to be decided upon before a recording session. With the advent of digital methods and signal processing, it is easy to record from all electrodes during a session, store on disk, and decide on which montage to show off-line. Moreover, the number of channels in modern EEG recordings can be as high as 256, giving superior temporal resolution and spatial resolution that rivals magnetic resonance imaging.

The EEG described so far is a spontaneous activity of the brain, depending in a global way on the mental state of the subject. Responses to sensory stimuli, which may give more specific answers as to sensory processing, can also be recorded. However, since the EEG stems from a huge number of neurons, responses to specific stimuli are usually too weak to be recognized from a single record as they are buried in a background of usually rhythmic brain activity. This is why such evoked potentials, more generally dubbed event-related potentials (ERPs), only become visible after substantial signal averaging (the event is not necessarily a sensory stimulus: it can also be a motor action of the subject). Of course, the averaging can be triggered on a fixed point, such as the start of the stimulus (an acoustical beep, a light flash, a brief touching of the skin, etc.) or the command for an action. An example of signal averaging is shown in Fig. 7.11. Averaging some 64 to 256 sweeps usually reveals a significant electrical response. Size, latency, and other quantities may then be related to the processing of the stimulus in question by the brain.

Whereas rhythmic brain activity such as alpha and beta waves are relatively widely distributed, coming from large parts of the cortex, ERPs are often related to activity in specific brain regions.

Averaging in combination with current source density analysis (CSD analysis, Chapter 7) helps to determine the location from where ERPs are emanating. Two examples are shown in Fig. 8.14. In example A, subjects were asked to press a button (Go) as soon as they perceived a circle at the top-left and bottom-right corners of a computer screen and to do nothing if the circles appeared in the other two corners (NoGo). Both events elicited a positive ERP about 300–400 ms after the presentation of the stimuli. This positive going ERP is typical for this kind of perception tasks and is known as the P300 or P3 wave. Note that a prefrontal origin of a sink becomes apparent using CSD analysis, which cannot be discerned in the FP data. Binaural auditory stimuli induce a negative ERP with a delay of about 100 ms, called the N100 or N1. N1 field potentials are shown in B. CSD analysis reveals two sinks and two sources. Because the two hemispheres carry out different tasks, a phenomenon known as functional lateralization, the spatial asymmetries of the ERP patterns between hemispheres of the source in A and the two sinks in B are common.

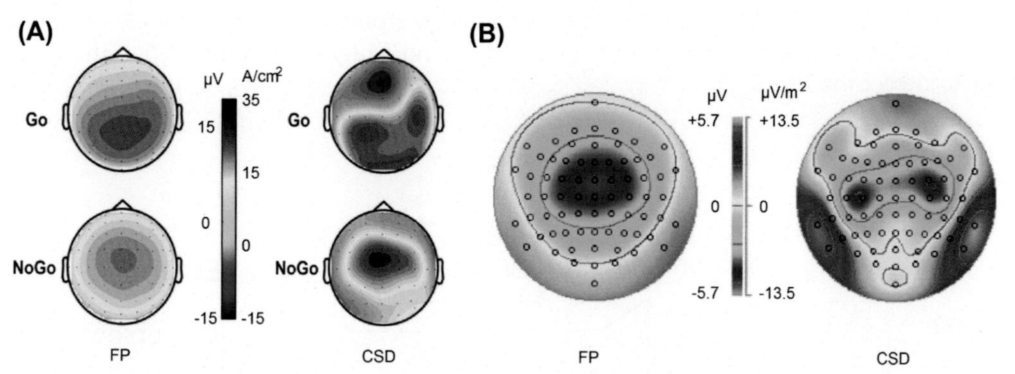

FIGURE 8.14 (A) EEG responses in a go and nogo task. If circles appeared at the top-right or bottom-left corners of a computer screen, a button had to be pressed (go). If the circles appeared at the top-left and bottom-right corners, nothing should be done. The field potentials (FP) on the left were recorded 380 ms following the stimulus presentation. The corresponding CSD on the right. Note the unconventional CSD dimension, A/cm^2. (B) ERPs in response to brief binaural tones (40 ms; 1000 Hz) averaged more than 127 presentations on the left and the corresponding CSD on the right. *(A) Adapted from Kamarajan C, Pandey AK, Chorlian DB, Porjesz B. The use of current source density as electrophysiological correlates in neuropsychiatric disorders: a review of human studies. Int J Psychophysiol 2015;**97**(3):310–22; Adapted from Tenke CE, Kayser J. Generator localization by current source density (CSD): implications of volume conduction and field closure at intracranial and scalp resolutions. Clin Neurophysiol 2012;**123**(12):2328–45 Both (A) and (B) permission Elsevier.*

OTHER SURFACE RECORDING TECHNIQUES

A number of other internal electrical activities of humans and other land animals can be recorded from the skin, most of them used again for clinical diagnosis. Although some of these signals stem from DC processes in the body rather than from action potentials, it is useful to treat them here together.

Examples are the electromyogram (EMG) from one or more muscles and the electronystagmogram, the recording of eye movements by electrodes at the temples. The multiunit recording shown in Fig. 9.1 is an example of an electromyogram.

The technique indicated with the term "lie detector" also falls in this category. It is actually a measurement of the resistance of the skin of the hand palm, which changes under the influence of the autonomic nervous system (i.e., the control of sweat glands). Although the term "lie detector" is a misnomer, the palmar skin resistance is a useful measurement, since it reflects the balance of (ortho)sympathetic and parasympathetic activities. It is thus a recording of covert behavior and as such useful in medical, psychological, and psychophysical research. Usually, skin resistance recording is combined with the recording of other physiological signals, such as the ECG, breathing rhythm, an EMG of postural muscles, and so on. The DC skin potential may also be used. The combination is called polygraphy and may give a better representation of the physiological state of the subject than each single quantity.

Note that polygraphy does not necessarily mean polyelectrode: The scheme in Fig. 8.15 describes a device we built for recording the ECG, EMG, skin resistance, and skin potential simultaneously with only two electrodes connected to the patient. The skin impedance is measured by supplying a sinusoidal current to an electrode on the hand palm.

The trick is done further by positioning the other electrodes on a neck muscle and segregating the signal components by electronic means: a DC preamplifier for the skin potential, a detector/amplifier (DET) to derive the skin impedance from the AC signal, a low-frequency band-pass filter (BP) and postamplifier for the ECG, and a high-pass filter (HP) with postamplifier to detect the EMG.

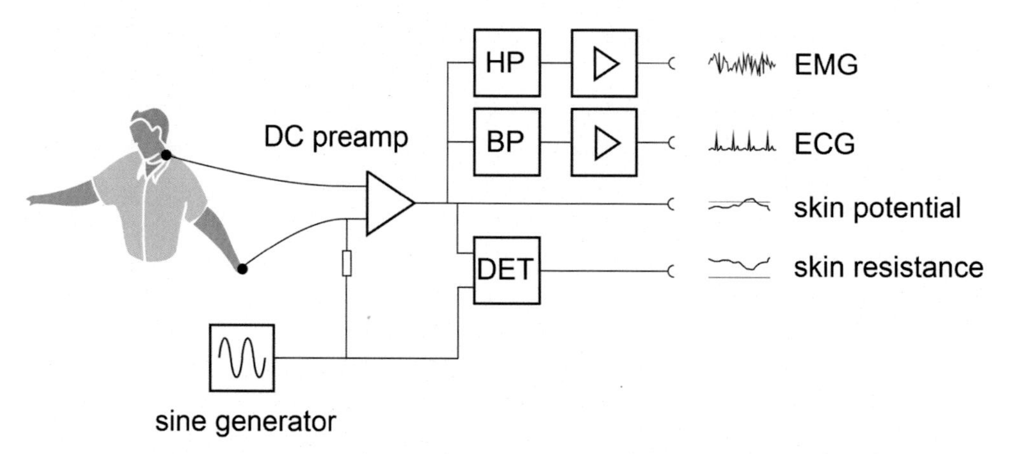

FIGURE 8.15 Simplified schematic of a polygraph using two electrodes. Ground wires are omitted. *HP*, high-pass filter; *BP*, band-pass filter; *DET*, detector (converts amplitude of AC into a DC).

The quantities skin impedance and electromyogram are processed with circuits like the one shown in Fig. 9.3. The resulting "polygrams" are usually interpreted directly from the chart records, i.e., in the time domain.

Two final remarks to skin surface recording:

1. All types of skin surface recording benefit from an electrically screened set-up. This is why the walls of ECG and EEG recording rooms are often covered with metal to form a large Faraday cage.
2. Often, human skin surface recordings are made with dedicated, medical-grade apparatus that fulfill the electrophysiological demands, including the safety aspects dealt with in the Appendix. Portable versions of ECG and EEG recorders are used by general practitioners to diagnose at the patient's bed. These apparatus have still better filters built in to yield clear results despite the lack of a screened room.

RECORDING OF SECRETORY EVENTS

During secretion, secretory vesicles fuse with the plasma membrane expanding its surface for a while before they are recycled. The changes in membrane surface are accompanied with changes in membrane conductance and membrane capacitance that, in mast cells, are in the order of 10 fF (1 fF $= 10^{-15}$ F) per event. Although the method described in the Chapter 4 can be used to get a rough estimate of the membrane capacitance, it is far too unprecise to resolve single secretory events. Instead, methods that use a small-amplitude sine wave voltage are often used. These methods rely on the electrical model as shown in Fig. 8.16.

In this model, R_s is the series resistance or the pipette resistance, G_s (=1/Rs) is the series conductance, R_m is the membrane resistance, $G_m(=1/R_m)$ is the membrane conductance,

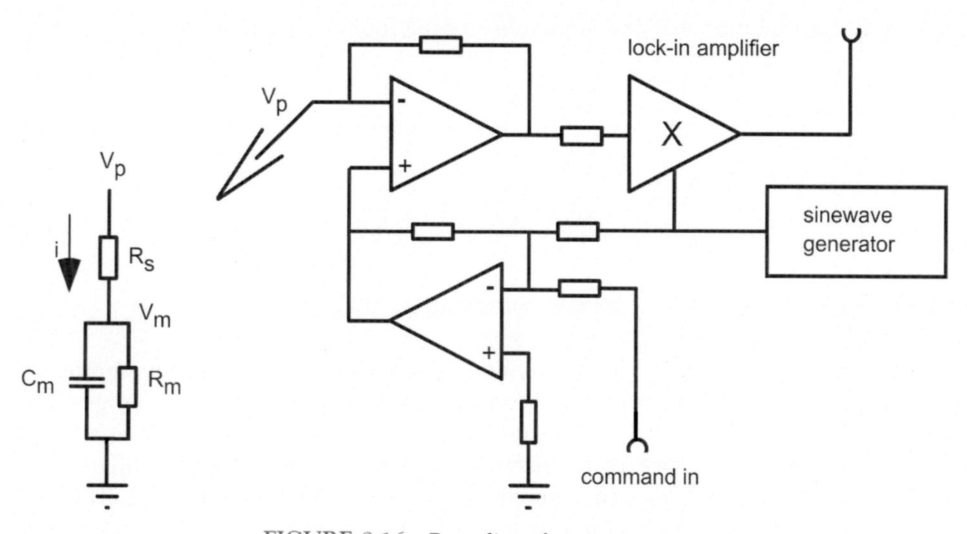

FIGURE 8.16 Recording of secretory events.

C_m is the membrane capacitance, V_p is the pipette potential or voltage-clamp potential, and V_m is the actual membrane potential. The membrane current under voltage-clamp is then:

$$i = V_m \cdot G_m + C_m \frac{dV_m}{dt}$$

which is equal to:

$$i = (V_p - V_m)G_s$$

and hence,

$$V_m = V_p - \frac{i}{G_s}$$

Substitution of V_m in the first equation gives:

$$V_p G_m G_s - i(G_m + G_s) + C_m \frac{d(V_p G_s - i)}{dt} = 0$$

Passing to complex notation with $V_p = e^{-j\omega t}$ and $i = I(\omega)\, e^{-j\omega t}$ gives:

$$G_m G_s e^{j\omega t} - (G_m + G_s)I(\omega)e^{-j\omega t} + C_m \frac{d(G_s e^{-j\omega t} - I(\omega)e^{-j\omega t})}{dt} = 0$$

and thus:

$$G_m G_s - (G_m + G_s)I(\omega) + j\omega C_m I(\omega) - j\omega G_s G_m = 0$$

Finally splitting $I(\omega)$ into real and imaginary parts: $I(\omega) = A + j\, B$ gives:

$$A = G_s \frac{G_m^2 + G_m G_s - \omega^2 C_m^2}{(G_m + G_s)^2 + \omega^2 C_m^2} \tag{8.1}$$

$$B = G_s \frac{\omega G_m G_s + 2\omega C_m}{(G_m + G_s)^2 + \omega^2 C_m^2} \tag{8.2}$$

The real part, A, vanishes for a certain combination of C_m and ω.

$A = 0$ for $G_m^2 + G_m G_s = \omega^2 C_m^2$

Hence, for a certain frequency of stimulation, small variations of the membrane capacitance can be measured without contamination by membrane conductance changes. In practice, the test frequency, ω, is fixed at a value between 1 and 5 KHz and Cm is adjusted by "cheating" with the capacitance compensation of the voltage-clamp amplifier to obtain $A = 0$. During this compensation, the imaginary output must be eliminated. This is done by chopping the voltage-clamp output current with a lock-in amplifier (this is an amplifier that inverts the signal during half the sine period) and averaging over ±10 periods. After compensation, the phase angle between sine wave stimulus and lock-in amplifier input is shifted 90 degrees to measure the imaginary part of the output current that now reflects changes in capacitance.[13,14] The present generation of computer-operated voltage-clamp amplifiers contain membrane capacitance measurement as a standard option.

If secretory granules are relatively large, such as in mast cells and chromaffin cells,[15] single events can be resolved appearing as minute stepwise increases in capacitance in Fig. 8.17. For other cell types, an alternative, second, way to measure release of substances from cells electrically is by reduction or oxidation of the substance released.

Molecules such as the neurotransmitters acetylcholine, adrenaline, and serotonin may be oxidized by an electrode held at 700 mV by a voltage-clamp circuit. Usually the electrode is made of platinum or carbon. In the latter case, a 5 μM carbon fiber is inserted in a glass or plastic pipette and then sealed with epoxy or by heat (in case of a plastic pipette). Often, after sealing the tip, the carbon fiber is broken off or polished to expose a clean surface. In addition, the tip is soaked in ethanol, butanol, or propanol for 15 min prior to experiment. Then the carbon electrode is brought very close to a cell and clamped at 700 mV. Each quantum released generates a spike of current ranging from 10 to 100 pA, depending on the distance between releasing site and electrode. Instead of using the classical voltage-clamp equipment, integrated circuits may be used (Fig. 8.18).

An experimental 10 × 10 array of platinum working electrodes on a chip and a single Ag/AgCl ground electrode has been developed to measure the transient release of adrenaline and noradrenaline from multiple chromaffin cells simultaneously.[17] The working electrodes are kept at 700 mV. Platinum has been chosen as the material for the working electrodes because they are almost perfect polarizable electrodes. This means that almost no DC current flows and that only capacitive currents pass (see Chapter 5 on polarized electrodes, Figs. 5.3 and 5.4). The chip also contains amplifiers and the logic circuits to relay the data from the 100 electrodes to a computer.

The design of the circuit, which is essentially the same as the two-electrode voltage-clamp approach, can be improved by using a three-electrode circuit as in Figs. 4.12 and 4.13.

Amperometric detection of other substances than catecholamines may be more difficult. For example, detection of insulin (by oxidation of sulfur bridges) requires special carbon electrodes on which a film of ruthenium oxide/cyanoruthenate is deposited and is operated at

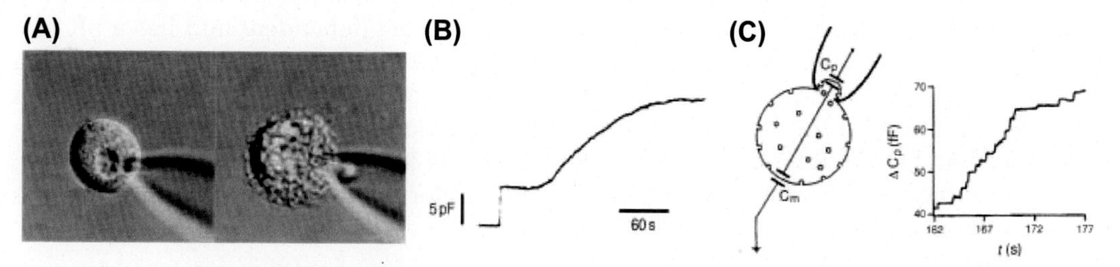

FIGURE 8.17 (A) Membrane capacitance of a mast cell was measured in whole-cell mode. (B) After intracellular dialysis with GTPγS, degranulation occurred resulting in an increase in membrane capacitance. (C) Single fusion events could be discerned when in cell-attached mode. *Adapted from Angleson JK, Betz WJ. Monitoring secretion in real time: capacitance, amperometry and fluorescence compared. Trends Neurosci 1997;**20**(7):281–87, Permission Elsevier.*

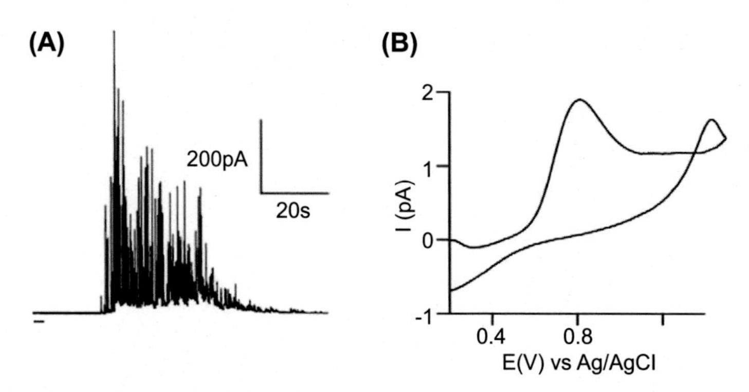

FIGURE 8.18 Oxidation of serotonin, released by mast cells, at the surface of a carbon electrode held at 700 mV. Each spike in (A) represents a secretory event. Time of stimulation is indicated by the line segment underneath the trace. The current to voltage curve in (B) was obtained by sweeping the working electrode potential starting at 0.2 V to and fro 1.4 V during an event. Maximum current was obtained at the oxidation potential for serotonin. *Adapted from Marquis BJ, Haynes CL. The effects of co-culture of fibroblasts on mast cell exocytotic release characteristics as evaluated by carbon-fiber microelectrode amperometry. Biophys Chem 2008;**137**(1):63–69, permission Elsevier.*

850 mV. These electrodes degrade rapidly and often other options are sought. One such an option is charging, for example, β cells (which secrete insulin) with serotonin, which is then cosecreted with insulin. A second option is coculture with a cell type that responds to the secreted substances. Hence, the second cell type serves as the detector.

A third option is to take advantage of the fact that the content of secretory vesicles is acidic (pH between 5 and 6). As we have seen in Chapter 2, ISFETs are sensitive to charged molecules in particular protons. When cells are grown on ISFETs, the release of protons onto the surface of the silicon layer separating the FET gate from the incubation medium changes the surface charge of the silicon layer thus affecting the source–drain current. Fig. 8.19 shows an example of this approach. Catecholamine secretion from chromaffin cells, grown on an array of ISFETs, can be elicited by a puff of $BaCl_2$. Release at the side of the cell in B was recorded by amperometry using a carbon fiber, while secretion at the base of the cell was recorded by one of the ISFETs. Even though the recording sites of FET and fiber were at some distance from one another, multiple coinciding events were detected (C). This correlation is probably due to events that occurred close to the bottom and the side of the cell where the carbon fiber was positioned. Hence, these events were therefore recorded by both methods.[18]

A fourth (optical) option also makes use of the acid nature of secretory vesicles. The fluorescence of enhanced green fluorescent protein (EGFP) is pH dependent and has a pK of 6 (meaning a half-maximum fluorescence at pH 6). PHluorin is a mutant variant having a more physiologically centered pK of 7. Transfection of cells with plasmids coding for a fusion protein of one of these two fluorophores and a vesicle-associated protein causes vesicles to become fluorescent. As the pH of the vesicles is low, fluorescence is equally low. Fluorescence increases largely upon fusion of the secretory vesicle with the plasma membrane since the pH then suddenly rises to 7.4. Secretion can thus be observed by fluorescence microscopy.

A fifth option is to measure secretion optically, using fluorescent dyes such as TMA-DPH (1-(4-trimethylammonium)-6-phenyl-1,3,5-hexatriene). TMA-DPH is a hydrophobic compound that can nevertheless be dissolved in water in micromolar concentrations. When it

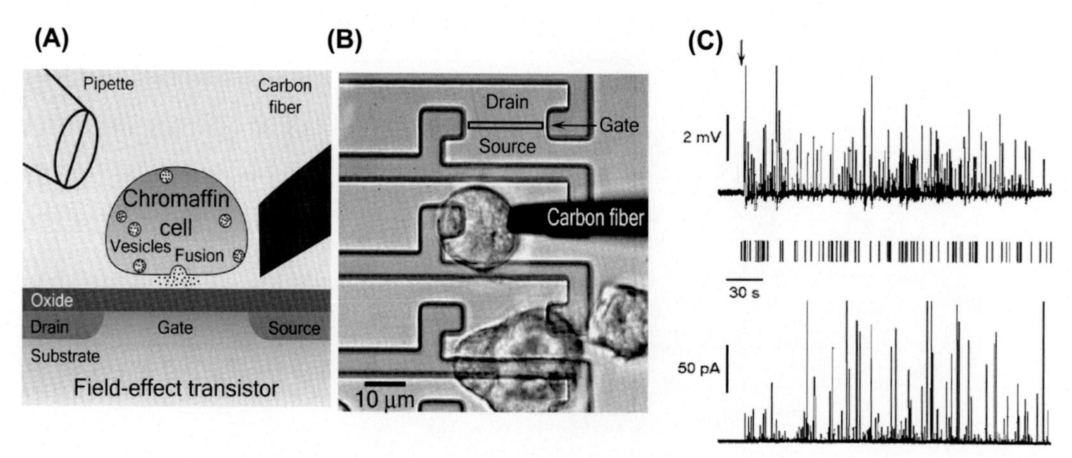

FIGURE 8.19 Chromaffin cell on a transistor. (A) Schematic cross section. Vesicles are released into the narrow extracellular space between cell and transistor. A carbon fiber is used for amperometry. Exocytosis is elicited by BaCl2 applied with a pipette. (B) Micrograph of silicon chip with linear transistor array. Source, drain, and gate (black frame) are marked on the upper transistor. A chromaffin cell covers about one-half of the open gate area. The cell is touched by a carbon fiber from the right. (C) Simultaneous detection of vesicle release by transistor recording and amperometry. Distribution of events in time for the transistor signal (top) and amperometry (bottom). The stimulation by a pulse of 5 mM BaCl2 is marked by an arrow. The band of vertical bars between the two records marks the events that are detected by the transistor and the carbon fiber within a time window of 5 ms. *Adapted from Lichtenberger J, Fromherz P. A cell-semiconductor synapse: transistor recording of vesicle release in chromaffin cells. Biophys J 2007;**92**(6):2262–68, permission Elsevier.*

comes in contact with the cell membrane, it rapidly dissolves in the lipid phase, at the same time becoming fluorescent (excitation 340 nm, emission 430 nm). When the cell is stimulated, secretory granules fuse with the membrane, take up TMA-DPH, and are subsequently internalized, making the cell interior fluorescent. Then the extracellular solution is replaced by a solution devoid of TMA-DPH, which also removes the dye from the extracellular leaflet of the plasma membrane, due to partitioning between aqueous and lipid phases. Then the cell may be stimulated again and now intracellular fluorescence decreases upon fusion of the secretory vesicles with the plasma membrane.[19,20]

RECORDING FROM BRAIN SLICES

The brain slice preparation is of inestimable value for the analysis of local neuronal circuitry such as those in the hippocampus and the cerebellum. The technique avoids certain disadvantages associated with in vivo recording from the CNS like the mechanical artifacts induced by the heartbeat and respiration and the need to sedate the animal. Multiple electrodes can be positioned in slices under visual control therewith assuring the location from which one is actually recording. In contrast, the *in vivo* placement of electrodes usually is carried out stereotactically (i.e., "blind" placement based on a coordinate system). The site of recording is labeled at the end of the experiment, and the animal's brain is then sectioned to check the recording position. The major challenge in the preparation of brain slices for

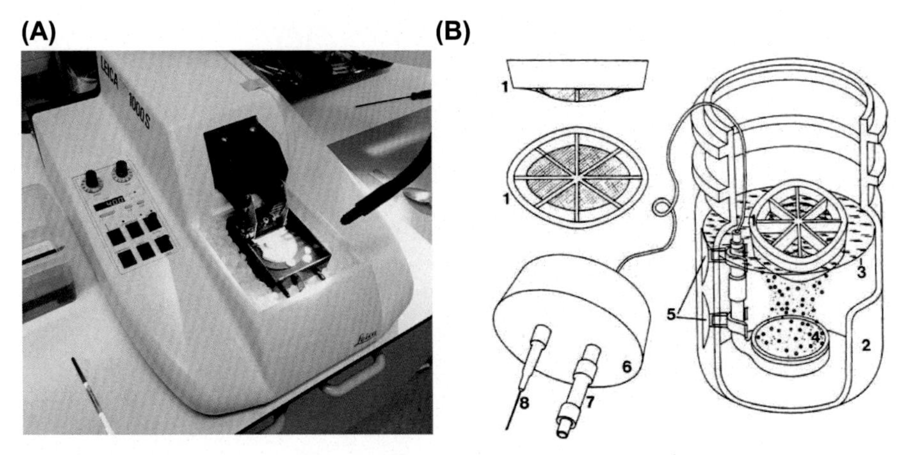

FIGURE 8.20 (A) Typical vibratome used to cut brain slices. (B) Drawing of the interface holding chamber: 1, holding chamber; 2, glass canning jar; 3, solution level; 4, gas dispersion ceramic disc; 5, suction cups; 6, rubber stopper; 7, gas inlet; 8, gas outlet. *Adapted from Krimer LS, Goldman-Rakic PS. An interface holding chamber for anatomical and physiological studies of living brain slices. J Neurosci Methods 1997;**75**(1):55–58, permission Elsevier.*

electrical recording is viability. Several excellent documents exist that describe the methods in detail.[21,22]

Usually young animals less than 3 weeks are used. Their brains are rapidly removed and immediately cooled in artificial cerebrospinal fluid (aCSF). A block of tissue is glued onto the stage of a vibratome (Fig. 8.20A) with superglue, and slices 100–300 μm are cut keeping everything cooled in ice-cold aCSF. The thickness of the slice is a compromise between the necessity to transmit light through it for visualization, the integrity of neuron projections, and maintenance of oxygen provision to the deeper layers of the slice. The slice is then transferred to a holding chamber (Fig. 8.20B) containing 32–35°C aCSF bubbled with 95%O_2/5%CO_2 carbogen gas. Different types of holding chambers are currently being used. The one shown in Fig. 8.20 keeps the slices close to the interface between fluid and carbogen gas above it, while others keep the slices well immersed.[22] Stable recordings can be made for several hours after preparation.

Once a number of slices have been cut and have been incubated in the holding chamber for at least half an hour, one of them is set up for recording on fixed microscope stage. The stage has to be fixed because the slice needs to be perfused with oxygenated room temperature aCSF, drug delivery should generally be possible, and at least two micromanipulators, one for recording and the other for stimulation, are required. The manipulation of each of these devices should not (mechanically) perturb the others. Hence, mounting all of these on a movable stage is hardly an option. Then essentially three options are open (Fig. 8.21). Option A is using an inverted microscope with an oil-immersion objective with high NA and long working distance. The condenser may be water immersible. Since the NA of the condenser should match the NA of the objective, immersion can hardly be avoided. One of the disadvantages of option A is that the bottom of the slice is the best visible. Therefore, the electrode has to traverse most of the height of the slice. This is OK for microelectrodes, but not well adapted for patch-clamp, because the patch electrode tip needs to stay clean. It is also the

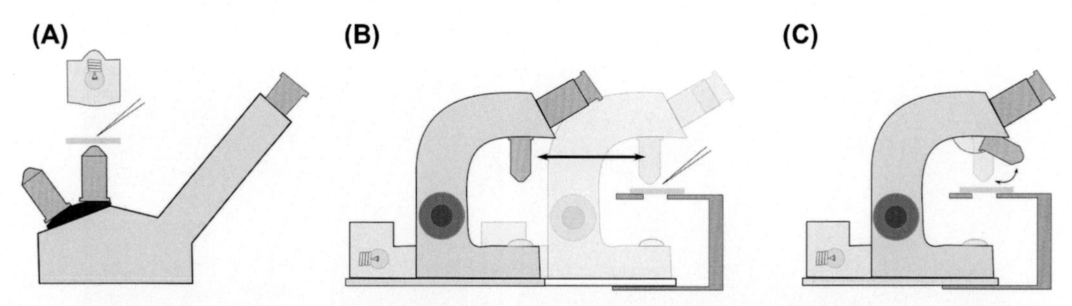

FIGURE 8.21 Three microscope configurations to record from slices: (A) inverted microscope, (B) sliding upright microscope with water-immersion objective, and (C) upright microscope with pivoting water immersion objective.

layer where oxygen supply is the lowest. Option B consists of an upright microscope. Contrary to most microscopes of this type, the objectives move for focusing and not the stage. It is mounted on a plate with ball bearings to move the entire microscope away from the stage for easy access. The high NA objective is water immersible. In option C, only the objective moves out of the way by pivoting it toward the experimenter.

A (little) problem concerning water-immersion objectives and condensers is the creation of a ground loop as soon as they are immersed into the experimental bath because both preparation and microscope are normally grounded separately. Insulation of the objective with nail polish usually solves this.

Visibility is a prime concern when approaching cells in a slice with electrodes. It is also very helpful for the selection of viable, undamaged cells. The illuminating light that traverses the slice is scattered, reducing visibility to such an extent that individual cells are difficult to discern. Visibility can be improved of course by cutting thinner slices, but then again damage to dendrites and circuits increases. Fortunately, scattering decreases with increasing wavelength, and for this reason, infrared illumination (>700 nm) is often used to visualize cells in slices (Fig. 8.22). A video camera and monitor are required because the human eye is insensitive to infrared radiation. The use of a monitor comes with a supplementary advantage in that image contrast can be enhanced thus setting out small differences in optical density. A further improvement is the use of differential interference contrast (DIC) microscopy. Most cells are essentially transparent objects that diffract rather than absorb light. Diffraction occurs because the cell interior has a somewhat higher optical density than the interstitial medium thereby retarding the progression of photons with respect to the interstice. This causes a small phase shift that DIC transforms into intensity contrast (Fig. 8.22C and D). The price to pay of using IR light instead of visible light is a reduced resolution: the resolution at 1000 nm wavelength is half that at 500 nm (i.e., 0.4 resp. 0.2 μm with an objective of NA 1.4).

For impalement of cells with microelectrodes, no more steps are required. However, for patch-clamping, the last obstacle to whole-cell access is the cleaning of the cell membrane. This can be done by following the method shown in Fig. 8.23, which has been developed in Sakmann's lab in the early 1990s.[22,26] The cell to patch is slowly approached maintaining a constant efflux of solution from the patch pipette. This can be achieved by including a three-way stopcock in the pressure tube that connects to the pipette holder. The formation of a sort of dimple can be seen when the pipette is close to the plasma membrane. Removal of the

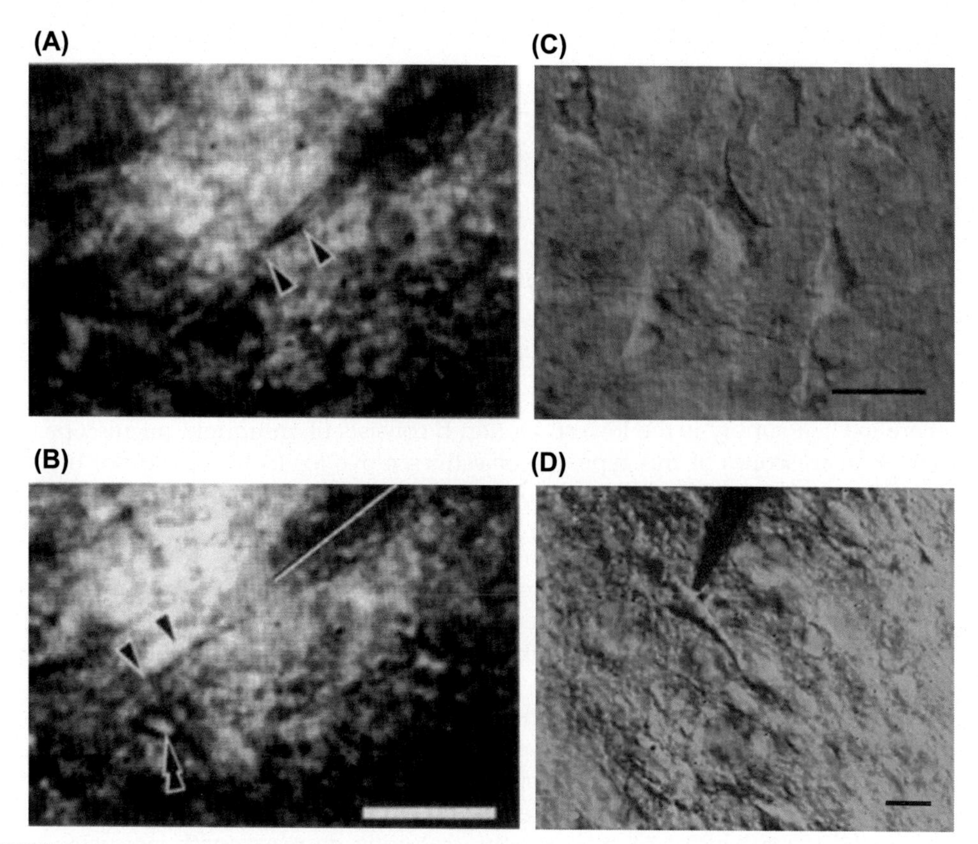

FIGURE 8.22 (A) Microelectrodes could be seen within the slice by refocusing along the shaft. The 400 μm thick rat hippocampal slice was observed with an inverted microscope illuminated by infrared light. A thicker portion of the microelectrode was in focus (between arrows). The main shaft extends to the upper right (dark shadow indicated by line in (B)). The tip (between arrows) was adjacent to a cell body (double arrowheads). Calibration bar: 30 μm. (C) A small cluster of neurons in lamina 2 of somatosensory cortex (40× objective, rat 15 days old). (D) An iontophoretic pipette touching a pyramidal neuron. The images in (C) and (D) were taken using an inverted differential interference contrast microscope lit by infrared radiation. *(B) Adapted from MacVicar BA. Infrared video microscopy to visualize neurons in the in vitro brain slice preparation. J Neurosci Methods 1984;**12**(2):133–39; Adapted from Dodt HU, Zieglgansberger W. Visualizing unstained neurons in living brain slices by infrared DIC-videomicroscopy. Brain Res 1990;**537**(1–2):333–36, permission Elsevier.*

pressure is sometimes enough to cause a seal, if not slight suction should do it. As always with patch-clamping, seal formation is determined by observing changes in pipette resistance. Next, a brief suction pulse should break the membrane under the patch pipette.

Field Potentials

The electric potentials measured by an extracellular electrode are called field potentials (FPs). The principles of field potentials and current source density (CSD) analysis were introduced in Chapter 7. The recording of field potentials is an often used method to measure

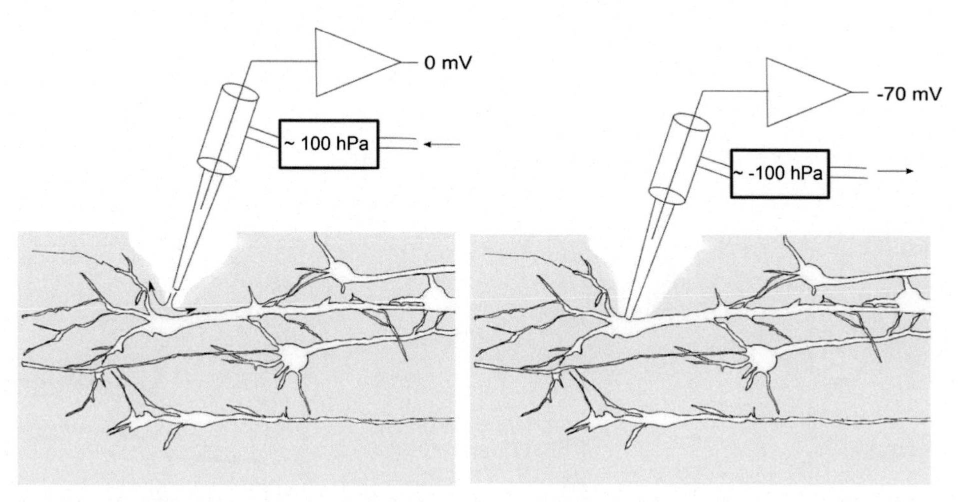

FIGURE 8.23 Hydrostatic pressure is maintained on the patch pipette while approaching a neuron, thus removing debris and extracellular matrix (left). After seal formation, short suction ruptures the plasma membrane under the pipette (right).

population spikes passing via nerve bundles or to evaluate synchronized synaptic activity in nerve tissue with parallel layout such as cortical structures. The oldest and most studied cortical structure is the hippocampus which forms part of the limbic system. It is highly plastic and the home of short time memory. The hippocampus gets input from the entorhinal cortex (EC), which relays information coming from higher structures, via the perforant path. The axons from EC layer III synapse onto *Cornu Ammonis* 1 (CA1) dendrites, while those of layer II contact CA3 and dentate gyrus cells (Fig. 8.24A). Mossy fibers from the dentate gyrus make synapse with CA3 cells. A bundle of axons from CA3 pyramidal cells, the Schaffer collaterals, project onto CA1 dendrites. When the tip of a carbon or glass electrode is placed in the CA1 region and Schaffer collaterals are stimulated near the CA3 region, field potentials of a few hundred microvolts can be recorded. These field potentials are due to the opening of neurotransmitter-operated channels in synapses. Depending on the pulse protocol used, facilitation, long-term potentiation, or long-term depression of synaptic transmission can be observed. The latter two phenomena affect the size and number of synaptic boutons and are associated with establishment and erasure of memory traces, respectively. The idea that synaptic weight constitutes memory has been originally proposed by Donald Hebb in 1949 and experimentally confirmed by Eric Kandel and others in the 1960s and 1970s.

CSD analysis is most easily applied whenever field potentials are recorded with multiple electrodes (>6) simultaneously. Multielectrode arrays (MEAs) can consist of active elements (CMOS) or metal electrodes deposited on glass. An example of the latter is shown in Fig. 8.25A. The electrodes consist of platinum deposited on glass and insulated by a layer of epoxy. Field potentials can then be recorded (Fig. 8.25B). A similar MEA, using gold electrodes, is used in Fig. 8.25C to record from the CA1 region of the hippocampus. One of the electrodes, the red dot in Fig. 8.25C, is used for stimulation of Schaffer collaterals. We just saw (Fig. 8.24) that stimulation of Schaffer collaterals provokes glutamate release and excitatory postsynaptic potentials (EPSPs) in CA1 pyramidal synapses. Cations enter the cell during

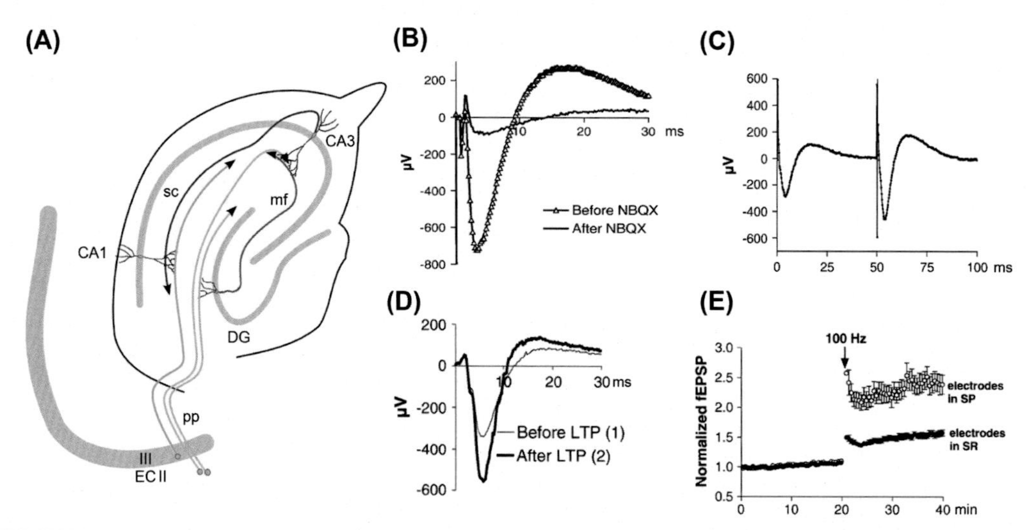

FIGURE 8.24 (A) Axons from the entorhinal cortex (EC) project onto dendrites of CA1 and CA3 pyramidal cells and onto neurons in the dentate gyrus (DG) of the hippocampus. DG granule cells send their projections (mossy fibers, mf) onto CA3 cells. Schaffer collaterals (sc) emanating from CA3 pyramidal cells make synapses with CA1 dendrites. (B) Field potentials were recorded in the CA1, and sc fibers were stimulated in CA3. Field potentials were suppressed by the AMPA/kainate receptor antagonist NBQX. (C) If the interval between two stimuli is short (50 ms), facilitation of synaptic transmission is observed. (D) Long-term p otentiation (LTP) may be induced by stimulation with a 100 Hz train of stimulations. (E) The potentiation observed seems larger when placing the field electrode in the striatum radiatum (where sc and mf contact pyramidal neurons) then in the striatum pyramidale (where the cell bodies lie, gray zones in (A)). *(B–E) Adapted from Steidl EM, Neveu E, Bertrand D, Buisson B. The adult rat hippocampal slice revisited with multi-electrode arrays. Brain Res 2006;**1096**(1):70–84, permission Elsevier.*

the EPSP, locally creating a current sink. Since cells need to stay close to electric neutrality, the same current necessarily leaves the neuron at some other place. Hence, every sink has its source somewhere and vice versa. The presence of a current sink in the striatum radiatum, where the Schaffer fibers synapse on CA1 dendrites, is strongly suggested by the negative field potential (blue) in the left panel of Fig. 8.25D. The image becomes more precise by applying CSD analysis (left panel of Fig. 8.25D). The first sign of current entering synapses occurs with a delay of about 2 ms. The sink it causes is accompanied by two sources, one of which is not apparent in the field potential image. The sink is then seen to spread upward to the pyramidal layer (see left arrow in F). At this point action potentials in the dendrite and axon hillock (green circle in F) may be generated, contributing to the current sink. Finally, inactivation of the glutamate-induced currents and activation of inhibitory postsynaptic K^+ currents carried by $GABA_b$ channels with slow kinetics terminate the sequence of events. An inversion of the sink into a source at about 10 ms poststimulus (horizontal arrow in F) is observed. Again, this source at $t > 10$ ms is accompanied by two sinks that are visible in the CSD panel but are hard to perceive in the FP panel. Of course, the sequence of events described above cannot be deduced just by inspection of the figures. Most of the conclusions have been obtained by CSD analysis in combination of systematic inhibition of synaptic current components with drugs specific for the different types of transmitter-operated ion channels.

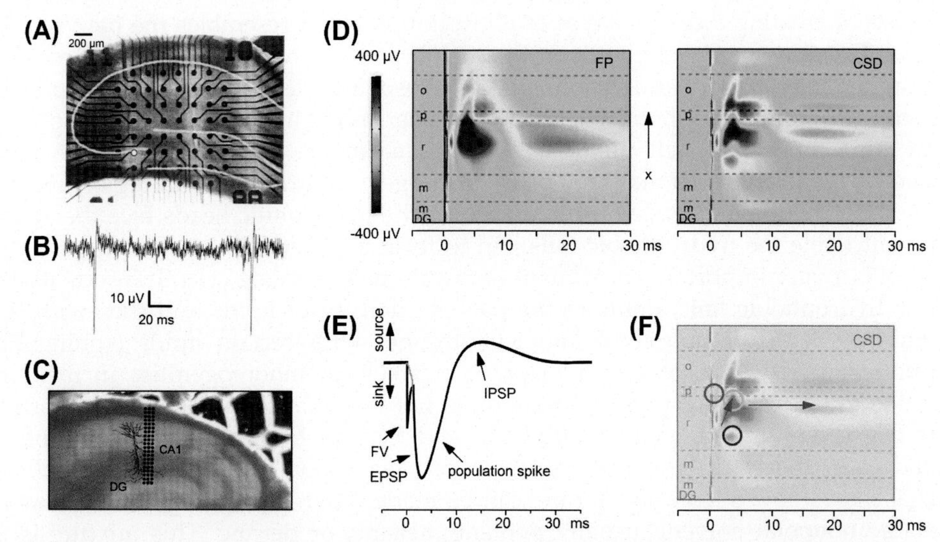

FIGURE 8.25 (A) A grid of 8 × 8 platinum electrodes on a chip can be used to record field potentials. The *white lines* sketch the layout of the hippocampus. The record in (B) shows spontaneous action potentials recorded with an electrode in the CA3 region, indicated by the white dot in (A). CSD analysis in the CA1 region of the hippocampus slice (C–F). (C) Three rows of 20 electrodes were placed as indicated on the left, only the potentials recorded by the middle column were used for figure (D). Distance between electrodes is 50 µm. The red electrode in the upper limit of the stratum radiatum, just below the stratum pyramidale, was used for stimulation. A neuron is drawn alongside the electrode array to appreciate the relation between cell morphology and electrode positions. (D) Field potential as a function of electrode position and time is shown in the left panel. CSD is on the right. Time 0 corresponds to time of stimulation. Note the sink at 1 ms at the pyramidal layer (blue circle) and the source (red circle) at 5 ms that have no counterpart in the left panel. Strata: o, oriens; p, pyramidale; r, radiatum; m, molecular (CA1 and DG). (E) Schematic drawing of CSD close to the pyramidal layer. *EPSP*, excitatory postsynaptic potential; *FV*, fiber volley; *IPSP*, inhibitory postsynaptic potential. *(A) and (B) Adapted from Heuschkel MO, Fejtl M, Raggenbass M, Bertrand D, Renaud P. A three-dimensional multi-electrode array for multi-site stimulation and recording in acute brain slices. J Neurosci Methods 2002;***114***(2):135–48, (C) and (D) adapted from Gholmieh G, Soussou W, Han M, et al. Custom-designed high-density conformal planar multielectrode arrays for brain slice electrophysiology. J Neurosci Methods 2006;***152***(1–2):116–29, permission Elsevier.*

Hence, the recording of field potentials in cortical structures and CSD analysis are great tools to figure out network connectivity and physiological function. The most striking example is perhaps the great advances that have been made in understanding the function of the hippocampus in memory, thanks to those techniques.

PLANAR LIPID BILAYERS

Before the invention of the patch-clamp, ion channels were studied by their reconstitution into artificial lipid membranes. This technique is still very useful to study the electrical properties of ion channels from membranes of organelles that are not easily accessible by patch pipettes such as the endoplasmatic reticulum or the mitochondrial membranes.

It consists of creating a double layer of phospholipids that resembles the native cellular membrane. The mayor phospholipids in the membranes of living cells are, in descending order of abundancy, phosphatidylcholine (PC), phosphatidylethanolamine (PE), phosphatidylinositol (PI), and phosphatidylserine (PS). Other lipids, among which the sphingolipids, are also present in cell membranes albeit at much lower quantities. All these constituents have in common that they possess a largely hydrophobic tail region and polar head group on the opposite side of the molecule. The hydrophilic heads face the aqueous environment while the hydrophobic tails join to form a water-impermeable double layer. Cholesterol is a very important constituent of membranes of living cells. It inserts itself between the hydrophobic tails, defining the packing density of lipids and therewith membrane fluidity. Because cholesterol interacts stronger with certain lipids (sphingolipids and saturated lipids) than others, it helps to create local inhomogenities in membrane composition known as lipid rafts. Cholesterol is not essential for the creation of artificial lipid double layers and is often omitted. PE and PC can be readily purified from egg yolk and are therefore popular construction materials for artificial membranes. Fig. 8.26A shows how such a membrane can be made. First, phospholipids are dissolved in a volatile nonpolar solvent, usually pentane, hexane, or decane. This mixture is then spread onto water where the polar heads insert into the water and the hydrophobic tails point in the air–solvent layer. After a few minutes, (most of) the solvent evaporates and a lipid monolayer remains at the water surface. Next, a thin hydrophobic sheet (usually Teflon, but sometimes Parafilm) containing a small aperture (10–200 μm) is lowered into the monolayer and the solution. The lipophilic tails adhere to the Teflon, dragging monolayers on each side of the sheet into the solution. The two monolayers meet to form a double layer when the aperture reaches the air–water interface. The final configuration is shown in Fig. 8.26B, where two compartments are separated by the artificial membrane. Agar bridges, containing 3M of KCL, connect to side chambers of which one is grounded with an Ag/AgCl electrode, while the other receives input from a voltage-clamp circuit.

After formation of the double layer with the method of Fig. 8.26, both cis and trans compartments contain solutions with identical ion compositions. One of the compartments needs to be refilled if one wishes to have different solutions on each side.

Rather than lowering the Teflon septum into the liquid, the septum can remain fixed and the liquids (with monolayers) on each side raised. The latter method also offers the possibility to have different monolayer compositions for the cis and trans sides (e.g., to have PS at the trans side as in the living cell's plasma membrane).

Ion channel proteins may be inserted into the double layer, once it has been formed, by adding detergent or liposomes containing these proteins to one of the compartments. Fusion of liposomes may be promoted by creating an osmotic gradient by having low osmolarity solution in the compartment (e.g., 10 mM HEPES). Lowering the fluid level below the aperture and raising it again also favors liposome integration into the bilayer.

Capacitance of Bilayers

The capacitance of a metal plate capacitor can be calculated by:

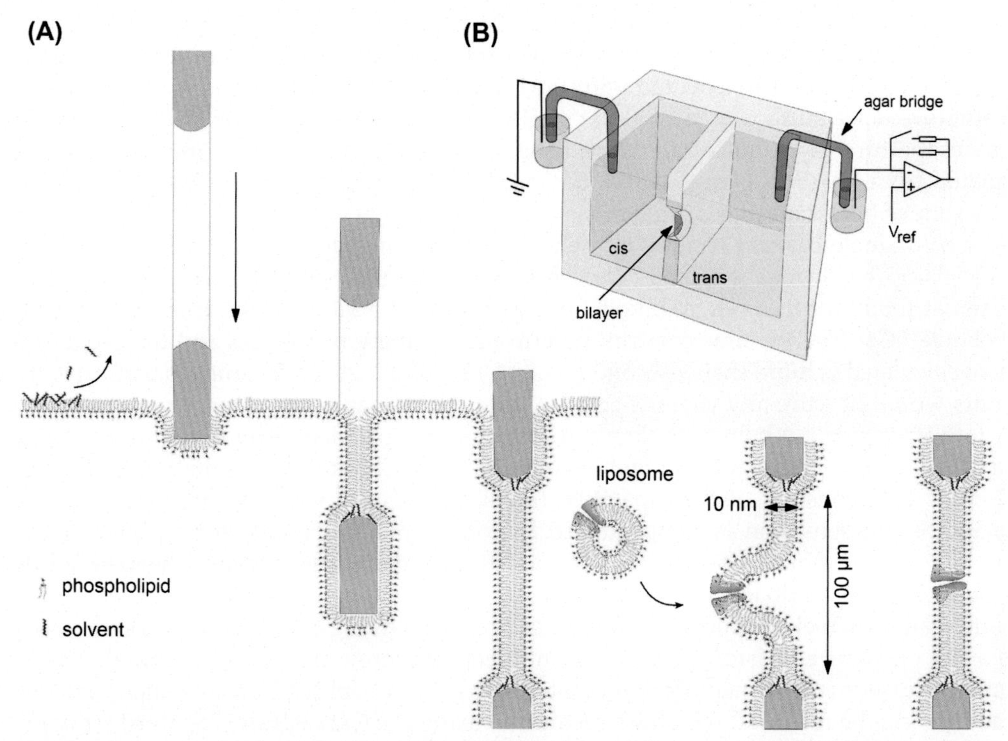

FIGURE 8.26 (A) A small quantity of phospholipid solution is spread onto an aqueous solution forming a monolayer. Then a hydrophobic septum with a small hole of typically 100 μm diameter is lowered into the solution thus forming a double layer over the aperture. Liposomes containing ion transporters or channels to be studied then may fuse with the artificial layer. (B) Two electrically isolated compartments separated by the artificial membrane are connected by agar bridges to chambers containing Ag/AgCl electrodes that voltage-clamp the double layer. Per convention, the cis compartment is at zero voltage. It is the "extracellular" side. Dimensions are not drawn to scale.

$$C = \varepsilon_r \varepsilon_0 \frac{A}{d} \tag{8.3}$$

This equation has already been mentioned in Chapter 1 (Eq. 1.1). Here C is the capacitance in farad; A, the surface area of the plates in m^2; d, the distance between the plates in m; ε_r, the dielectric constant (relative static permittivity) of the medium separating the two plates; and ε_0, the electric constant (= $8.86e^{-12}$ F/m). The thickness of a bilayer as determined by crystallography is about 10 nm, and the dielectric constant of phospholipids is close to 3. The capacitance of a square centimeter of membrane can be calculated with these figures and amounts to 0.27 $\mu F/cm^2$. This value is about four times lower than what has been experimentally determined (i.e., 0.9–1 $\mu F/cm^2$). It has been argued that the effective "electric" distance between the aqueous phases is smaller than estimated because water molecules may penetrate the bilayer farther than presumed. Whatever the physical reasons behind the discrepancy, we should assume capacity of 0.9 $\mu F/cm^2$ for phospholipid bilayers.

Now that we know all this, let us calculate the capacitance of the bilayer set-up. Typical dimensions for bilayer septa are: 1 cm diameter Teflon membrane of 25 μm thickness having a 100 μm diameter aperture covered with the bilayer. The dielectric constant of Teflon is 2.1. The aperture has a surface of $\pi * (50e^{-6})^2 = 7.85e^{-9}\ m^2 = 7.85e^{-5}\ cm^2$. With 0.9 $\mu F/cm^2$ this gives an aperture capacitance of 70.7 pF. The capacitance of the Teflon membrane can be calculated from Eq. (8.3), giving $C = 8.86e^{12} * 2.1 * \pi * (0.5e^{-2})^2/25e^{-6} = 58.4$ pF.

The sum of both capacitances for this example is C = 129 pF.

To record single channel activity in the pA range of currents, the feedback resistor of the voltage-clamp head-stage is usually chosen between 10 and 50 GΩ, converting the unitary currents at the input to a few millivolts at the output. The bilayer resistance ($1M\Omega/cm^2$) is also about 10GΩ. The RC time constant determined by these resistances and the capacitance we just calculated is more than a second (10GΩ * 129 pF = 1.29 s). This means that capacitive currents would obscure any channel activity for a considerable time after a voltage step.

A few strategies are being employed to counter this problem. The first and most practiced is capacitance compensation that we mentioned in Chapter 4. This method only works well if the time constant of the capacitive transient does not vary in time. Unfortunately, capacitance compensation is compromised by ion channels opening during the transient, thus changing membrane resistance. The second strategy is the use of a head-stage that has two or more feedback resistors, between which the computer program controlling the amp can switch electronically (schematically shown in Fig. 8.26B). Just before a voltage step a low resistance resistor is switched on (creating therewith a short RC time). The low resistance resistor is switched off immediately after the capacitive transient such that channel activity can be recorded with high resolution using the GΩ resistor. The third strategy is to reduce aperture and Teflon membrane diameters. They are somewhat more difficult to make in the lab, but Teflon partitions with aperture sizes down to 10 μm diameter are commercially available.

A simple method to eliminate the problem altogether has been found in the "tip-dip" technique using standard patch-clamp equipment (Fig. 8.27).

Tip-dip glass pipettes may be drawn from common hematocrit glass capillaries and coated with Sylgard elastomer to reduce stray capacitance as is usual for patch-clamping. They have a somewhat larger tip (2–4 μm diameter) than patch pipettes, which are about 1 μm. A

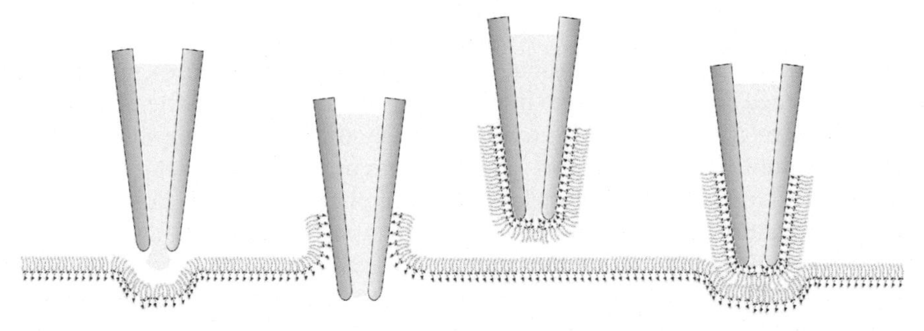

FIGURE 8.27 To obtain a double lipid layer with the tip-dip method, a patch pipette traverses a phospholipid monolayer a few times. During the first passage, liquid is squirted out of the pipette to prevent clogging of the tip. Polar head groups attach to the glass. Dimensions are not drawn to scale.

bilayer is formed by passing the pipette through the water–air interface until a GΩ pipette resistance is obtained. Bilayers having different composition for the inside and outside layers can be made by switching Petri dishes with different monolayers between dips.

AUTOMATION

Ever since several prescription drugs needed to be withdrawn from the market because of their adverse effect on cardiac rhythm, both the Food and Drug Administration (FDA) and the European Medicines Agency (EMA) recommend strongly to determine whether new candidate drugs are liable to cause cardiac *torsade de pointes* and sudden death due to a prolongation of the repolarization period after a ventricular contraction (long QT interval). Since the most common mechanism of QT interval prolongation, as we have seen, is due to the inhibition of the hERG channel, most efforts of the pharmaceutical industry in early safety screening are concentrated on this channel. Many hundreds to thousands of compounds enter each year in the pipeline of an average pharmaceutical firm, most of which do not make it to the first clinical trials for many different reasons. On average, only 1 in every 1000 chemical compounds survives preclinical testing and 1 in every 5000 candidate compounds finally gets FDA and/or EMA approval. This gives a rough idea of the number of molecules that needs to be tested for their effect on hERG or, in fact, any other ion channel in case molecules are being developed for the ion channel in question as target.

VOLTAGE-CLAMP

The need for high-throughput patch-clamp recording for preclinical testing became pressing around the turn of the century for the reasons mentioned above. New automated electrophysiology techniques had to be developed to replace the manual approach. Solutions to this problem have been sought from different perspectives.

The **first**, and least effective one when considering throughput, is the partial replacement of the experimenter by a robotized system. Commercially available set-ups following this approach use cells in suspension (Fig. 8.28A). The set-ups in question still require human

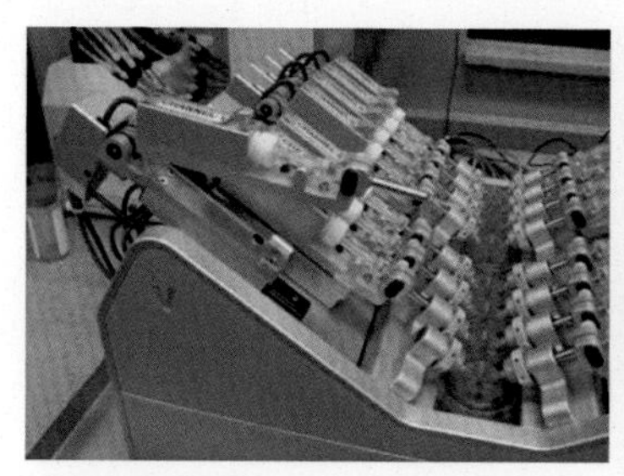
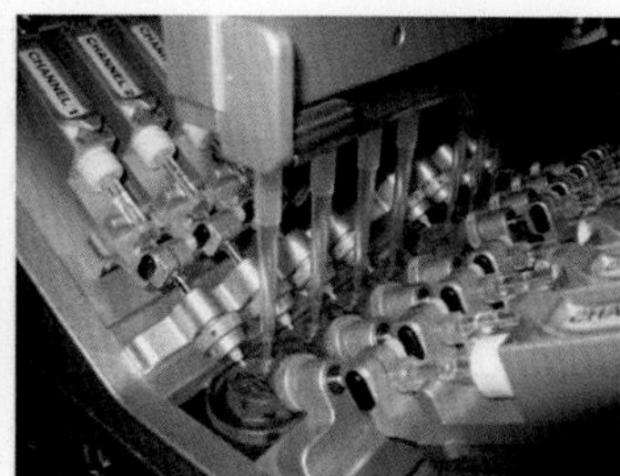

FIGURE 8.28 The semiautomated OpusXpress system from Molecular Devices for two-electrode voltage-clamp of eight *Xenopus* oocytes in parallel. One electrode has been lifted in the leftmost photograph such that the glass electrode may be inserted. A pipetting robot takes care of the application of experimental solutions in the middle and rightmost photographs. *Adapted from Papke RL, Stokes C. Working with OpusXpress: methods for high volume oocyte experiments. Methods 2010;**51**(1):121–33, permission Elsevier.*

intervention. A single person may be able to serve three set-ups that each clamp just a single cell. A similar approach has been successfully applied to blind patch-clamping of neurons in vivo.[30]

Xenopus oocytes are very practical for the expression of plasma membrane channel proteins and transporter proteins carrying charged molecules. They are large and easy to manipulate. Oocytes are voltage-clamped using two glass electrodes (two-electrode voltage-clamp, TEVC). Because their intracellular potential is negative with respect to the exterior, a robot can easily detect impalement by a negative going voltage jump. The first (semi)automated system for oocyte TEVC that became commercially available is OpusXpress system shown in Fig. 8.28. It can clamp and record from 8 oocytes in parallel and it takes care of adding tests substances to the chambers. The system increases throughput of a person attending the device by roughly 20 times if compared to using a manual setup. Roboocyte, developed by Multi Channel Systems, records from oocytes in a 96 well plate sequentially. The firm also sells an automated cDNA/mRNA injection robot. Little is known about its performance, except that currents could be recorded from 50% of the oocytes according to one report.[31]

The **second** solution to voltage-clamp automation that has been found is an adaptation of the planar lipid bilayer technique. In the bilayer technique, a lipid bilayer is spread over a small hole punctured in a Teflon sheet after which channel proteins solubilized in detergent are fused with the bilayer. Electrodes on each side voltage-clamp the artificial membrane. The aperture in the Teflon is usually a few hundred microns. The relatively large membrane surface brings along a large capacitance that reduces temporal resolution. By using borosilicate glass rather than Teflon and by etching the glass with hydrofluoric acid at a spot that was beforehand fragilized by penetration with a single high-energy gold ion (produced by a linear accelerator), a 1 μm perforation could be made.[33] This method effectively reduces the size of the aperture to that of a classical patch-clamp pipette tip. A cell suspension can be added on top of the glass slide and a gigaseal is made with one of the cells that happen to be close by after suction (Fig. 8.29D). Because of the planar layout, the series resistance that can be

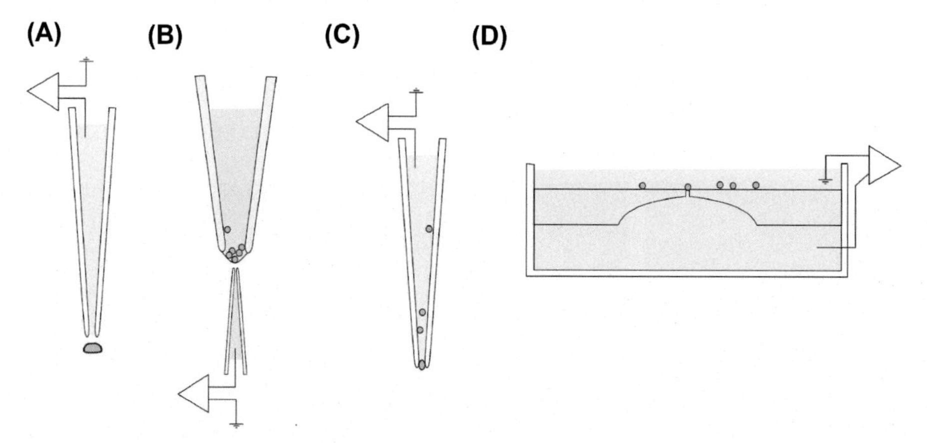

FIGURE 8.29 (A) In classical manual patch-clamping, a pipette is brought to a cell by an experimenter using a microscope for visual control. (B) Automated patch-clamp methods may bring the meniscus of a larger pipette containing a suspension of cells in contact with a patch pipette, (C) flush cells through a patch pipette until a seal is formed, or (D) suck a cell into a pore in a 200 μm thick glass plate.

obtained is lower when compared to the classical patch electrode (~200 kΩ vs. a few MΩ, respectively).

Another method to make a planar patch-clamp device makes use of a high-energy gallium ion beam to "drill" a hole in a SiO2 layer coated with Cr/Pt electrodes and insulated by a so-called low temperature SiO2 layer.[34]

A big advantage is that the planar method does not require a microscope. Seal formation can be monitored and hence automated, by measuring impedance.[35] Several instruments taking advantage of this planar chip technology are on the market, which clamp many cells in parallel (containing 48, 96, or even 384 patch-clamp amplifiers). The numbers correspond to the familiar culture plate formats containing 48, 96, or 384 wells. It should be mentioned here that not all seal initiations lead to a successful recording. The manufacturers mention a 75% success rate.[36]

The **third** approach takes conventional patch pipettes as a starting point. However, rather than placing the pipette against a cell, the cells are flushed inside the pipette and sucked to the tip, where they form a gigaseal with a 60% success rate according to at least one of the manufacturers (Fig. 8.29C). One of the systems that is available on the market requires very little human intervention. It makes ~200 recordings a day, largely enough to establish a dose-effect curve, for example, a task that takes a human experimenter about a week.

The success of these systems in high-throughput screening and the ease of use of automated devices have prompted a few manufacturers to propose single-well variants that could be useful for students or science labs whose primary interest and expertise is not necessarily electrophysiology (Fig. 8.30).

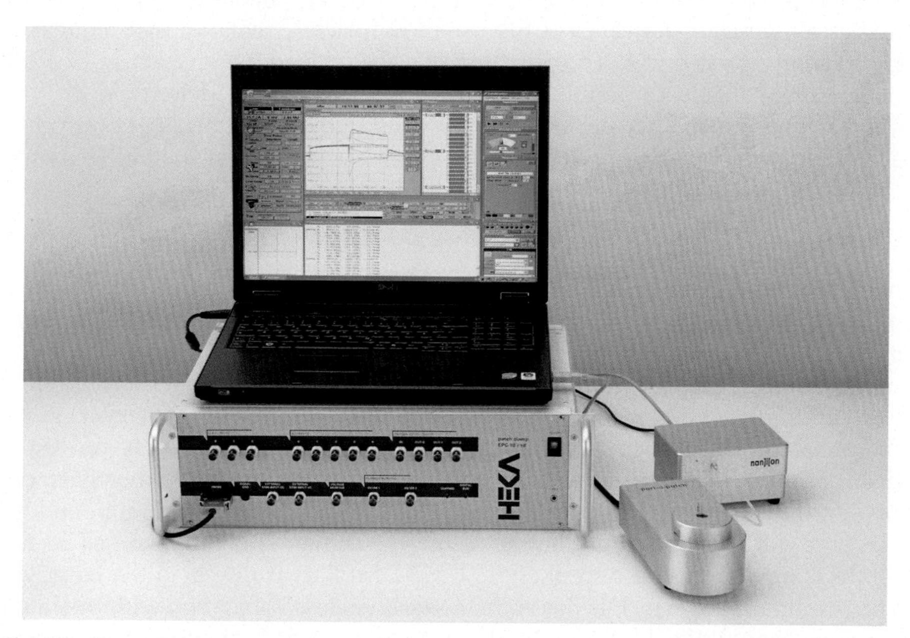

FIGURE 8.30 Port-a-Patch is a miniature, single-well, automated patch-clamp device manufactured by Nanion Technologies. The voltage-clamp amplifier is by HEKA Electronics. *Courtesy Nanion Technologies.*

Pros and Cons of Substitute Methods to Screen Ion Channel Activity

Before automated patch-clamping had been developed, other high-throughput screening methods were employed and still are, since the equipment for automated electrophysiology is prohibitively expensive at the time this is written. This is not only so as an initial investment but also because of the specific (patented) consumables that the equipment requires.

ALTERNATIVE ESSAYS

Ligand-Binding Assays

A medical compound often fixes itself to an ion channel to affect its function. Ligand-binding assays may be used to test compounds that displace a radio-labeled ligand from a known binding site. This method can be used to find compounds with improved affinity for a given binding site but does not help to find new binding sites nor gives it a clue about its functional effect. The drug may be an agonist or an antagonist or just sitting there doing nothing. Even so, ED_{50}s (half the maximum effective doses) determined with binding assays usually correspond reasonably well to those found with patch-clamping.

Ion Flux Assays

Ion flux based essays use radio-labeled ions such as $^{22}Na^{+}$, $^{45}Ca^{2+}$, and $^{86}Rb^{+}$, to follow the passage of ions into or out of cells through Na^{+}, Ca^{2+}, and K^{+} channels, respectively. Like for the binding essay, a major drawback is the use of radioactive material that requires special handling and safety procedures. To overcome this inconvenience, an isotope-free Rb^{+} test has been developed that uses atomic absorption spectroscopy to determine Rb^{+} content. This method is now much used in industry for tests with candidate drugs on cells (often from the Chinese Hamster Ovary cell line, CHO) that stably express the hERG channel.

Voltage-Sensitive Dyes

The membrane potential of a cell is determined by ion gradients and the conductances for these ions. The conductance of an ion species may be indirectly inferred from the change in membrane potential. Fluorescent voltage-sensitive dyes can do this job. Cells are grown in multiwell plates, incubated with the dye for a certain time and subjected to a test procedure during or after which fluorescence is read with a standard plate fluorimeter. However, in order to obtain interpretable data, the conductance under study should be the only one or at least the dominant conductance present in the cell, which limits this approach mostly to cell lines overexpressing the ion channel in question. Most of the dyes also distribute in endoplasmic membranes and mitochondria, thus contaminating the signal. This problem has been solved by adding the coumarin-tagged phospholipid, CC2-DMPE, that inserts in the outer leaflet of the plasma membrane, but which is not voltage-sensitive, in combination with the fluorescent voltage-sensitive oxonol dye, $DisBAC_4(3)$. The two form a pair in fluorescence resonance energy transfer (FRET)-based voltage sensing (Fig. 8.31). $DisBAC_4$ concentrates at the external

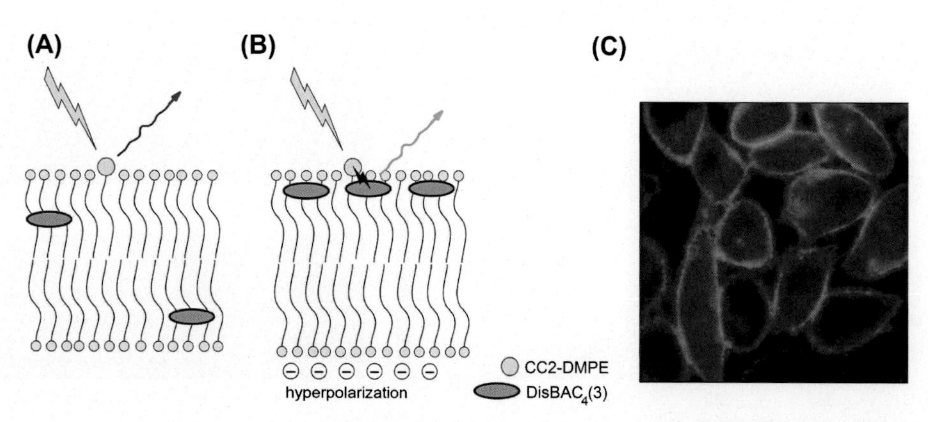

FIGURE 8.31 (A) and (B) Schematic drawing of the plasma membrane with a synthetic phospholipid inserted in the outer leaflet. (A) and (C) The polar coumarin head of the phospholipid fluoresces in the blue if excited with ultraviolet light. (B) When the anionic dye, $DisBAC_4(3)$, is forced close to CC2 by a potential gradient, energy is transferred to the dye which then emits in the green. DMPE = 1,2-Dimyristoyl-sn-glycero-3-phosphoethanolamine, DiSBAC = Diethylthiobarbituric Acid. *(C) From Gonzalez JE, Oades K, Leychkis Y, Harootunian A, Negulescu PA. Cell-based assays and instrumentation for screening ion-channel targets. Drug Discov Today 1999;***4***(9):431–39, permission Elsevier press.*

face of the membrane upon hyperpolarization. The coumarin dye normally emits a blue photon when excited by near ultraviolet light, but if it is in close vicinity of the oxonol, its energy is transferred to the oxonol molecule instead, upon which the latter emits a green photon.

An evident problem of measuring potential changes as a surrogate for ion permeation is that not all change in potential implies activity of the ion channel under study. It may also be the result of an indirect, nonspecific effect (e.g., metabolic, toxic, ion-pump modulation).

Ion-Sensitive Dyes

The presence of ions can also be detected directly by fluorescent probes. Probes for zinc, sodium, magnesium, chloride, and calcium are available but, except for Ca^{2+} probes, their specificity and sensitivity leaves to be desired. Transmembrane ion currents are often not large enough to appreciably change the intracellular ion concentration and they inactivate with time. Calcium indicators are usually fluorescent variations on the aminopolycarboxylic Ca^{2+} chelator BAPTA, itself having a structure close to EGTA. Fura-2 and Fluo-3 are popular examples (Fig. 8.32). As for most intracellular fluorophores, both water-soluble and lipophilic esterified variants are available. The ester forms, because of their lipophilic character, easily permeate the plasma membrane. Once in the cell, they are hydrolyzed by intracellular esterases, thereby adopting their impermeable state. This way, the Ca^{2+} probes accumulate intracellularly until the probe is removed from the medium. A major difficulty of this technique is buffering of intracellular Ca^{2+} by the fluorescent chelator. The incubation time must be chosen such that enough probe enters the cell to obtain a reasonable signal-to-noise ratio without fixing the Ca^{2+} concentration.

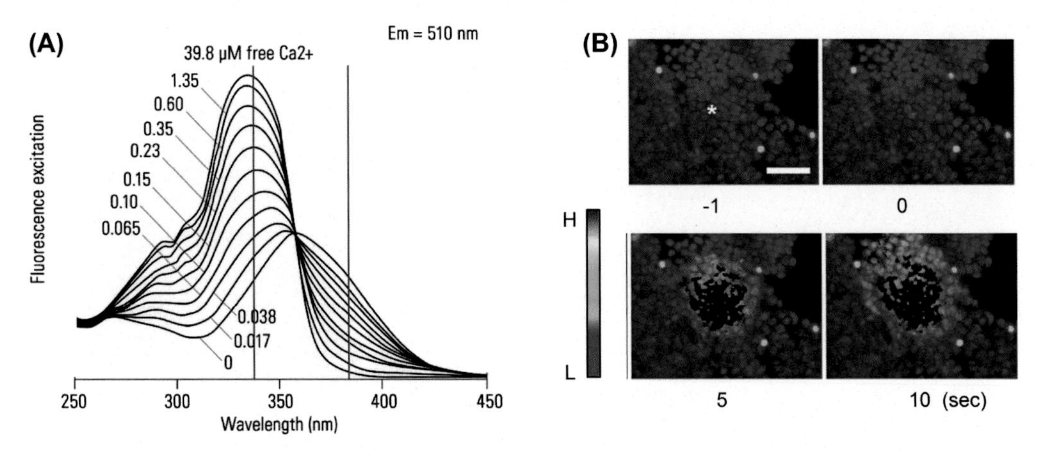

FIGURE 8.32 (A) The shape of the excitation spectrum of Fura-2 depends on the Ca^{2+} concentration. (B) Emission at 510 nm increases with Ca^{2+} concentration if excited at 340 nm and decreases if excited at 380 nm. Salivary gland cells in culture were locally stimulated mechanically (asterisk), resulting in a spreading Ca^{2+} wave. False colors indicate the ratio of emission at 510 nm for the excitation wavelengths 340/380. de Cheveigne A, Parra LC. Joint decorrelation, a versatile tool for multichannel data analysis. Neuroimage. 2014;98:487–505.[38] *Ryu SY, Peixoto PM, Won JH, Yule DI, Kinnally KW. Extracellular ATP and P2Y2 receptors mediate intercellular Ca(2+) waves induced by mechanical stimulation in submandibular gland cells: role of mitochondrial regulation of store operated Ca(2+) entry. Cell Calcium 2010;47(1):65–76. Permission Elsevier press.*

Thallium permeates K^+ channels and is at the base of a fluorescent potassium channel assay developed by Molecular Probes using a thallium-sensitive dye. Cells are first incubated with the esterified form of the fluorescent probe. The remaining extracellular dye is either washed out or quenched with an impermeable molecule after the incubation period.[39] Next, thallium is added to the medium. Thallium entering the cells combines with the dye, thus increasing fluorescence. Other similar methods are being proposed, but, per definition, they all suffer from the same flaw: false-positive and false-negative events.[40] This is demonstrated by the example in the Table 8.1. It compares the IC_{50}s (half-maximum inhibitory concentrations) and MDDs (Minimum Detection Doses) of five compounds known to inhibit the hERG channel obtained with different methods. As can be seen, Sertindole and Cisapride would probably considered to be safe according to the tests with the voltage-sensitive fluorescent dyes, DiBAC4(3) and FMD, while the Rb^+ efflux test would suggest that Cisapride and Terfenadine are OK. Although this example only shows a few cases for which false-negative results were obtained with techniques other than patch-clamp, these techniques may also produce the opposite effect, false-positive results. The latter experimental error would then lead to the rejection of a safe compound.

In conclusion, patch-clamp remains the superior technique for the screening of drug effects on ion channels, albeit expensive in man hours or equipment. It should be noted here that automated patch-clamp tends to shift the IC_{50} concentrations to higher values with respect to manual patch-clamp results.[41] Today, new compounds are often prescreened using one of the techniques other than electrophysiology, such as Rb^+ flux in the case of K-channel tests, followed by patch-clamp checking the promising compounds only.

TABLE 8.1 IC_{50} Values and Minimum Detection Doses (MDDs) of Different Screening Methods

Compound	Patch-Clamp IC_{50} (nM/L)	Rb^+ Efflux IC_{50} (nM/L)	DiBAC4(3) MDD (nM/L)	FMD dye MDD (nM/L)
Dofetilide	10	69	1000	100
Sertindole	14	352	10,000	10,000
Cisapride	45	1500	10,000	10,000
Terfenadine	56	1800	1000	1000
Astemizole	6	59	1000	100

Data taken from Tang W, Kang J, Wu X, et al. Development and evaluation of high throughput functional assay methods for HERG potassium channel. J Biomol Screen *2001;**6**(5):325–31.*

IMPEDANCE AND METABOLISM

The biochip technology has spawned other applications that lend themselves for automation and possibly high-throughput implementations. Cell migration, growth, and adhesion can be followed by measuring (complex) impedance (see Appendix on complex numbers and section Recording of Secretory Events) when the cells are grown on a grid of electrodes. Malignant cells generally display a tendency for proliferation, migration, increased glucose consumption, and lactate rejection. Migration is accompanied by a loss of adhesion, glucose consumption, and proliferation, which go hand in hand with O_2 consumption and CO_2 production. Both CO_2 and lactate production acidify the extracellular medium. Because many anticancer drugs aim to reduce growth and migration (infiltration and metastasis) of malignant tissue, industry is interested in automated methods to measure impedance, O_2 levels, and pH. Several impedance measuring devices are on the market that all work based on the principles that we discussed in the section on secretion. One company has integrated ISFETs to measure pH, miniaturized Clark probes to measure O_2, and an impedance electrode into a single chip (Fig. 8.33). Note that the Clark sensor in A has the same structure

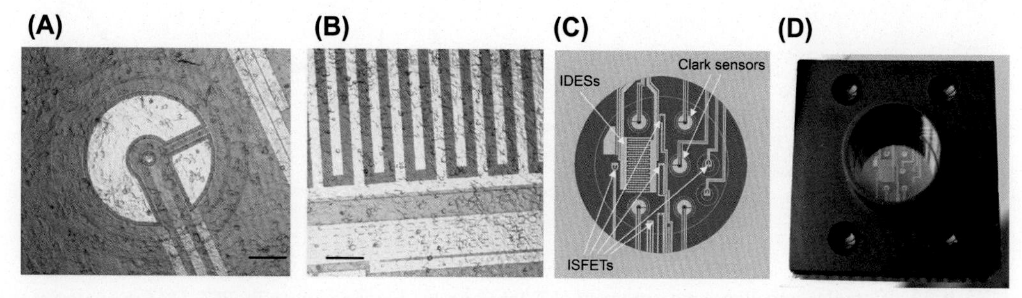

FIGURE 8.33 (A) A three-electrode Clark-type sensor covered with a O_2-permeable membrane. (B) Impedimetric interdigitated electrode structure (IDES) measuring complex impedance. (C) Chip layout. (D) Photo of Bionas metabolic chip SC1000. The well diameter is just over 6 mm. The Bionas Discovery 2500 system may contain up to 6 of these units. Scale bar in (A) and (B) is 200 μm. *Adapted from Ceriotti L, Kob A, Drechsler S, et al. Online monitoring of BALB/3T3 metabolism and adhesion with multiparametric chip-based system. Anal Biochem 2007;**371**(1):92–104., permission Elsevier.*

as in Fig. 4.11, i.e., a small working electrode in the center, a reference electrode to the right, and a larger counter electrode to the left.

Cell adhesion and migration can also be monitored by (arrays of) BioFETs. These transistors are a variation on the ISFET theme. Whereas the gate of the ISFET is covered by a metal-oxide, SiO2, or Si3N4 layer, whose dielectric properties depend on protons and other ions present in the bath, the gate of the BioFET is bare SiO_2. A charged double layer over the gate is created by a potential between gate and reference electrode in the bath (Fig. 8.34A), thus forming an electric double layer having mainly capacitance. The complex impedance of this layer changes in the presence of a cell occupying the gate electrode (Fig. 8.34B). The change is most appreciable in the range between 200 kHz and 1 MHz.

OPTIMIZING SCREENING DATA ACQUISITION

At this point let us discuss the use of the receiver operating characteristic (ROC) to estimate errors in decision-making (accept or reject a screening result) and ways to minimize those errors. The principle of the ROC is explained in Appendix. Drawing a ROC curve can be applied to questions such as "To what extent is method X well adapted to decide whether drugs cause *torsade de pointes*?" and "Is method Y better than method X?". In what follows 55 drugs, of which 32 had been shown to cause *torsade de pointes* were tested on voltage-sensitive hERG K^+ channels, Cav1.2 L-type Ca^{2+} channels, and Nav1.5 Na^+ channels expressed in HEK 293 or CHO cells.[44] The Ca^{2+} and Na^+ channels depolarize, while the K^+ channel repolarizes cardiac myocytes. For each channel type, the negative log ratio of the IC_{50} and the effective free therapeutic plasma concentration (ETPC) was taken as the parameter of interest (PI). Hence, $PI = -\log(IC_{50}/ETPC)$. This was done for all 55 compounds using automated patch-clamping. Then ROC curves were made considering the PIs for each

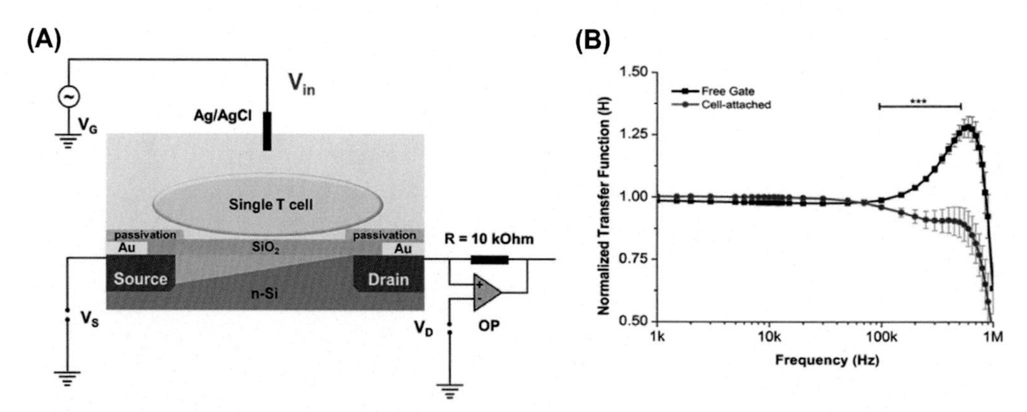

FIGURE 8.34 Complex impedance measurement of T lymphocytes crawling over a BioFET. (A) Schematics of the recording configuration. (B) Frequency-dependent impedance in the absence and the presence of a cell adhering to the open gate of the FET. *From Law JK, Susloparova A, Vu XT, et al. Human T cells monitored by impedance spectrometry using field-effect transistor arrays: a novel tool for single-cell adhesion and migration studies. Biosens Bioelectron 2015;**67**:170–76, permission Elsevier.*

channel separately or by making linear combinations of PIs of multiple channels. It was thus found that the IC_{50} data obtained with the Nav1.5 preparation had no predictive value whatsoever. The AUC (area under the curve) for the hERG data was 0.77. This improved considerably if Cav1.2 PIs were linearly combined with the hERG data (AUC = 0.93). From this result, one should conclude that testing for adverse effects on hERG only is not enough, but that drug effects on the cardiac L-type channel should be considered as well. Accuracy values between 0.7 and 0.8 were found when estimating the performance of automated patch-clamp screening of multiple channels (hERG, Nav1.5, and Cav1.2) in predicting QT shortening or prolongation in *ex vivo* heart preparations.[45] This finding indicates that even combining data obtained with automated patch-clamp of multiple ion channels is not entirely fail-safe. Part of the problem is due to quantification errors when parameter estimations are based on only a few experiments. This is illustrated in Fig. 8.35 where results of a large number of automated patch-clamp recordings obtained with inhibition of the hERG channel by Cisapride, a positive control in hERG screening tests, are pooled.[46] In each of the independent experiments, the IC50 and the slope of the Hill equation fitting the data points making up dose-effect curves were estimated. The Hill equation was originally developed as a model for ligand binding to a receptor (Fig. 8.35A). IC50 in (B) and (C) corresponds to Km in (A). It should be clear that the estimated IC50 and slope cannot be accurately determined by taking only two or three samples randomly from the distributions in (B) and (C).

The other part of the problem likely is that, even if dose-effect parameters are obtained with high precision, their impact on cardiac physiology still needs to be estimated. Hence, the prediction depends also on the quality of the model employed.

It is recommended for these reasons that new compounds that have passed the *in vitro* screens are finally tested on live animals equipped with telemetric ECG devices.

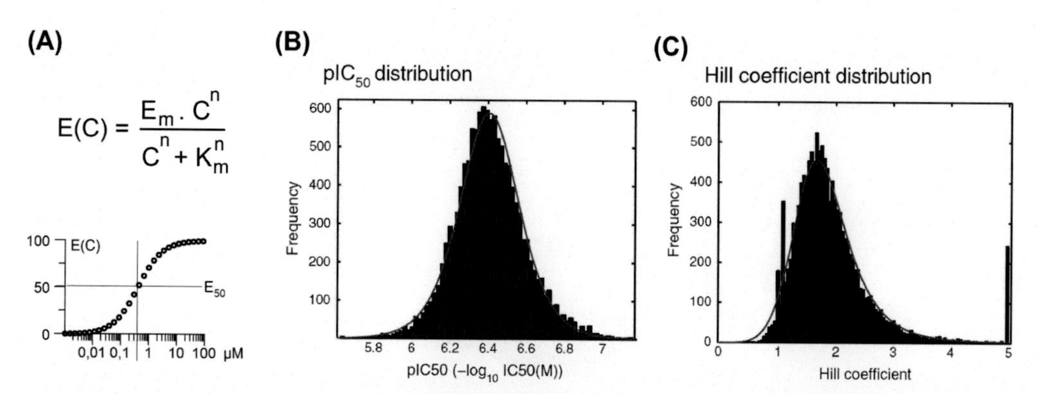

FIGURE 8.35 The Hill equation of ligand binding. C is concentration of the ligand, Em is maximum binding, Km is equilibrium constant, and n is the Hill coefficient. The histogram (A) is a simulation with Em = 100, n = 1, and Km = 0.4 μM. Experimental histograms of (B) pIC50 values (i.e., the negative logarithm of the IC50) and (C) Hill coefficients, from an IonWorks Quattro Cisapride hERG assay. N = 12,638 independent runs. Solid red lines denote Gaussian curve fit.[46] *Elkins RC, Davies MR, Brough SJ, et al. Variability in high-throughput ion-channel screening data and consequences for cardiac safety assessment. J Pharmacol Toxicol Methods 2013;68(1):112–22. (B) and (C) permission Elsevier.*

TELEMETRY

Telemetry of physiological data is often necessary if the data is to be collected from animals. Animals tend to be stressed if they are restrained by the experimenter, a condition surely influencing heart rate, blood pressure, secretion, and other physiological parameters of interest. Telemetry is also a solution if a phenomenon occurs rarely such as seizures in epileptic patients. A standard 20 min. EEG recording is normal in 50% of epilepsy patients, which makes it a dubious diagnostic tool. A 48h recording from ambulatory patients raises the reliability to 95%.[47]

Telemetric devices transmit data by means of electromagnetic waves in the radio frequency spectrum, i.e., between 5 MHz and 300 GHz, the higher limit being just below the microwave frequency range. Low-frequency radio signals can transmit over long distances (a few times the diameter of the Earth or even over interplanetary distances), while signals in the gigahertz range are rapidly attenuated due to absorption by water and particles in the air. FM radio emits at about 90 MHz, Bluetooth at about 2.4 GHz, and Wi-Fi both at ~2.4 GHz and at ~5 GHz, to give a few examples. These examples were not taken randomly because they are of practical use in animal and patient monitoring.

Since FM radio waves travel relatively far, they are ideal to follow the movement of birds, bats, and other volatiles. Actually, a license-free radio band, centered at 446 MHz, is normally used to prevent interference with FM radio stations. Transmitters weighing less than half a gram can be glued onto these animals in order to localize them. Weight necessarily increases if transmission of extra data such as temperature, respiration, and the like are required, mostly because the batteries need to be heavier (see Fig. 8.36).

Bluetooth and Wi-Fi solutions are currently used for short range telemetry in dogs and patients. The animal or the patient wears a jacket or shoulder bag containing the device that transmits multichannel ECG, temperature, blood pressure, and/or EEG data to a near-by Bluetooth receiver (e.g., a smartphone) or Wi-Fi modem.

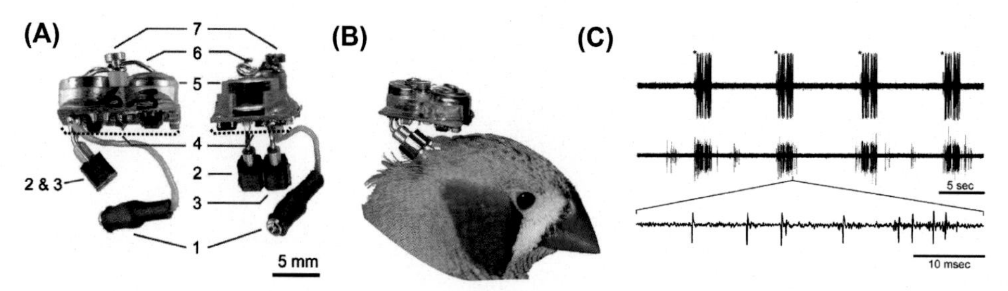

FIGURE 8.36 (A) Left side: Side view of the telemetric device. Right side: Rear view of the device. 1: Connector for recording electrode, attached via a shielded wire to the transmitter. 2: Connector for reference electrode. 3: Dummy connector for support. 4: Components on printed circuit board. 5: Batteries. 6: Spring wire connecting batteries in series. 7: Screw. (B) Zebra finch with a transmitter. (C) Repeated stimulation with this bird's own song (upper trace) causes repeated activation of the unit that was recorded (lower trace). Asterisks (*) denote the start of each stimulus. *Adapted from Schregardus DS, Pieneman AW, Ter Maat A, Jansen RF, Brouwer TJ, Gahr ML. A lightweight telemetry system for recording neuronal activity in freely behaving small animals. J Neurosci Methods 2006;***155***(1):62–71, permission Elsevier.*

References

1. Rakotoarivelo C, Petite D, Lambard S, et al. Receptors to steroid hormones and aromatase are expressed by cultured motoneurons but not by glial cells derived from rat embryo spinal cord. *Neuroendocrinology* 2004;**80**(5):284–97.
2. Hodgkin AL, Huxley AF. A quantitative description of membrane current and its application to conduction and excitation in nerve. *J Physiol* 1952;**117**(4):500–44.
3. Milnes JT, Witchel HJ, Leaney JL, Leishman DJ, Hancox JC. Investigating dynamic protocol-dependence of hERG potassium channel inhibition at 37 degrees C: Cisapride versus dofetilide. *J Pharmacol Toxicol Methods* 2010;**61**(2):178–91.
4. Chaves AA, Zingaro GJ, Yordy MA, et al. A highly sensitive canine telemetry model for detection of QT interval prolongation: studies with moxifloxacin, haloperidol and MK-499. *J Pharmacol Toxicol Methods* 2007;**56**(2):103–14.
5. Gowda RM, Khan IA, Wilbur SL, Vasavada BC, Sacchi TJ. Torsade de pointes: the clinical considerations. *Int J Cardiol* 2004;**96**(1):1–6.
6. Redfern WS, Carlsson L, Davis AS, et al. Relationships between preclinical cardiac electrophysiology, clinical QT interval prolongation and torsade de pointes for a broad range of drugs: evidence for a provisional safety margin in drug development. *Cardiovasc Res* 2003;**58**(1):32–45.
7. Ramon y Cajal S. *Comparative study of the sensory areas of the human cortex*. 1899.
8. Creutzfeldt OD, Watanabe S, Lux HD. Relations between EEG phenomena and potentials of single cortical cells. I. Evoked responses after thalamic and erpicortical stimulation. *Electroencephalogr Clin Neurophysiol* 1966;**20**(1):1–18.
9. Berger H. Über das Eletrenkephalogramm des Menschen. *Arch Psych Nerverkr* 1929;**87**:527–70.
10. de Cheveigne A, Parra LC. Joint decorrelation, a versatile tool for multichannel data analysis. *Neuroimage* 2014;**98**:487–505.
11. Kamarajan C, Pandey AK, Chorlian DB, Porjesz B. The use of current source density as electrophysiological correlates in neuropsychiatric disorders: a review of human studies. *Int J Psychophysiol* 2015;**97**(3):310–22.
12. Tenke CE, Kayser J. Generator localization by current source density (CSD): implications of volume conduction and field closure at intracranial and scalp resolutions. *Clin Neurophysiol* 2012;**123**(12):2328–45.
13. Neher E, Marty A. Discrete changes of cell membrane capacitance observed under conditions of enhanced secretion in bovine adrenal chromaffin cells. *Proc Natl Acad Sci USA*. 1982;**79**(21):6712–6.
14. Lindau M, Neher E. Patch-clamp techniques for time-resolved capacitance measurements in single cells. *Pflugers Archiv Eur J Physiol* 1988;**411**(2):137–46.
15. Angleson JK, Betz WJ. Monitoring secretion in real time: capacitance, amperometry and fluorescence compared. *Trends Neurosci* 1997;**20**(7):281–7.
16. Marquis BJ, Haynes CL. The effects of co-culture of fibroblasts on mast cell exocytotic release characteristics as evaluated by carbon-fiber microelectrode amperometry. *Biophys Chem* 2008;**137**(1):63–9.
17. Kim BN, Herbst AD, Kim SJ, Minch BA, Lindau M. Parallel recording of neurotransmitters release from chromaffin cells using a 10x10 CMOS IC potentiostat array with on-chip working electrodes. *Biosens Bioelectron* 2013;**41**:736–44.
18. Lichtenberger J, Fromherz P. A cell-semiconductor synapse: transistor recording of vesicle release in chromaffin cells. *Biophys J* 2007;**92**(6):2262–8.
19. Zhou Z, Misler S. Amperometric detection of quantal secretion from patch-clamped rat pancreatic beta-cells. *J Biol Chem* 1996;**271**(1):270–7.
20. Illinger D, Kuhry JG. The kinetic aspects of intracellular fluorescence labeling with TMA-DPH support the maturation model for endocytosis in L929 cells. *J Cell Biol* 1994;**125**(4):783–94.
21. Edwards FA, Konnerth A, Sakmann B, Takahashi T. A thin slice preparation for patch clamp recordings from neurones of the mammalian central nervous system. *Pflugers Arch* 1989;**414**(5):600–12.
22. Sakmann B, Neher E. *Single-channel recording*. 2 ed. New York: Plenum press; 1995.
23. Krimer LS, Goldman-Rakic PS. An interface holding chamber for anatomical and physiological studies of living brain slices. *J Neurosci Methods* 1997;**75**(1):55–8.
24. MacVicar BA. Infrared video microscopy to visualize neurons in the in vitro brain slice preparation. *J Neurosci Methods* 1984;**12**(2):133–9.
25. Dodt HU, Zieglgansberger W. Visualizing unstained neurons in living brain slices by infrared DIC-videomicroscopy. *Brain Res* 1990;**537**(1–2):333–6.

26. Stuart GJ, Dodt HU, Sakmann B. Patch-clamp recordings from the soma and dendrites of neurons in brain slices using infrared video microscopy. *Pflugers Archiv Eur J Physiol* 1993;**423**(5–6):511–8.
27. Steidl EM, Neveu E, Bertrand D, Buisson B. The adult rat hippocampal slice revisited with multi-electrode arrays. *Brain Res* 2006;**1096**(1):70–84.
28. Heuschkel MO, Fejtl M, Raggenbass M, Bertrand D, Renaud P. A three-dimensional multi-electrode array for multi-site stimulation and recording in acute brain slices. *J Neurosci Methods* 2002;**114**(2):135–48.
29. Gholmieh G, Soussou W, Han M, et al. Custom-designed high-density conformal planar multielectrode arrays for brain slice electrophysiology. *J Neurosci Methods* 2006;**152**(1–2):116–29.
30. Kodandaramaiah SB, Franzesi GT, Chow BY, Boyden ES, Forest CR. Automated whole-cell patch-clamp electrophysiology of neurons in vivo. *Nat Methods* 2012;**9**(6):585–7.
31. Pehl U, Leisgen C, Gampe K, Guenther E. Automated higher-throughput compound screening on ion channel targets based on the *Xenopus laevis* oocyte expression system. *Assay Drug Dev Technol* 2004;**2**(5):515–24.
32. Papke RL, Stokes C. Working with OpusXpress: methods for high volume oocyte experiments. *Methods* 2010;**51**(1):121–33.
33. Fertig N, Meyer C, Blick RH, Trautmann C, Behrends JC. Microstructured glass chip for ion-channel electrophysiology. *Phys Rev E Stat Nonlin Soft Matter Phys* 2001;**64**(4 Pt 1):040901.
34. Li S, Lin L. A single cell electrophysiological analysis device with embedded electrode. *Sensor Actuators A* 2006;**134**:20–6.
35. Fertig N, Blick RH, Behrends JC. Whole cell patch clamp recording performed on a planar glass chip. *Biophys J* 2002;**82**(6):3056–62.
36. Wood C, Williams C, Waldron GJ. Patch clamping by numbers. *Drug Discov Today* 2004;**9**(10):434–41.
37. Gonzalez JE, Oades K, Leychkis Y, Harootunian A, Negulescu PA. Cell-based assays and instrumentation for screening ion-channel targets. *Drug Discov Today* 1999;**4**(9):431–9.
38. Ryu SY, Peixoto PM, Won JH, Yule DI, Kinnally KW. Extracellular ATP and P2Y2 receptors mediate intercellular Ca(2+) waves induced by mechanical stimulation in submandibular gland cells: role of mitochondrial regulation of store operated Ca(2+) entry. *Cell Calcium* 2010;**47**(1):65–76.
39. Titus SA, Beacham D, Shahane SA, et al. A new homogeneous high-throughput screening assay for profiling compound activity on the human ether-a-go-go-related gene channel. *Anal Biochem* 2009; **394**(1):30–8.
40. Tang W, Kang J, Wu X, et al. Development and evaluation of high throughput functional assay methods for HERG potassium channel. *J Biomol Screen* 2001;**6**(5):325–31.
41. Dunlop J, Bowlby M, Peri R, Vasilyev D, Arias R. High-throughput electrophysiology: an emerging paradigm for ion-channel screening and physiology. *Nat Rev Drug Discov* 2008;**7**(4):358–68.
42. Ceriotti L, Kob A, Drechsler S, et al. Online monitoring of BALB/3T3 metabolism and adhesion with multiparametric chip-based system. *Anal Biochem* 2007;**371**(1):92–104.
43. Law JK, Susloparova A, Vu XT, et al. Human T cells monitored by impedance spectrometry using field-effect transistor arrays: a novel tool for single-cell adhesion and migration studies. *Biosens Bioelectron* 2015;**67**:170–6.
44. Kramer J, Obejero-Paz CA, Myatt G, et al. MICE models: superior to the HERG model in predicting Torsade de Pointes. *Sci Rep* 2013;**3**:2100.
45. Beattie KA, Luscombe C, Williams G, et al. Evaluation of an in silico cardiac safety assay: using ion channel screening data to predict QT interval changes in the rabbit ventricular wedge. *J Pharmacol Toxicol Methods* 2013;**68**(1):88–96.
46. Elkins RC, Davies MR, Brough SJ, et al. Variability in high-throughput ion-channel screening data and consequences for cardiac safety assessment. *J Pharmacol Toxicol Methods* 2013;**68**(1):112–22.
47. Lawley A, Evans S, Manfredonia F, Cavanna AE. The role of outpatient ambulatory electroencephalography in the diagnosis and management of adults with epilepsy or nonepileptic attack disorder: a systematic literature review. *Epilepsy Behav* 2015;**53**:26–30.
48. Schregardus DS, Pieneman AW, Ter Maat A, Jansen RF, Brouwer TJ, Gahr ML. A lightweight telemetry system for recording neuronal activity in freely behaving small animals. *J Neurosci Methods* 2006;**155**(1):62–71.

CHAPTER

9

Analysis of Electrophysiological Signals

OUTLINE

Introduction to Electrophysiological Methods and Instrumentation, Second Edition
https://doi.org/10.1016/B978-0-12-814210-3.00009-8

Information contained in electrophysiological records requires shaping into meaningful numbers or graphs. In rare occasions, a short stretch of data suffices to demonstrate a point to be made, but in most cases, some quantity has to be determined, averaged over multiple records, and its variation between experimental conditions evaluated. In this chapter, we introduce a series of methods to extract information from electrophysiological data. The chapter can be roughly divided into two halves: the first dealing with action potential series and the second with voltage-clamp records. In both cases, the analysis usually consists of three parts: (1) detection of events, (2) measurement of a parameter, and (3) fit to a model. Not rarely, measurement and eventually detection already presupposes a model.

ANALYSIS OF ACTION POTENTIAL SIGNALS

Except for unicellular species such as amoebae and other protists, all animals need to convert neural signals into trains of action potentials to send them through long nerve fibers. It is tempting to call the spike code a digital code, to segregate it from the clearly analogue potential variations that arise in sense organs, neuronal cell bodies, and so on. Indeed, given the regenerative mechanism that restores the amplitude of an action potential everywhere along the nerve fiber, the amplitude may be considered a binary signal: At any moment, there is either a spike or no spike.

However, most series of action potentials are continuous in time. The information being transmitted remains "analog" in the sense that it may be encoded in spike density (spike frequency modulation). The physiologically important parameter may also reside in the temporal delay or coherence with respect to the electrical activity of other neurons converging onto the same target neuron. In any case, continuously varying spike intervals contain the information of interest.

Before tackling the analysis of spike signals, the distinction must be made between single unit and gross activity. In most cases, one records the activity of a single nerve fiber with a suitably small electrode. The fiber may be connected to a sensory cell or a neuron and is usually called a unit, hence single-unit recording. In other cases, especially with extracellular recording, the signal contains more than one spike train. A multiunit signal, which contains a small number of spike trains, is very confusing and often of little use. In some cases, the

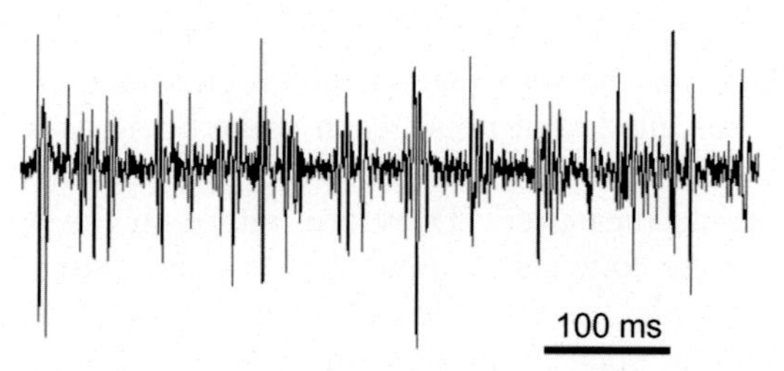

FIGURE 9.1 A multiunit recording from a small muscle (ball of the thumb).

spike trains can be segregated by their respective amplitudes. If the fibers have different distances to the recording electrodes, their spike trains will have different amplitudes (Fig. 9.1) and so can be segregated by a computer algorithm.

If one records from a whole nerve or muscle, the number of units contributing to the signal is so large that they cannot be segregated. In fact, the resultant, so-called gross-activity signal behaves as a form of noise.

POPULATION SPIKE AND GROSS ACTIVITY

Gross activity is the term for the joint activity of tens, hundreds, or even thousands of fibers present in a single nerve or muscle. In some special cases, the activity (firing) of all fibers is approximately synchronous, for instance, if the whole nerve or muscle is stimulated. This is often the case in clinical applications, where the functioning of nerves and muscles after a trauma can be assessed by recording the activity elicited by pulses from external (i.e., transcutaneous) stimulation. In this case, the summed spike signals may be recognizable as a relatively large spike, as if it were a single unit. This is called a population spike, or compound action potential (CAP): see Fig. 9.2. The sharpness of the resultant spike is determined by the degree of synchrony. If the different fibers differ in conduction velocity, the spikes in parallel fibers tend to diverge so that the best signal is obtained close to the source (stimulus electrode).

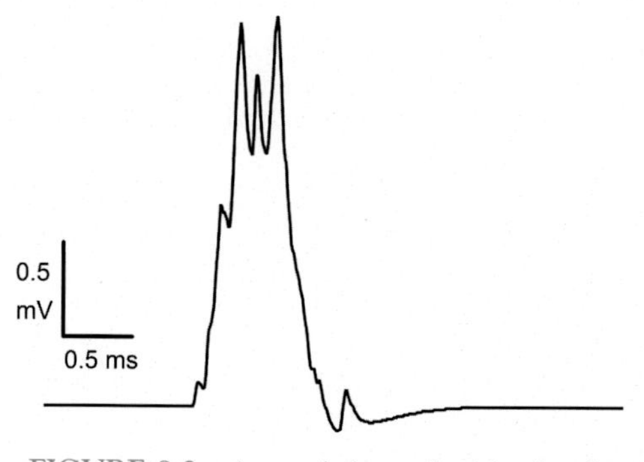

FIGURE 9.2 A population spike (simulated).

Under more natural circumstances, however, spike trains in parallel fibers tend to be independent. In this case, a random, noiselike signal called gross activity arises. More precise, it has the properties of the so-called shot noise, each spike being a "shot" (see the section on noise in Chapter 3). The amplitude of this noise is a good measure of the average spike activity in the nerve or muscle measured. Therefore, although the spike frequencies cannot be recorded directly, the spike code can be analyzed to a fair extent. The desired amplitude can be found by a very simple circuit, called a diode pump. It consists of a diode and a low-pass filter. The output of such a circuit shows the average spike activity in a whole nerve (Fig. 9.3), provided the input signal is strong enough to overcome the (Schottky) diode threshold voltage of 0.2–0.7 V (see Chapter 3, diodes). Alternatively, computer algorithms performing essentially the same operation may be used with signals of arbitrary amplitude.

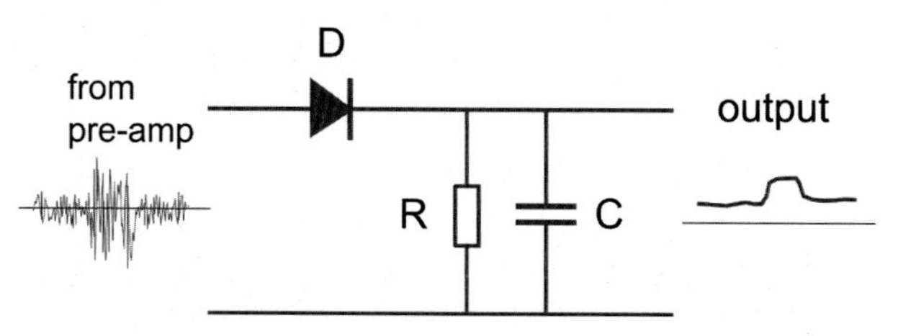

FIGURE 9.3 Gross-activity analysis by a "diode pump."

The functions of some hitherto unknown sense organs have been found by gross-activity recording, which can be performed on intact animals (Fig. 9.4).

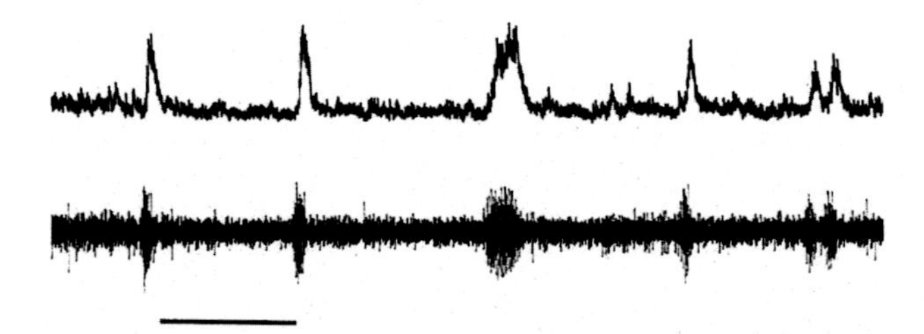

FIGURE 9.4 Gross-activity recording of a nerve that proved to innervate a chemoreceptor in the skin of a fish. Bottom trace: gross spike activity; top trace: "diode pump" running average activity. Bar 1 s. *Courtesy Peters RC, van Steenderen GW, Kotrschal K. A chemoreceptive function for the anterior dorsal fin in rocklings (Gaidropsarus and Ciliata: Teleostei: Gadidae): electrophysiological evidence.* J Mar Biol Assoc UK *1987;***67***(4):819–23.*

SINGLE-UNIT ACTIVITY

The simplest situation to illustrate the problem of analyzing action potential series is the example shown in Fig. 7.1: a sense organ picking up some quantity from the environment and encoding it in the form of an action potential series. The encoding of light flashes by a photoreceptor and its subsequent encoding, as a spike train is only the beginning of a long journey of that signal through several parts of the peripheral nervous system and the brain. During that journey, the form of the signal will alternate many times between an analog potential (receptor potential, postsynaptic potential) at the points where the signal must be processed and a spike train in the long-distance transport sections.

The intrinsic properties of the spike code determine the fate of any neural signal, but in addition, they will determine which analysis methods are suitable for physiologists to hunt down the functioning and the function of the studied neurons.

UNCERTAINTY AND AMBIGUITY IN SPIKE SERIES

Formally, a series of action potentials is a signal continuous in time. However, for most analyses of neural signals, the duration of each spike can be neglected, and only the timing of the pulses carries information. Moreover, most spike trains are irregular, making statistical analysis necessary. Therefore, a time series of action potentials is considered to be a stochastic point process. A point process because it exists only at certain moments in time, and stochastic because fluctuations are conspicuous. As we will see later on, the fluctuations may even form an essential part of the code.

A typical spike train might look like Fig. 9.5.

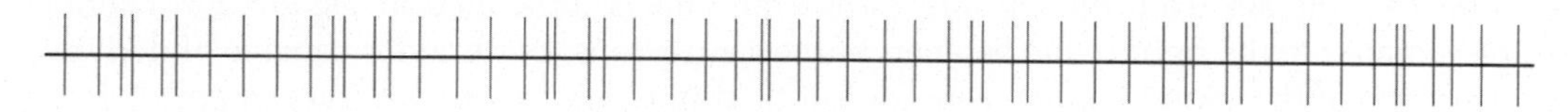

FIGURE 9.5 A spike train.

This signal can be considered as a regular firing frequency modulated by noise. Both components may be characterized by well-known statistical measures such as mean and standard deviation. In the nervous systems, such a signal might represent the activity of a sense organ, the contraction force of a muscle fiber or the activity of a neuron in the brain. Apart from the fluctuations, the spike train will be modified and varied in time. In a sense organ, this might stem from a changing stimulus (any important quantity in the environment); in a motor circuit controlling a muscle, it stems from instructions from the brain to change the activity of that particular muscle fiber (or motor unit).

Fig. 9.6 shows a spike train modified by a change in stimulus/activity level.

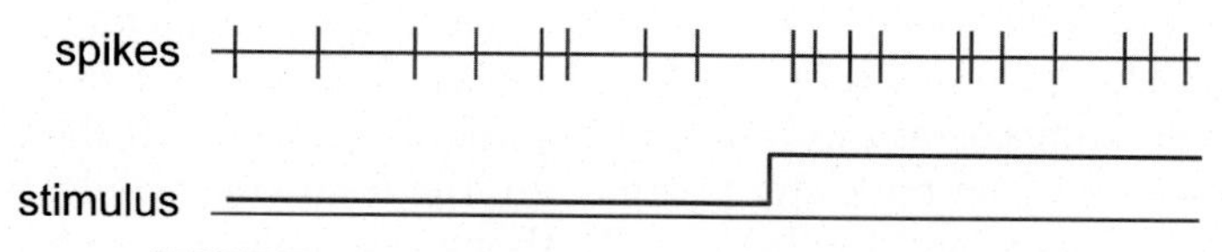

FIGURE 9.6 Uncertainty at a change in activity level.

Note that the spike code per se introduces an uncertainty, since a sudden change in activity will only be reflected in the spike train at the moment of the next spike. Between two spikes, the signal is uninformative. The ubiquitous fluctuations cause a second type of

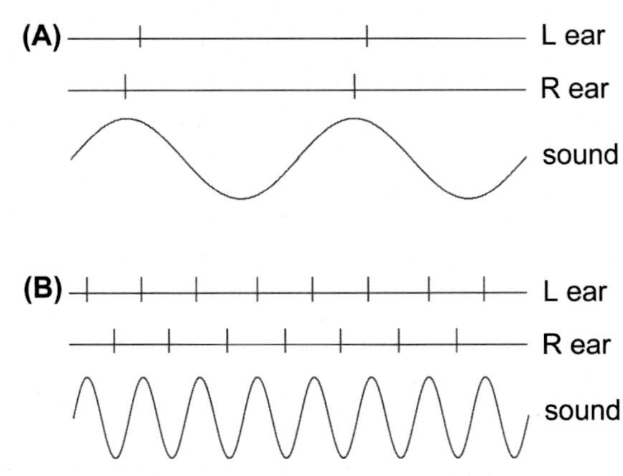

FIGURE 9.7 Ambiguity in directional hearing. (A) Timing difference allowing perception of the direction of the sound (i.e., from right to left). (B) Ambiguity leading to a diffuse sound sensation.

uncertainty: The central nervous system must perform some kind of statistical analysis to find out whether the change has a relevant cause (such as a change in ambient light intensity) or is a mere fluctuation.

In addition to uncertainty, a spike signal may be ambiguous. An example of this ambiguity occurs in directional hearing, which depends partially on timing differences between the sound-evoked spikes from the left and right ears. This is illustrated in Fig. 9.7 for a sound source situated to the right of the listener. In most situations, our brain is capable of deriving the direction of the sound source from the subtle differences in arrival times of spikes from our left and right ears. This is shown in Fig. 9.7A. With neuronal circuits, the short delay between spikes from the right and left ears can be translated into a sensation of direction and be distinguished from the much longer delay left–right. At a certain frequency the two intervals become equal (shown in Fig. 9.7B) and the perceived sound direction is getting indeterminate: The sound seems to stem from a broad front.

In sense organs, two main types of activity can be distinguished: silent and spontaneously active. Some sensory cells and fibers show spontaneous activity, i.e., in the absence of a stimulus, a more or less regular spike train is visible. The spontaneous rate is modulated by the sensory input. Other sense organs are silent at rest and only fire when a certain threshold is crossed. These two cases are illustrated in Fig. 9.8.

The fluctuations in spike timings can be very large and occur in virtually every nerve fiber. It may seem purely detrimental to the value of spike signals in encoding real-world quantities. However, it is not always true: The fluctuations may serve to overcome certain disadvantages of the point process code. Suppose, for instance, that a neuron from the ear fires in response to a sound stimulus at a certain (not too high) frequency. If the spike code were entirely deterministic (i.e., without any fluctuation), the following would happen. At very low intensities, no spikes would be fired at all. Above threshold, a single spike per period would appear. A further increase would not influence the spike signal, until the point where the sound amplitude reached a second threshold, above which two spikes per period would appear. Hence, a strictly regular spike train would code amplitude in a digital (discrete) way. More subtle amplitude changes would go unnoticed. Enter the noise: because of the

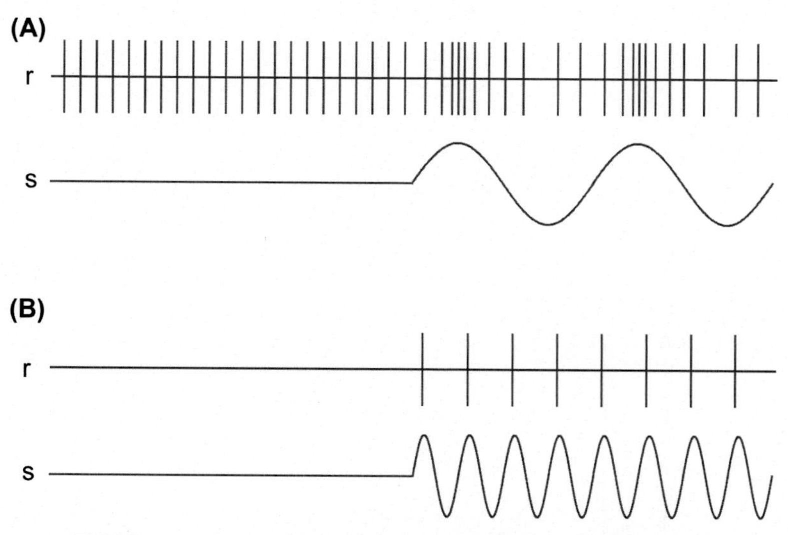

FIGURE 9.8 (A) Spontaneously active versus (B) silent nerve cells.

ubiquitous fluctuations, the situation is very different. At extremely low sound intensities, again no spikes will appear. However, even at relatively low intensities, a spike will appear occasionally. The stronger the sound, the higher the *probability* that a spike will be fired, until it is about one spike per period. At that point, an occasional second spike per period will occur, and so on. This is shown in Fig. 9.9. In this case, the amplitude is sensed in a continuous, thresholdless way, *provided the CNS is allowed to average the spike signal*, either in time or over a number of parallel fibers. In reality, both ways of averaging occur.

Whatever the part played by the fluctuations, electrophysiologists investigating a spike signal face the task of separating the deterministic changes from the stochastic ones. Fortunately, there exists a plethora of methods to extract relevant information on the functioning of neurons and neuronal networks.

The basic ones will be treated below.

INTERVAL HISTOGRAM

A number of spike pattern analysis methods are based on the interval time of spikes, the spike interval, or, more precisely, the interspike interval (formally, the duration of the action potential is also a time interval, the intraspike interval). The most compact description of a (large) number of interval values is of course the classical statistical set of moments. The mean, also known as the first moment, is the single most important characteristic. This describes the deterministic content in the signal, but gives no clue as to the scatter, or degree of (un-)reliability of the spike train. Therefore, we need at least one other measure such as the standard deviation. This is the most used (and misused) measure of scatter. It is called the second central moment formally because it reflects the weighted (root mean square)

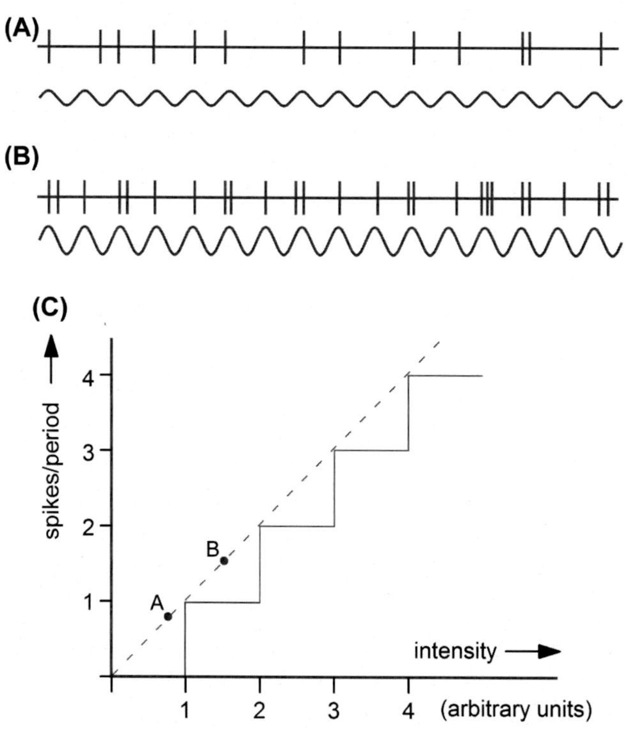

FIGURE 9.9 The advantage of fluctuations in a spike signal. (A) Sound intensity where about 0.8 sp/period arise. (B) A stronger sound yields about 1.6 sp/period. (C) The input/output curves with random fluctuations (*straight line*) and without (*staircase*) fluctuations.

deviation of the data from the mean. Of course, this holds only for data series that follow approximately a normal distribution. If a distribution of interspike intervals is markedly skewed or peaked (i.e., a distribution that differs significantly from the normal, or Gaussian, distribution), one might need further moments, such as the third central moment, called skew for short, and the fourth central moment or kurtosis (peakedness). Details and equations can be found in most books on statistical methods.

Unfortunately, spike interval distributions are often skewed, which have several reasons and depend on the spike generating mechanism. The two extremes may illustrate this: On one end, a perfectly regular spike train, entirely determined by a pacemaker mechanism, and at the other end, a highly irregular, entirely random one. The first case is approximated by, e.g., the cardiac pacemaker and by some sensory receptor cells and fibers. A perfect pacemaker would have a very unusual, "spike"-shaped interval histogram: all values in one bin. In real-world pacemakers, however, there is always some noise, which stems from the random opening of ion channels, a fluctuating number of synaptic vesicles and other molecular processes. This causes the interspike intervals to fluctuate around a certain mean value, often giving rise to an approximately Gaussian distribution. This situation, best described as "largely deterministic plus slightly random," is shown in Fig. 9.10, trace 1. Finally, one can

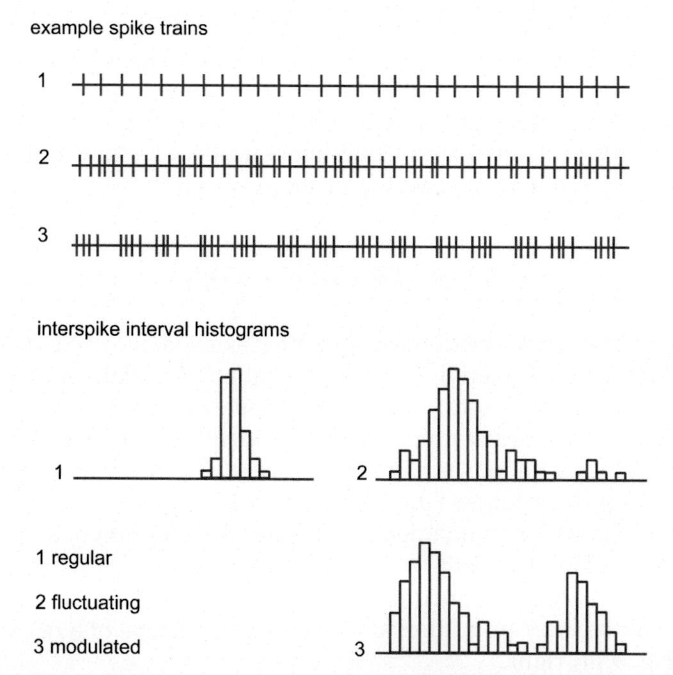

FIGURE 9.10 Schematic representation of sample spike trains and their respective interval histograms: (1) a fairly regular series with a narrow, approximately normally distributed interval distribution; (2) a more irregular spike train with a skewed histogram; and (3) a spike train modulated by a deterministic signal (sinusoid) and its histogram (discussed in the text).

imagine an entirely random interval series, which is again approximately the case with some real-world neurons.

The reader may wonder why action potentials, arising successively in one and the same physical structure (a neuron), can be found to be largely independent from one another. In general, any physical structure will have some degree of inertia, or memory. Indeed, in most neurons, the degree of depolarization determines the overall firing rate or firing probability, and it will always take some time to change the membrane potential and the firing rate. Depending on the duration of this dependence, successive spikes may show more or less interdependency, the degree of which can be assessed by statistical inference.

However, the high degree of independence of successive spikes in some neurons can be explained by two facts. First, far more ion channels are open during an action potential than in between the spikes. In other words, each spike causes a brief shunting of the local membrane potential, a kind of "reset" pulse erasing the "memory" of past situations to a certain degree. The degree of shunting is determined by the shape of the neuron soma and spike-initiating zone and by the number and distribution of voltage-sensitive ion channels. This partial shunting, or partial reset, explains most irregular spike firing. Second, interspike interval times are often longer than the membrane time constants involved so that the electrical "memory" is very short.

Most neurons show a compromise between the extremes of a pacemaker, having a regular spike rate, and a noisy, virtually independent spike generator. Fig. 9.10, trace 2, shows a spike train recorded from a sensory neuron and its interval histogram in the absence of stimulation. Contrary to trace 1, this represents the situation "largely random plus slightly deterministic." The histogram is skewed with a tail at the long interval side. The third trace shows the effect of a periodical, i.e., deterministic modulation of the spike rate.

POISSON PROCESSES

To explain why spike interval distributions are often skewed, we need to delve a bit deeper into the statistical principles underlying spike generation. At relatively long-spike interval times and/or a high degree of shunting, subsequent spike occurrences are virtually independent. In statistical theory, a series of mutually independent random events forms a relatively simple and well-known case known as a Poisson process. The number of events per time unit (i.e., the rate) of such a process follows the Poisson distribution. This is a skewed distribution, well known from most books on fundamental statistics. A histogram of a Poisson-distributed spike series is shown in Fig. 9.11, left.

However, spike analysis is performed usually at the single interval level. In a Poisson process, the histogram of the interspike interval lengths has the shape of a negative exponential. This is shown in Fig. 9.11, right.

The exponential distribution has only one parameter: the mean frequency, usually called λ. The amount of scatter, reflected in the standard deviation (the second central moment), is $\sqrt{\lambda}$.

Thus, in principle, one could expect the interval histogram of some spike series to follow this exponential distribution. Most interval histograms made from real spike trains, however, have different shapes. This is shown in Fig. 9.12. Graph A shows the exponential distribution discussed above. A first modification would be caused by the refractory period, which is inherent in the neuronal spike generating mechanism. This would cause the distribution to shift to the right (graph B).

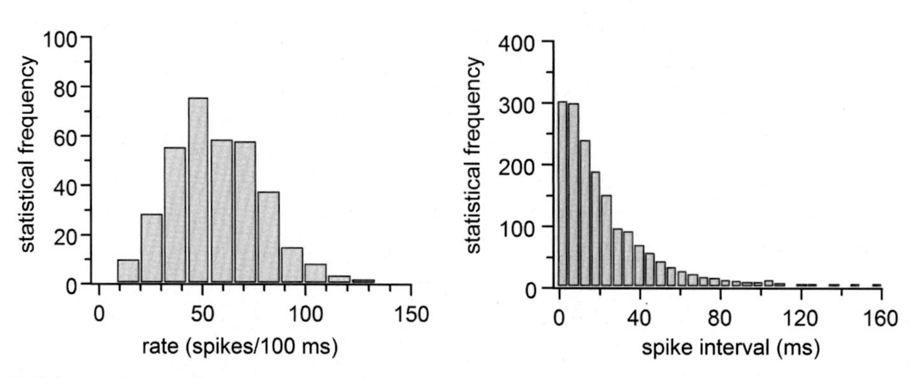

FIGURE 9.11 Left: Rate histogram of a Poisson-statistics spike train. Right: Interspike interval histogram of a simulated Poisson (i.e., independent) spike interval series.

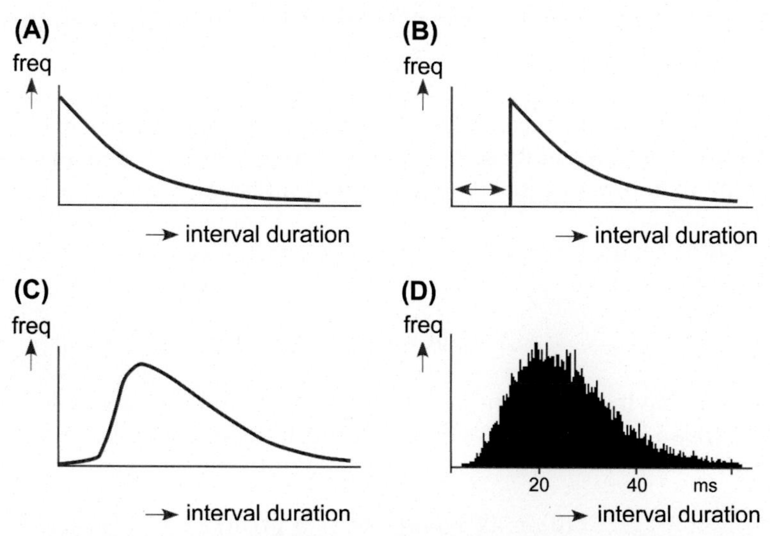

FIGURE 9.12 Theoretical interval distributions: (A) exponential distribution; (B) the same, but taking the refractory period into account; (C) a gamma distribution; and (D) a spike train from a sensory receptor fiber (ampullary electroreceptor of a catfish[2]).

A more fundamental consideration is that the spikes may indeed be mutually independent, but that in addition, each spike is caused by a process in which a number of also independent events are necessary to trigger one spike. One might think of small postsynaptic potentials, where a spike is only elicited after, say, three to four psps in a short time span. In this case, the distribution followed is a so-called gamma distribution, which derives its name from the Γ (gamma) function, which is part of its mathematical description.

A gamma distribution, shifted by the aforementioned refractory period, is shown in Fig. 9.12C. Fig. 9.12D, finally, is an interval histogram made from neuronal data (single-unit recording of spontaneous activity of an ampullary electroreceptor organ).

THE GAMMA DISTRIBUTION

A gamma distribution has two parameters, called λ and r. The lambda is a scale parameter, reflecting the time scale (mean spike rate). Gamma distributions with different λs have essentially the same shape, only stretched more or less on the time (interval duration) axis. The parameter r is called the shape parameter. The example in Fig. 9.12 C has $r = 4$, which means that four independent events are needed to cause each spike. Thus, the overall shape of a gamma distribution is determined by r. For $r = 1$, the gamma distribution reduces to the exponential distribution treated earlier and so is skewed strongly. For $r \rightarrow \infty$, the gamma distribution approaches the Gaussian ("normal") distribution and so gets more and more symmetrical. Most interval histograms can be fitted with a gamma distribution. By the way, this does not mean necessarily that the underlying cause is clear.

THE MATHEMATICS OF RANDOM POINT PROCESSES

Finally, the shape of interval histograms can be treated in a more general way.

Consider a process that generates random events at an average rate of μ events per second. Further, suppose that the probability of an event occurring is independent of time. Divide an epoch of duration t in m bins of equal lengths, with m sufficiently large to prevent the occurrence of more than 1 event per bin. Then the process is a homogeneous Poisson point process. If μ is a function of time, the process is an inhomogeneous Poisson process that we will not discuss here. The bin size is:

$$\delta t = t/m$$

The probability of having one event per bin (P(1)) is $\mu\delta t$ and the probability of having no event in a bin (P(0)) is $1 - \mu\delta t$. It follows that the probability of having no event in the epoch of duration t is:

$$P(0) = (1 - \mu\delta t)^m = (1 - \mu t/m)^m$$

Taking the limit of P(0) for increasing m gives:

$$P(0) = \lim_{m \to \infty} (1 - \mu t/m)^m = e^{-\mu t}$$

The probability to get 1 event in any of the bins is:

$$P(1) = m\mu\delta t(1 - \mu\delta t)^{m-1} = \mu t(1 - \mu t/m)^{m-1}$$

Taking the limit of P(1) for increasing m:

$$P(1) = \lim_{m \to \infty} \mu t(1 - \mu t/m)^{m-1} = \mu t e^{-\mu t}$$

The time it takes before the first event occurs is called the waiting time and its associated probability density function is called the waiting time distribution, W. Suppose the event occurs in the last, mth bin, the other bins 1 through $m - 1$ being empty, then the probability,

$$\delta P(1) = \mu\delta t(1 - \mu t/m)^{m-1}$$

or:

$$\delta P(1)/\delta t = \mu(1 - \mu t/m)^{m-1}$$

Taking the limit of $\delta P(1)/\delta t$ for increasing m gives the probability density:

$$W(1) = \mu \cdot e^{-\mu t} \tag{9.1}$$

Hence, the event intervals (e.g., spike intervals) have an exponential probability density distribution; the intervals are said to be Poisson distributed. From this equation, the mean duration between two successive events can be obtained:

$$t_{mean} = \int_0^{\infty} t\mu e^{-\mu t} dt = 1/\mu$$

which, as may be expected, is 1 over the event rate, and the variance, $\sigma^2 = \text{mean}(t^2) - (t_{mean})^2$:

$$\sigma^2 = \int_0^{\infty} t^2 \mu e^{-\mu t} dt - \frac{1}{\mu^2} = \frac{1}{\mu^2}$$

The ratio of the standard deviation to the mean interval is called the coefficient of variation:

$$c_v = \sigma/t_{mean} = 1$$

The coefficient of variation characterizes the variability in the event intervals. It is a distinguishing feature of a homogenous Poisson process that the coefficient of variation equals one.

The probability of the occurrence of n events, with the nth event in the last, mth, bin, and $n \ll m$ is:

$$\delta P(n) = \mu \delta t (1 - \mu t/m)^{m-n} (m\mu\delta t)^{n-1} \Big/ (n-1)!$$

and hence,

$$\delta P(n)/\delta t = \mu (1 - \mu t/m)^{m-n} (m\mu\delta t)^{n-1} \Big/ (n-1)!$$

Here the factor $m^{n-1}/(n-1)!$ relates to the distribution of $n - 1$ events in m bins. Now, after taking the limit for increasing m, the probability density W(n) is obtained:

$$W(n) = \lim_{m \to \infty} \mu \frac{(1 - \mu t/m)^{m-n} \cdot (m\mu\delta t)^{n-1}}{(n-1)!} = \frac{\mu^n t^{n-1} e^{-\mu t}}{(n-1)!} \tag{9.2}$$

This waiting time distribution, which describes the waiting time until the nth event, is also known as the standard gamma distribution. It is easy to verify that W(n) reduces to Eq. (9.1) for $n = 1$.

When fitting a function to experimental data, it is often desirable to have rational (floating point) parameters to fit rather than integers. This can be achieved for Eq. (6.8), by replacing the (n − 1)! faculty with the gamma function Γ(n), which is a kind of faculty for rational numbers:

$$W(n) = \frac{\mu^n t^{n-1} e^{-\mu t}}{\Gamma(n)} \tag{9.3}$$

where n may now be floating point.

The interpretation of Eqs. (9.2) and (9.3) is that each spike in a train may be caused by a process in which a number of independent events are necessary to trigger one spike. One might think of small postsynaptic potentials (psps) where a spike is only elicited after, say, three or four psps in a short time span.

MARKOV CHAINS

The assumption in Eq. (10.2) is that all events leading to the final observable nth event have, in a probabilistic sense, identical intervals, i.e., the rate constants all equal μ. If we wish to drop that restriction, we could think of the problem in the following way:

$$S_1 \underset{\mu 1}{\rightarrow} S_2 \underset{\mu 2}{\rightarrow} \ldots \rightarrow S_{n-1} \underset{\mu(n-1)}{\rightarrow} S_n \tag{9.4}$$

Starting from state 1, S_1, the system evolves through different states until it reaches the observable state S_n. The transitions between states are Poisson distributed, but now having different rate constants, μ_i. Such a series is called a Markov chain.

The mathematics of Markov chains is explained in detail in Appendix.

TIME SERIES ANALYSIS: SPIKE RATE, INTERVAL SERIES, AND INSTANTANEOUS FREQUENCY

Apparently, histograms show several interesting aspects of neural functioning, but they lack one fundamental quantity: time. Therefore, we will focus now on the analysis of spike trains in time.

SPIKE FREQUENCY OR RATE

The most intuitive way of analyzing a time series of events is to measure the frequency, i.e., the number of events in a predetermined time interval. This is shown in Fig. 9.13. An

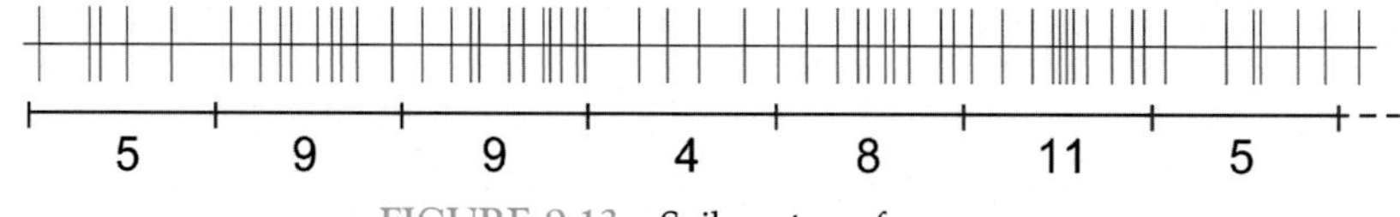

FIGURE 9.13 Spike rate or frequency.

irregular spike train is divided into a number of fixed-time intervals. The number of spikes counted in each interval is a measure of the spike frequency, also called the spike rate. The distinction between rate and frequency may need an explanation. The well-known notion of a frequency stems from the harmonic (sinusoidal) motion, in which the motion (or any other physical quantity) is fluent. If the frequency is doubled, so are all aspects of the signal, such as speed and acceleration. In the frequency domain, the whole spectrum is shifted by a factor of two. In a spike train, however, the duration of the individual events is always the same and so is independent of the number of events per time unit. In this case, it is better to call the number of events per time unit a rate rather than a frequency.

To amplify neural spikes, an amplifier must have its *frequency* bandwidth centered at about 1 kHz, irrespective of the spike *rate*. Even one spike per hour would still be a short pulse. Obviously, there must be differences in the frequency domain too, but these are far smaller than in the abovementioned case of a sinusoid. Since both frequency and rate are expressed in the quantity s^{-1} or Hz, it has been proposed to use the name Adrian for the unit of spike rate, in honor of the famous pioneer electrophysiologist and Nobel Prize winner Lord Edgar Adrian. The name did never catch on, and since the frequency content of spike trains is known well enough by the people recording them, using the term spike frequency poses no problem in practice. The most common expression is "spikes per second," abbreviated "sp/s."

The series of values shown in Fig. 9.13 demonstrates the disadvantages of spike rate as a measure of neuronal activity. If the measuring interval is taken rather short, as in the figure, the number of spikes per measurement is very small, yielding a rather crude measure of the signal. If the measuring window is taken longer, say one or a few seconds, the measured rates are more precise (say between zero and a few hundred spikes per second), but now the time resolution is unacceptably low. Even a snail could not survive when informed about its environment only once per couple of seconds.

A number of problems in neural coding arise from the fact that the frequency of the "carrier" (the average spike rate) is not much higher than the frequency with which an organism needs to be informed about its environment. The neural code resembles the principle of frequency modulation (FM) used in radio transmissions, but the orders of magnitude are very different. An FM radio station transmitting at, say, 100 MHz codes for speech and music, which contains frequencies up to about 20 kHz. So, from the carrier wave's viewpoint, the frequency needs to change only very slowly. This is called a rate code, and many nerve cells do approximately the same, although at very different time/frequency scales. In human hearing, for example, the spike rate from cochlear nerve fibers may be identical to the frequency of the sound wave, say 400 sp/s at 400 Hz. Comparing this with the duration of the shortest syllables (or musical notes), about 20 ms, shows that the spike frequency needs to change every few spikes. Since there is not enough time to determine the spike rate, it is better to call such a code a time code, or interval code, rather than a rate code. Each individual spike (or spike interval) may contain relevant information about the signal.

This has consequences for the way in which spike trains must be analyzed to infer the relevant neural information from them.

INTERVAL SERIES AND INSTANTANEOUS FREQUENCY

To obtain a more detailed record of the processes coded by spike signals, we will evaluate each individually occurring spike. In the early days of electrophysiology, the spike interval times had to be measured by hand from chart recordings or photographs of spike trains. This was a tedious task, since getting interesting results implies the processing of thousands of spikes. Nowadays, we will perform the same task with the aid of a computer. Mark the formulation of the previous sentence: man is still in control, or has to be. This means that, although we are glad that we do not need to process the bulk data, we still need to check the computer algorithms that recognize spikes from an input signal and count the proper interval. To this end, it's a good idea to analyze a small sample by hand and compare it with the computer analysis of the same spike train.

As an illustration, a very short sample is shown in Fig. 9.14. The spike signal, the interspike interval series (t_i), and its inverse ($1/t_i$), the instantaneous frequency, are shown in the bottom, middle, and upper trace, respectively. If this spike train is analyzed in real-time, each spike interval time is known only at the occurrence of the next spike. Therefore, both measures of spike activity are undetermined during the first interval, i.e., in the time between spike #1 and spike #2 (hatched areas). In the next interval, the value of the first interval (or its inverse) is plotted, and so on. In summary, both interval series and instantaneous rate are lagging for the duration of one spike interval. Obviously, if the spike train is analyzed later, i.e., after its complete acquisition, this delay does not arise. The figure shows that the

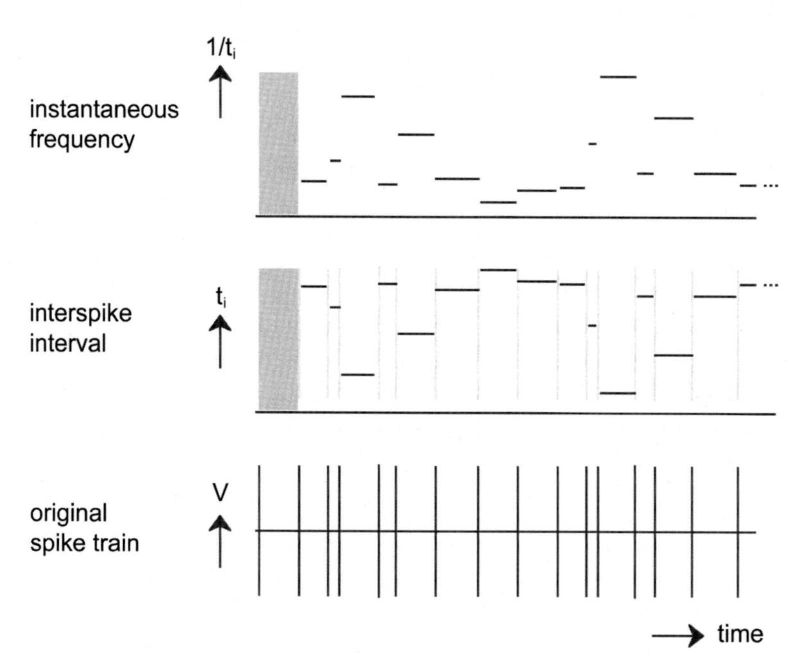

FIGURE 9.14 From bottom to top: a small sample of a spike train, a plot of the interspike interval times, and its inverse, the instantaneous frequency.

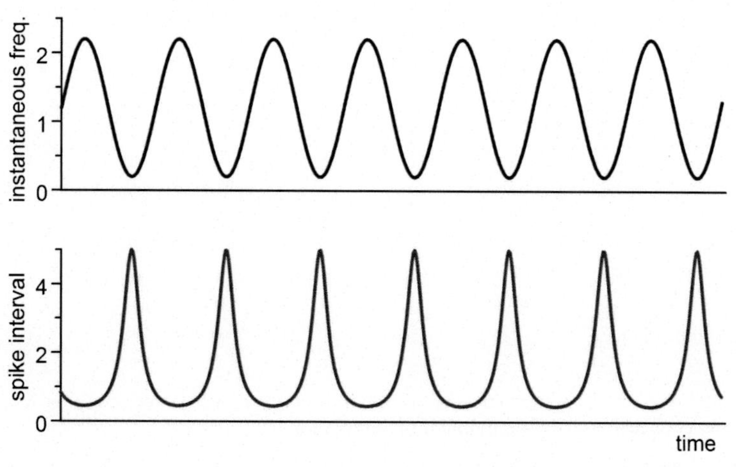

FIGURE 9.15 A sinusoidal curve (upper trace) and its inverse (lower trace).

interval series and the instantaneous frequency series are approximate mirror images of one another. Small intervals yield large rate values, whereas long intervals yield small rate values. This is the general picture for fairly small changes in spike rate or interval, but it does not hold for large excursions of the spike rate. The inverse of a sine function is also a periodical signal but a much distorted one.

This distortion can be illustrated with an analogue (continuous) example. Fig 9.14 shows a sine wave; $\sin(t) + 1.2$ and its inverse; $1/(\sin(t) + 1.2)$.

Although in principle interval time and instantaneous frequency carry the same information, the choice of one over the other is not arbitrary. In the nervous system, the spike rate is often approximately proportional to a physiological quantity such as the strength of a signal from the environment (eye, ear, etc.) or the position of our own head, limbs, etc. (via proprioceptors). Therefore, spike rate is arguably a better measure than interval time. If there is hardly any activity, the intervals are getting very long. If activity ceases altogether, the interval time tends to infinity. This is not in keeping with our intuition, in which a dead neuron is not infinitely active. If, in the example of Fig. 9.15, the amplitude of the sinusoid (upper trace) would be increased only slightly, the spike frequency would touch the zero line, and hence the inverse would grow to infinity. The bottom trace is that, although spike data is collected as a series of interspike interval times, the conversion of this series into an instantaneous frequency series yields a better record of the neurophysiological activity studied.

DOT DISPLAY

A simple method to display the activity of a neuron, called a dot display or raster display, consists of plotting each spike as a dot on a computer screen or on chart paper and using our own visual system as a pattern analyzer. Fig. 9.16 shows a dot display of the response of some neurons to a repeated stimulus. Each horizontal trace plots all spikes that occurred shortly before and after a stimulus (such as an electric pulse, a light flash, and a sound click).

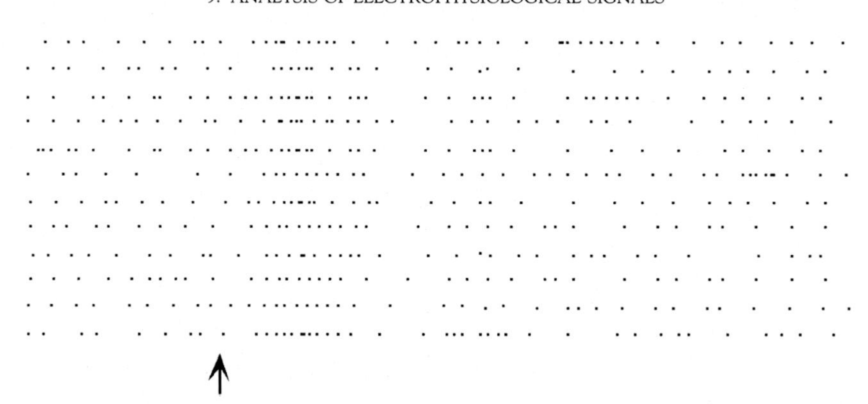

FIGURE 9.16 A dot display of a hypothetical neuron that responds to a repeated stimulus, in addition to strong fluctuations in spontaneous activity. *Arrow*: moment of stimulation.

Each spike train shows relatively large random fluctuations, but by repeating the stimulus a large number of times and lining up all traces vertically, a pattern is seen to arise. In the example given, the fluctuations cause bursts of relatively closely spaced spikes as well as gaps to occur spontaneously in each trace. In addition, however, the higher activity directly after the stimulus, and a gap somewhat later, is seen to occur in all traces.

This type of display can give a researcher a good impression of the degree in which a neuron responds to a certain stimulus. An automated version of the dot display is the so-called poststimulus time histogram, or PSTH for short.

STIMULUS-RESPONSE CHARACTERISTICS: THE PSTH

Traditionally, the PSTH plots a frequency histogram (here frequency means the *statistical* frequency) of spike occurrences in a number of bins at regular times, starting from a brief stimulus. However, in sensory physiology, the stimulus is often a continuous waveform, such as a sound wave or a modulated light intensity. In this case, the stimulus is always present, and the PSTH is dimensioned so that one period of the stimulus is rendered. Such a plot is called a peristimulus time histogram, having the advantage that the traditional abbreviation PSTH can be kept. The principle is illustrated in Fig. 9.17.

A neuron is stimulated with an appropriate stimulus signal. This may be either a short pulse (such as a light flash to a photoreceptor, a sound click to a hearing organ, an electrical pulse to an interneuron, etc.) or a continuous waveform (such as a sinusoidal or compound sound wave for a hearing organ, a periodical head rotation for a semicircular canal organ, etc.). In Fig. 9.17, a sinusoidal signal is used to get a peristimulus time histogram. The spike train is recorded repeatedly, starting from a fixed point in the stimulus waveform (sweeps 1 through n in the figure). At relatively short time scales, only one or a few spikes per sweep will occur, yielding a crude histogram at first. After many sweeps, however, a detailed histogram reveals the average response of the organ to the stimulus. Note that in the PSTH, the T stands for the times of occurrence of the spikes with respect to a fixed trigger point in the stimulus signal. Interspike interval times play no role here.

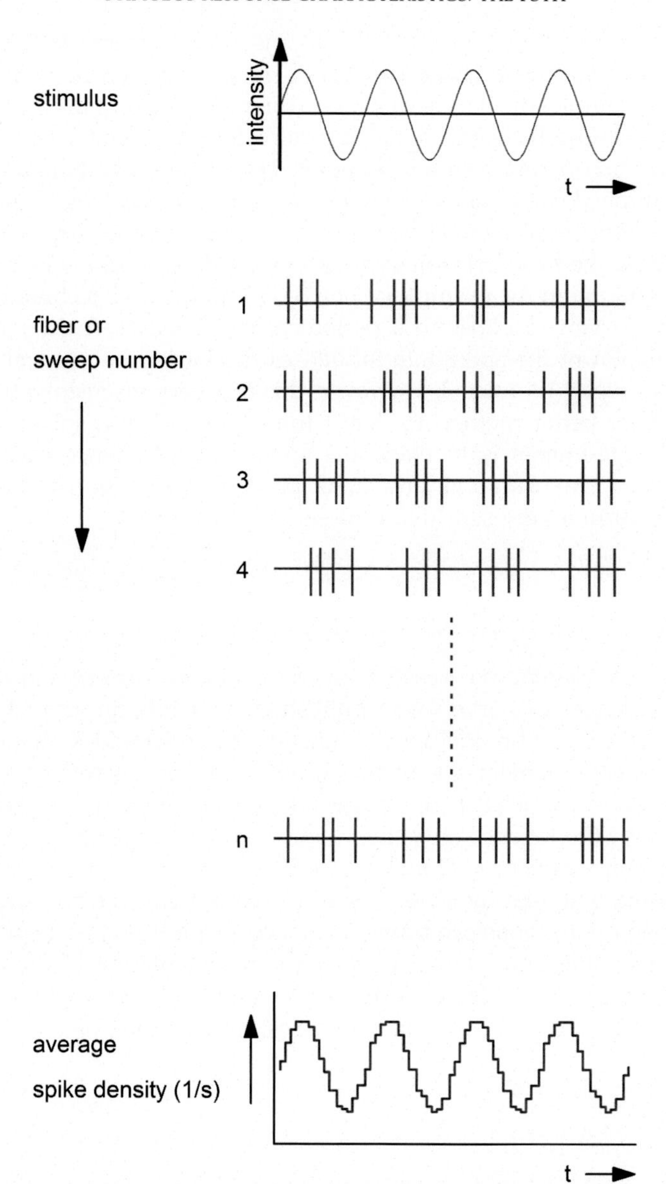

FIGURE 9.17 Principle of poststimulus time histogram determination (see text for details).

In addition to illustrating the way in which physiologists can probe the response of a single nerve fiber, Fig. 9.17 may serve to illustrate the way in which the nervous system is able to process data reliably, given the fluctuating spike trains shown. In this case, the traces 1 through n represent the simultaneous responses of a number of parallel fibers. Although electrophysiologists record usually from one or a very limited number of fibers simultaneously, the nervous system can rely on a massively parallel information stream. In most sense organs, such as the eye, the cochlea (hearing organ), the organs of equilibrium, and many skin sensors for touch, temperature, and so on, many parallel fibers represent virtually the same quantity. In the labyrinth, for example, the nerve from a semicircular canal organ consists of a few hundred fibers, all representing the same signal: a head rotation around one axis. It is plausible that nature performs an operation similar to the PSTH-scheme by letting a number of fibers from the nerve mentioned converge onto a small number of brain cells. Combining the postsynaptic potentials from these parallel fibers averages the spike occurrence times in a way similar to the one described for PSTH determination, yielding a reliable measure of the original input (a head rotation in the example). The PSTH is a sensitive technique to find a small but nevertheless important response to a periodical input signal. If the response is strong enough, the modulation of the spike rate influences the interspike interval histogram. This is the case shown in Fig. 9.10, trace 3. However, this is a very insensitive indicator of spike rate modulation. A far better method to detect the properties of a spike train that is used frequently during experiments is to make the spikes audible. Since each spike causes a distinct click of about 1 ms duration, any small audio amplifier and loudspeaker will do. Our ears plus brain form a very sensitive time series processor!

TERMINOLOGY: THE HODGKIN AND HUXLEY CHANNEL

In 1952, a paper "A quantitative description of membrane current and its application to conduction and excitation in nerve" was published by Hodgkin and Huxley[3] in which they presented a mathematical model for the functioning of Na and K channels in the nerve axon. Although the existence of ion channels had not been demonstrated yet at that time, the terminology that was introduced in this paper to describe membrane permeability is still used today to describe the kinetics of (voltage-sensitive) ion channels.

According to their model (the H&H model), a channel contains a pore that is kept closed at the resting membrane potential by a number of activation gates. Upon depolarization, the activation gates open up and ions can traverse the pore until it is closed again by a (number of) inactivation gate(s). Fig. 9.18 shows a schematic representation of the voltage-sensitive Na^+ channel according to the H&H model. During the brief moment that all gates are open simultaneously, an inward Na^+ current passes through the pore.

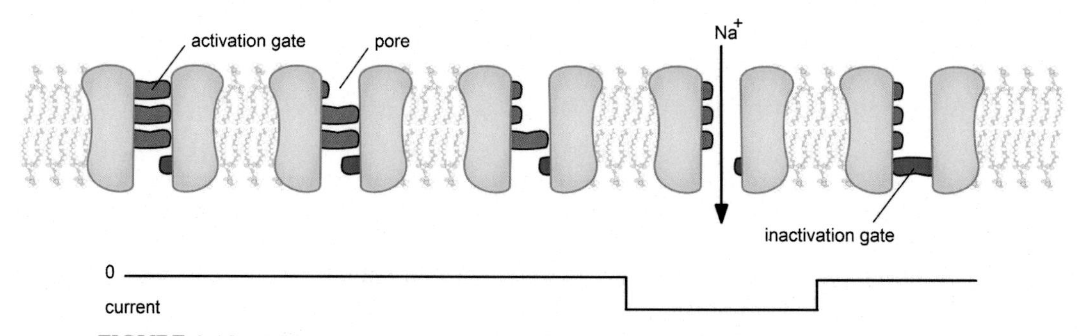

FIGURE 9.18 Schematic indicating the different states of a voltage-sensitive sodium channel.

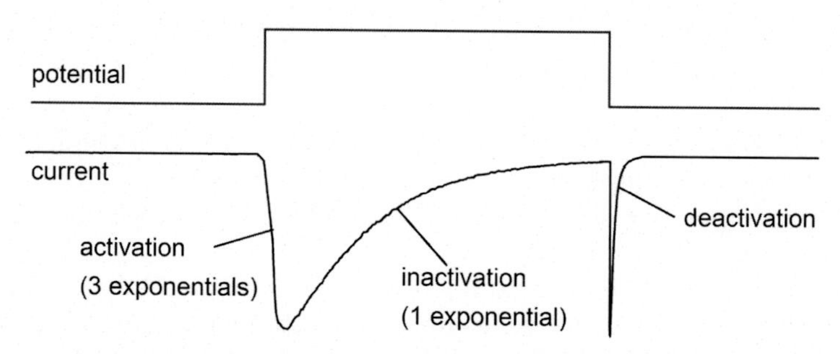

FIGURE 9.19 The resultant, macroscopic, current through many ion channels.

The gates are supposed to operate independently of one another and to open and close stochastically with Poisson-distributed life times. Hence the kinetics of each gate may be considered as a monomolecular chemical reaction. The energy to move the gates is furnished by the transmembrane electric field. In the H&H theory, the opening of a single gate requires the translocation of a single elementary charge across the membrane.

Due to the stochastic nature of the movement of the gates, the macroscopic current measured from a large population of Na^+ channels is not square as shown for a single channel in Fig. 9.18, but a function of 4 exponentials (Fig. 9.19).

Upon return to the resting membrane potential, a tail current may be observed that decays rapidly due to the closing of the activation gates (deactivation).

We know since its cloning that the voltage-gated Na^+ channel α subunit consists of 4 repeats having 6 transmembrane segments. Each repeat has its own voltage sensor, so a model with 4 activation gates might currently seem more appropriate than the 3 originally proposed in the H&H model.

ANALYSIS OF MACROSCOPIC (WHOLE-CELL) CURRENTS

The first and most important step in the analysis of patch-clamp experiments is taken before the actual experiment: its design. What do we want to measure and how do we eliminate currents and channels we are not interested in? What data do we have to feed to our analysis program such that it can deal with the question that we ask?

As an example, let us suppose we wish to determine the voltage dependence of activation of the low-threshold (L-type) calcium channel.

To eliminate voltage-dependent sodium currents from many preparations, tetrodotoxin (TTX) can be applied extracellularly. Unfortunately, not all voltage-dependent Na^+ currents are sensitive to this toxin, but fortunately, they are not sensitive to either inhibitors of the L-type Ca^{2+} channel such as the dihydropyridines (e.g., nifedipine) or diphenylalkylamines (e.g., verapamil). Potassium currents may be eliminated by replacing extracellular K^+ by tetraethylammonium (TEA) ions and by replacing intracellular KCl by a mixture of CsCl and TEACl (K^+ channels are poorly conductive for CsCl, but since they are usually much more numerous than Ca^{2+} channels, a small cesium current may contaminate the records if 10 mM of TEACl is not added to the pipette solution). N-type Ca^{2+} channels are inhibited by omega-conotoxin; P/Q-type Ca^{2+} channels are inhibited by omega-agatoxin. There is no inhibitor for the T-type Ca^{2+} channel, but its voltage dependence and kinetics are different from the L-type channel. There is no inhibitor for R-type Ca^{2+} channels either.

The contribution of voltage-insensitive ("leak") currents may be subtracted from the records after measuring their amplitude at potentials at which the Ca^{2+} channel is inactive. Subsequent linear extrapolation should then give an estimation of their contribution at other potentials.

The contribution of currents that are still left may be eliminated by recording once in the absence of L-type Ca^{2+} blockers and recording a second time in the presence of L-type Ca^{2+} blockers. Subtraction of the records then yields the true L-type current. Often, extracellular calcium is raised from 2 to 10 mM in order to increase the size of calcium currents. As L-type Ca^{2+} channel inactivation is due to binding of Ca^{2+} ions to an intracellular domain of the channel protein, increasing calcium currents increases inactivation, but Ca^{2+} may be replaced by Ba^{2+}, which binds poorly to this site, thus reducing inactivation. However, the substitution of Ca^{2+} for Ba^{2+} also shifts the voltage dependence of calcium channel activation. Relatively large currents carried by Ca^{2+} channels may be obtained by eliminating all extracellular calcium and recording the currents in 140 mM NaCl. In that case, the L-type channel current is carried by Na^{+} ions, but again, the voltage dependence of channel activation is changed.

THE CURRENT TO VOLTAGE (I/V) CURVE

By presenting a succession of square voltage pulses of increasing amplitude to a voltage-clamped cell and recording the current responses, an I/V curve is obtained. For each trace, the peak (maximum) current is measured and plotted as a function of the voltage of the square pulse. The following I/V curve (Fig. 9.20) was obtained in conditions to measure calcium currents in the absence of extracellular Ca^{2+}, hence in extracellular 140 mM NaCl, 0 Ca, and TTX 100 nM and intracellular 25 mM NaCl, 100 mM CsCl, 10 mM TEACl, and 2.5 mM EGTA:

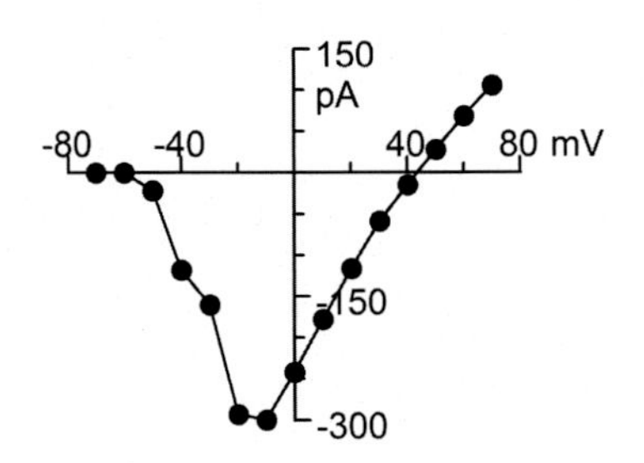

FIGURE 9.20 I/V curve of a calcium current.

The I/V curve is the result of the activation of both T- and L-type channels. The same series of pulses given after inhibition of L-type channels by nifedipine gives the I/V curve for the T-type channel (Fig. 9.21, left). Then subtracting this curve from the first curve gives the I/V curve for the L-type channel (right).

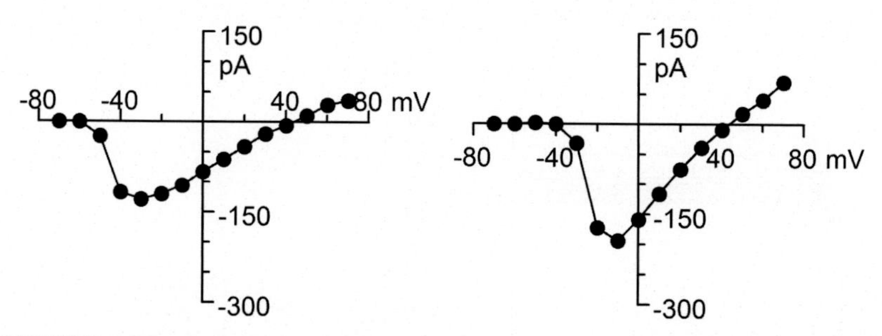

FIGURE 9.21 Segregated I/V curves of the type T (left) and type L (right) channel.

The reversal potential, which is the potential where the current changes polarity, is 43.5 mV in this example and close to the equilibrium potential for sodium. As the whole-cell current is proportional to both the fraction of open channels and the driving force (the difference between reversal potential and actual potential), the fractional conductance C/C_{max} can be obtained by dividing the data points in the graphs above by the driving force and normalizing with respect to the maximum conductance. This gives, in Fig. 9.22, the so-called activation curves for T-type channels (open circles) and L-type channels (closed circles):

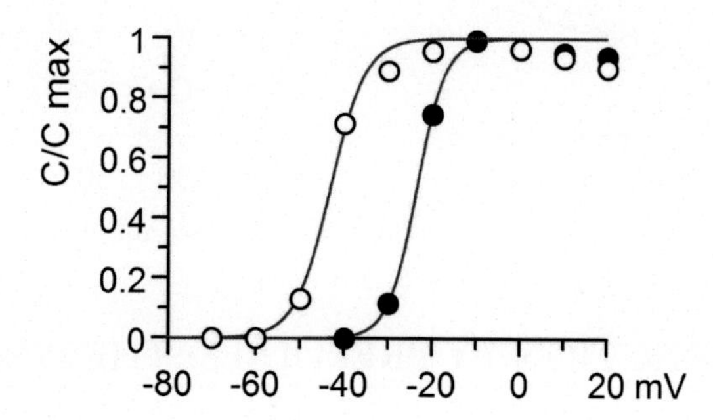

FIGURE 9.22 Activation curves of T and L types of calcium channels.

LEAK SUBTRACTION BY EXTRAPOLATION

The following is an example to illustrate how this method works (Figs. 9.23 and 9.24):

The family of K^+ currents on the left of Fig. 9.23 were recorded in response to a series of square voltage pulses of increasing amplitude (hence, a similar I/V protocol as in the preceding section). The figure to the right shows the resulting I/V curve. A priori knowledge of the K^+ channel in question (an outward rectifier) says that it should not open below −60 mV. Hence the first five data points represent a linear leakage current. Linear regression of these five points gives a leak conductance of 2.2 nS (Fig. 9.24, left), and after subtraction of the leak, the curve to the right is obtained.

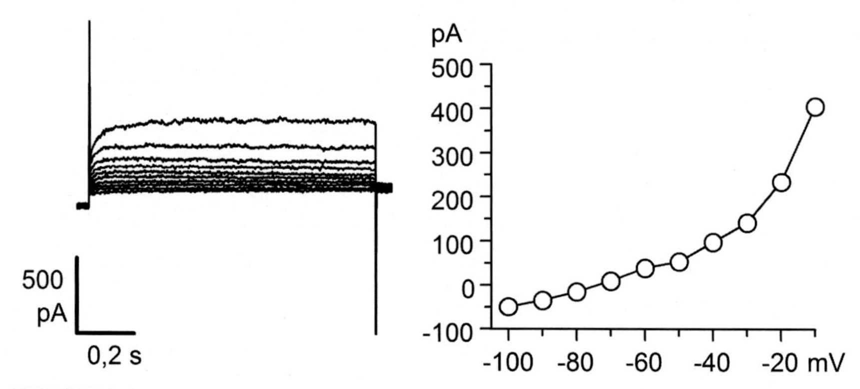

FIGURE 9.23 Potassium currents (left) and the I/V curve (right) before leak subtraction.

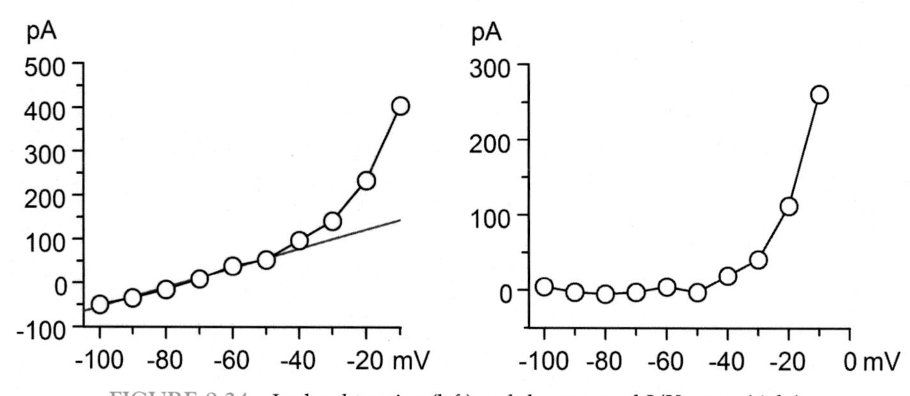

FIGURE 9.24 Leak subtraction (left) and the corrected I/V curve (right).

LEAK SUBTRACTION BY PREPULSES: THE P/N METHOD

This method is used to eliminate linear leak and (most of) the capacitive currents.

In the preceding paragraphs, the I/V protocol consisted of a series of square depolarizing voltage pulses (Fig. 9.25):

In the P/N (one positive over several negative pulses) protocol, each depolarizing pulse of amplitude v is preceded by a number, n, of hyperpolarizing pulses of size $-v/n$. The responses to the $n + 1$ voltage jumps are summed with the result that all linear responses cancel out and the nonlinear response of the voltage-sensitive current of interest remains. This is shown in Fig. 9.26 for $n = 2$.

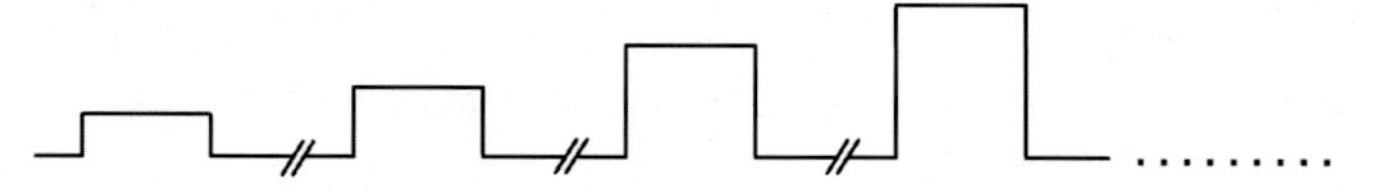

FIGURE 9.25 Series of pulses used in I/V curve measurement.

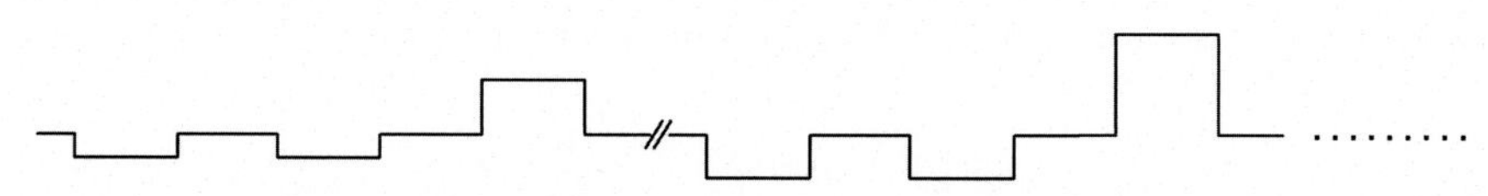

FIGURE 9.26 Pulse series used in the P/N protocol, for n = 2.

Of course, the resting potential must be chosen such that it lies well outside the range of the voltage-sensitivity of all ion channels in the cell.

Capacitive currents of opposite polarity are elicited by the hyperpolarizing and the depolarizing pulses and tend to cancel. However, they do not quite so because the relaxation time constant of the capacitive current is a function of the membrane resistance, which is high during the hyperpolarizing pulses but gradually decreases during the depolarizing pulse due to the opening of voltage-sensitive channels. Similarly, upon return to the resting potential, a tail current may change the relaxation time constant of the capacitive current. Hence, the P/N protocol cannot altogether compensate for a poorly set clamp amplifier.

NOISE ANALYSIS: ESTIMATING THE SINGLE-CHANNEL CONDUCTANCE FROM WHOLE-CELL OR LARGE PATCH RECORDINGS

Part of the noise recorded in the whole-cell configuration is due to the random opening and closing of ion channels. This noise contains information from which the single-channel conductance and the total number of channels can be obtained. Consider a cell that contains predominantly a population of noninactivating voltage-dependent channels (e.g., noninactivating outwardly rectifying K^+ channels). If the probability of channel opening, p, is low (e.g., at hyperpolarized potentials), then the mean current generated by the population of channels is low and the random fluctuations around the mean current are small. At the other extreme, if all channels are open all the time (p = 1), the mean current is maximal and the random fluctuations around the mean are again small. Half-way between these conditions, at p = 0.5, the mean current is half maximal and the random fluctuations are the largest, since now channels open and close all the time. Hence plotting the variance, s(p), of the macroscopic current as a function of the mean current, m(p), gives a humplike graph (Fig. 9.27).

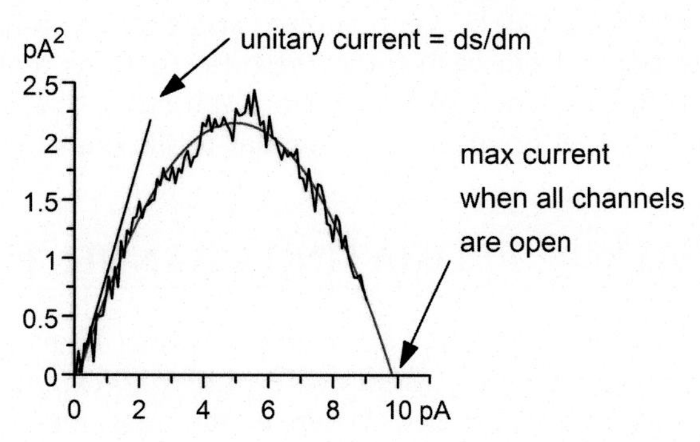

FIGURE 9.27 Variance of the macroscopic current as a function of the fraction of open channels. Fitted parabola in red.

According to Neher and Stevens,[4] the relation between variance (s(p)) and mean (m(p)) is:

$$s(p) = i \cdot m(p) - \frac{m^2(p)}{N}$$

with p, the probability of channel opening; i, the unitary current; and N, the total number of channels. Differentiation of s with respect to m gives:

$$ds/dm = i - 2 \cdot m/N$$

Hence at zero mean current ($p = 0$): $ds/dm = i$. Fitting the histogram with a parabola:

$$s(x) = a + b \cdot (x - x_0)^2$$

gives two equations: at $x = 0$: $ds/dx = -2.b.x_0 = i$ and at the intersection of the parabola with the $y = s(x) = 0$ axis: $x = N.i$. Therefore, fitting the histogram with a parabola yields both the unitary current and the total number of channels.

Often, the variance at zero mean current is not zero due to amplifier and thermal membrane noise. In such a case, the variance at zero mean current gives an estimate of the background noise that should be subtracted (vectorially) from the data points before fitting the curve with a parabola. Hence, if the background variance is b, then the actual channel noise variance, a, for each data point is obtained as follows:

$$a = (\sqrt{s} - \sqrt{b})^2$$

where s is the observed variance.

Above, stationary channel activity was assumed, but the method can also be applied to transiently active channels, like the voltage-sensitive Na^+ channel. The only thing that is important is to obtain variance and mean by varying the open channel probability. During a voltage jump from a hyperpolarized to a depolarized potential, the Na^+ currents starts with all channels closed ($m = 0$, $s = 0$). During the activation phase of the current, the probability of channel opening increases ($m > 0$, $s > 0$) and then decreases again when inactivation sets in. If one presents many identical voltage jumps to the cell, the macroscopic current of each of the responses will be very similar except for the random fluctuations. By subtracting the mean response from each trace, these fluctuations remain and they can be used to estimate the variance at each phase of the mean current. Plotting these estimates as a function of the mean current gives a figure similar to the one above.

NOISE ANALYSIS: ESTIMATING CHANNEL KINETICS

Before the advent of single-channel voltage-clamp, channel kinetics could be estimated either by fitting transient macroscopic currents with (multiple) exponential functions or by spectral analysis of steady-state currents. Whereas the first method works well with voltage-sensitive currents, the second is often required in those cases that transient currents

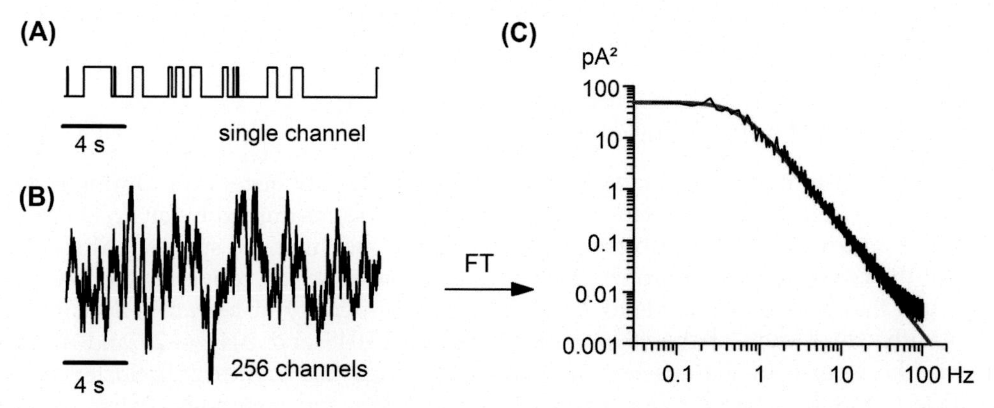

FIGURE 9.28 Simulated single channel (A), macroscopic membrane current (B), and the power spectrum of the latter (C). The Lorentzian fit gives $k = 3.67\ s^{-1}$, corresponding to a -3 dB point at $3.67/2\pi = 0.58$ Hz.

cannot be (easily) obtained, such as for constitutively active or ligand-gated channels. Consider a hypothetical cell containing constitutively active channels that flip between open and closed states with the following kinetics:

$$C \underset{k2=3}{\overset{k1=1}{\rightleftharpoons}} O$$

If it were possible to force all channels in the closed state (C) and then have them relax to equilibrium, a macroscopic current with a relaxation rate constant of $k1 + k2 = 4/s$ would have been obtained. Since this is not possible, we have to use the fact that the cell only contains a limited number of channels that produce measurable current noise. The general idea behind the theory, which has been developed by DeFelice and others in the 1970s,[5] is that the noise fluctuations represent statistical deviations from the mean that subsequently relax to equilibrium with, in this case, a rate constant of $k1 + k2$. As we have seen in Eq. 7.6, exponential decay in the time domain translates into a Lorentzian function in the frequency domain. Hence, by taking the power spectrum of the channel noise and fitting the spectrum with a Lorentzian function, the relaxation rate constant may be found. In Fig. 9.28, 20 s sweeps of the activity of 256 channels were simulated, and the mean power spectrum was obtained and fitted with a single Lorentzian function, yielding an estimated rate constant of $3.67\ s^{-1}$.

ANALYSIS OF MICROSCOPIC (UNITARY) CURRENTS

The analysis of the unitary current (or single-channel current) usually involves four steps, i.e.,

1. estimation of the unitary current,
2. detection of opening and closing events,

3. determination of the number of channels in the patch, and
4. measurement of dwell times.

Estimation of the Unitary Current

The unitary current may be obtained from the current-density histogram. During acquisition, the output voltage of the patch-clamp amplifier has been sampled by an A/D (analog-to-digital) converter with a certain precision, say 12 bit, such that the span of values at the output of the converter ranges from −2048 to +2048. Suppose the converter has been calibrated such that 2048 corresponds to 20 pA. The current-density histogram reflects the frequency with which each of the 4096 values between −2048 and 2048 (−20 and +20 pA) occurs in the record. In the absence of channel openings, the values will cluster around 0 pA. Due to random charge movements in both amplifier and membrane patch, the peak at 0 pA has a bell shape obeying a Gaussian distribution. If the patch contains a single channel that opens and closes regularly, then the current-density function will show an additional peak. The two peaks are separated by a distance (in pA) that corresponds to the unitary current. If more channels of the same type are present, then a series of equidistant peaks will appear in the current-density histogram (Fig. 9.29).

The distance of the peaks can of course be measured by hand, but it is better to fit a number of equidistant Gaussians to the histogram, the difference of the means giving the unitary current (7.0 pA in this example). Fitting gives the system offset at the same time (note that the peak at 0 pA is not quite centered on 0). This offset needs to be subtracted from the records

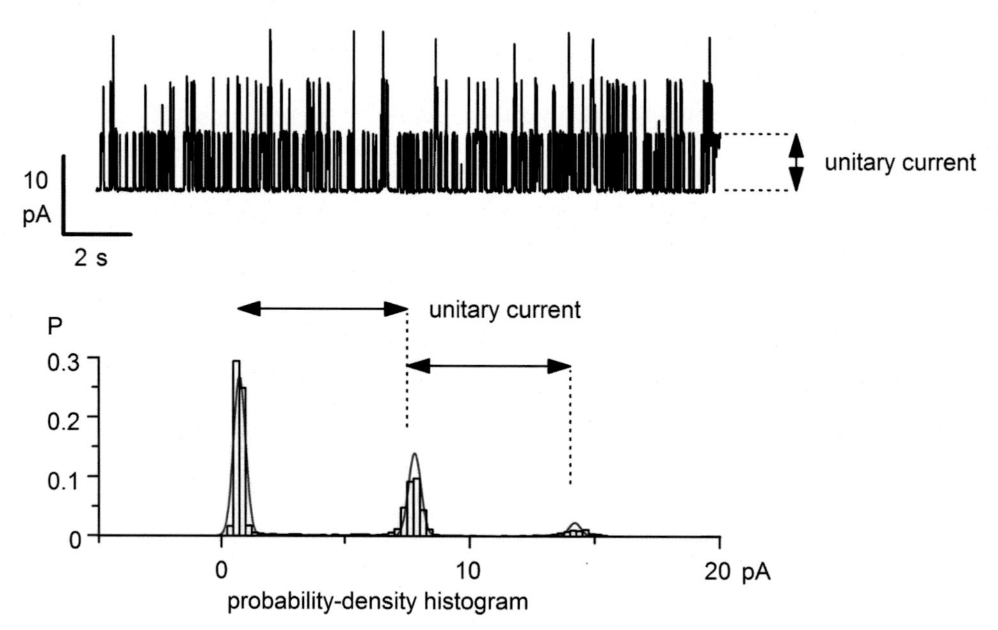

FIGURE 9.29 Single-channel currents (upper trace) and the probability density histogram (bottom trace). Fitted Gaussians in red.

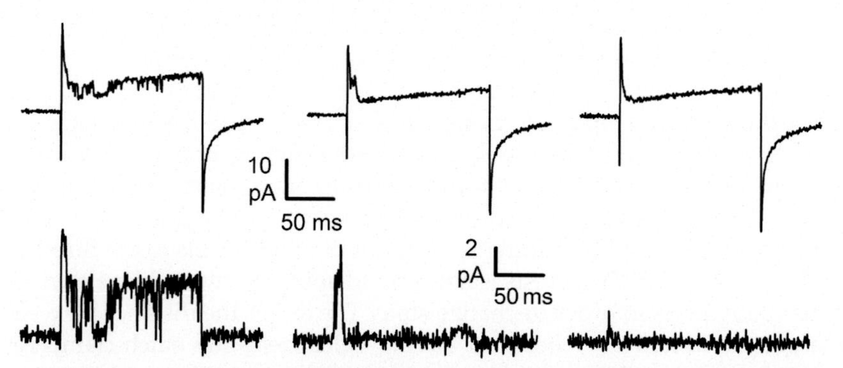

FIGURE 9.30 Capacitive current peaks that hamper channel current measurement (upper trace). Subtraction of traces without channel openings from those with openings removes most of the artifacts (bottom trace).

before detection of opening and closing events may be carried out. The example given above concerns a relatively large voltage-insensitive K^+ channel without overlap between the bell-shaped curves and without baseline fluctuations. This condition is not always met, and especially if channel conductance is low (<1 pS), baseline drift and fluctuations need to be removed by subtraction of a polynomial or a cubic spline.

A special case of baseline fluctuation is the one caused by changes in the clamp voltage, e.g., when stimulating the opening of voltage-dependent channels by a jump in potential. The voltage jump generates a capacitive current, which, if not carefully compensated with the amplifier settings, tends to contaminate the current record, as shown in Fig. 9.30.

The first two records show channel openings during a depolarizing pulse, while the third stimulation did not elicit openings. The fact that some records do not contain openings can be used to eliminate the stimulus artifacts. By taking the average of the traces devoid of channel openings and subtracting the average from each individual trace, most of the capacitive current is removed.

Detection of Opening and Closing Events

Once all artifacts are removed, baseline fluctuations subtracted, and the value of the unitary current determined, the detection of opening and closing events can take place.

The simplest way to detect channel openings is to set thresholds at $(n + 0.5) \times i$, where i is the unitary current and n, an integer greater than or equal to 0 (Fig. 9.31). Then, values above

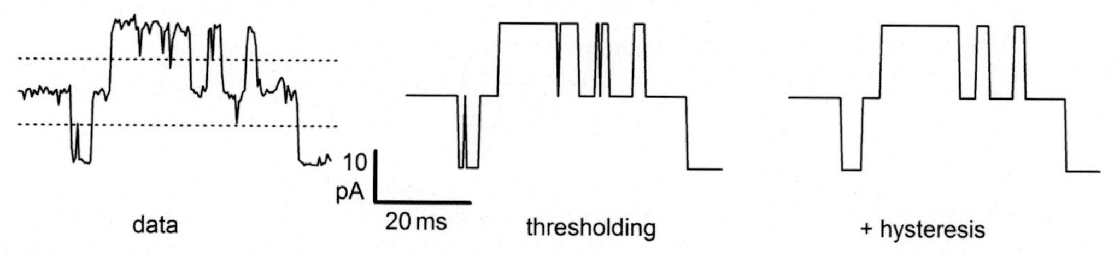

FIGURE 9.31 Detection of channel openings by thresholding, using hysteresis to eliminate brief threshold crossings.

threshold n correspond to n openings (the dotted lines in the figure below indicate thresholds for $n = 1$ and $n = 2$; the middle trace shows the result of thresholding).

As can be seen from the first two traces, short spurious excursions from the current state, barely crossing the threshold, may lead to detection of events. Although it is not shown here, this phenomenon is a nuisance when background noise is more important or if the record is less filtered. In those cases, each channel transition from one state to the next may be accompanied by a few rapid crossings of the threshold before the channel settles at the next state. This situation can be remedied by adding some amount of hysteresis to the thresholds (rightmost trace). A threshold with hysteresis is a threshold split into two sublevels with the signal having to cross both levels in order to change state. The larger the distance between the sublevels, the larger the hysteresis. After thus having "idealized" the patch-clamp records, the times between channel transitions can be measured to create dwell-time histograms.

Estimation of the Number of Channels in a Patch

When multiple channels are present in a membrane patch, the number of channels needs to be known in order to carry out correctly the kinetic analysis using dwell-time histograms. A conservative estimation is given by counting the maximum number of levels that occur in the patch-clamp records. However, especially if open channel probability is low, the rare occasions when all channels are open simultaneously might have been missed. The binomial analysis discussed below gives a second estimate.

Suppose the patch contains N channels. All channels are identical and have a probability of being open of p. The probability of being closed is $1 - p = q$.

For $N = 2$, the probability the both channels are open is p^2, the probability that one is open and one closed equals $2.p.q$ and both closed is q^2.

For $N = 3$, this gives p^3, $3.p^2.q$, $3.p.q^2$, q^3 for three open, two open and one closed, one open and two closed, and three closed, respectively. The coefficients of the terms can be found by expanding "Pascal's triangle":

$$
\begin{array}{ccccccccc}
 & & & & 1 & & & & \\
 & & & 1 & & 1 & & & \\
 & & 1 & & 2 & & 1 & & \\
 & 1 & & 3 & & 3 & & 1 & \\
1 & & 4 & & 6 & & 4 & & 1
\end{array}
$$

The numbers in each new row are the sum of the two numbers just above, adding 1s at the borders. The coefficients may also be calculated using the equation:

$$c_k = \frac{N!}{k! \cdot (N - k)!}$$

In the following stretch of data, at least three channels are present as judged by the number of levels in the trace at the left and the number of peaks in the current-density histogram at the right (Fig. 9.32).

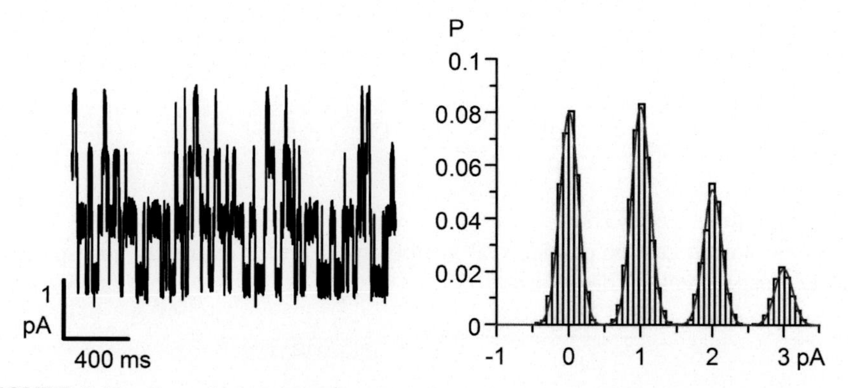

FIGURE 9.32 Example current with at least three channels. Fitted Gaussian functions in red.

Now, two unknowns need to be solved: N, the total number of channels, and p, the probability of opening. We first calculate the product of N and p:

$$N \cdot p = (S1 + 2 \cdot S_2 + 3 \cdot S_3)/St$$

Here S_1 is the surface of the peak in the current-density histogram corresponding to one open channel, S_2 is the surface corresponding to two open channels, S_3 is the surface for three open channels, and St is the total surface of the histogram. In the case of this example, the current-density histogram could be fitted by four equidistant Gaussians with identical variance, meaning that all peaks are isomorphic. Therefore, it is not necessary to calculate the surface of each peak. It suffices to take the square of the amplitude of each peak as a measure of its surface and hence:

$$N \cdot p = \left(A_1^2 + 2 \cdot A_2^2 + 3.A_3^3\right) / \left(A_0^2 + A_1^2 + A_2^2 + A_3^2\right)$$

In the current example, $A_0 = 0.080072$, $A_1 = 0.081944$, $A_2 = 0.050924$, and $A_3 = 0.021246$ giving N.p = 0.8197086. If N were 3 then p would be 0.8197086/3 = 0.273236.

With the hypothetical N and p, the expected surfaces ($E_0 \ldots E_3$) of the current-density histogram can be calculated using the approach discussed above.

If p = 0.273236, then q = 1 − p = 0.726763.

The expected probabilities and the observed probabilities, with $O_i = A_i^2 / \sum (A_i^2)$ for N = 3, are:

$E_0 = q^3$	= 0.383866	$O_0 = 0.396483$	$(E_0 - O_0)^2/E_0$	=0.00041470
$E_1 = 3 \cdot q^2 \cdot p$	= 0.432958	$O_1 = 0.415238$	$(E_1 - O_1)^2/E_1$	=0.00072522
$E_2 = 3 \cdot q \cdot p^2$	= 0.162776	$O_2 = 0.160364$	$(E_2 - O_2)^2/E_2$	=0.00003573
$E_3 = p^3$	= 0.020399	$O_3 = 0.027913$	$(E_3 - O_3)^2/E_3$	=0.00276807
Σ	1	1		0.00394372

The sum of Chi^2 deviations between expected and observed surfaces ($O_0 \ldots O_3$) gives a measure of error (dE):

$$dE = \sum_i (E_i - O_i)^2 / E_i$$

which for $N = 3$ equals $3.943723e^{-3}$.

The same calculation can be carried out supposing 4 and 5 channels with p is 0.2049271 and 0.1639417, respectively. This results in:

$N = 3$	$dE = 3.943723e^{-3}$
$N = 4$	$dE = 6.825411e^{-5}$
$N = 5$	$dE = 1.234697e^{-3}$

Hence it is most likely that the patch contains 4 channels.

Measurement of Dwell Times

As its name, at least partially, suggests, a dwell-time histogram describes how often and how long a signal spends in a certain state n. Fig. 9.33 shows an example for a patch containing a single K^+ channel for $n = 0$ (channel closed) and $n = 1$ (channel open).

From both histograms, it is clear that short dwell times are more abundant than longer ones. In general, channel kinetics are considered to behave like radioactive decay or monomolecular chemical reactions. This implies that transition probabilities are independent of time and that decay follows an exponential time course. The above histograms are fairly well fitted by single exponential distributions, which give the rate constants (in s^{-1}) for a simple monomolecular model of the behavior of the channel (with $C =$ closed state and $O =$ open state):

$$C \underset{k2 = 399.5}{\overset{k1 = 64.5}{\rightleftharpoons}} O$$

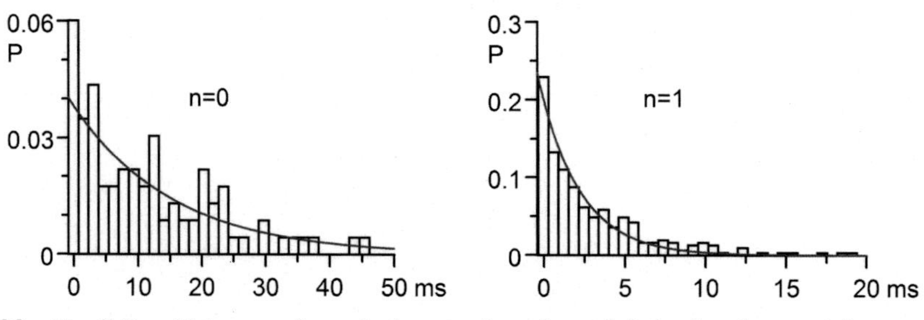

FIGURE 9.33 Dwell-time histograms for a single potassium channel. Left: closed state; right: open state. Fitted exponentials in red.

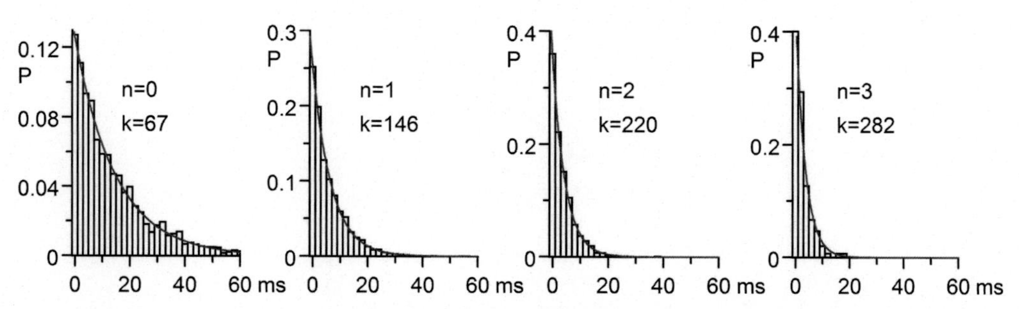

FIGURE 9.34 Dwell-time histograms for a patch with three channels. Fitted exponential functions in red.

If the patch contains more than one channel, the analysis is somewhat more complicated. In the following example, the patch contained 3 channels, resulting in 4 dwell-time histograms (one for all channels closed, one for 1 channel open ... etc.). Each of the histograms is fitted with a single exponential (Fig. 9.34).

Suppose that the 3 channels are identical and each can be either in the closed state (C) or the open state (O). If all channels are closed, then the apparent rate constant leading from $n = 0$ to $n = 1$ is $3 \cdot k_1$. Similarly, if all channels are open, then the apparent rate constant from $n = 3$ to $n = 2$ is $3 \cdot k_2$. If only 1 channel is open, then the apparent rate constant is $2 \cdot k_1 + k_2$, since 2 channels are available for opening and 1 for closing. For 2 open channels that is $k_1 + 2 \cdot k_2$. In general, with N, the total number of channels in the patch, and n, the current state:

$$k(n) = (N - n) \cdot k_1 + n \cdot k_2 \left(s^{-1}\right) \tag{9.5}$$

With the data of the figure above, we get four equations:

$$3 \cdot k_1 = 67$$

$$2 \cdot k_1 + k_2 = 146$$

$$k_1 + 2 \cdot k_2 = 220$$

$$3 \cdot k_2 = 282$$

It is of course easy to use the first and fourth equation to find the rate constants k_1 and k_2, but then not all available data would have been used. Moreover, often the probability of having all channels open simultaneously is low with the consequence that the associated dwell-time histogram (the fourth equation) is of poor quality and has to be ignored. The relation between state n and apparent rate constant k is a linear one, and therefore we can find estimates for $k(n = 0)$ and $k(n = 3)$ using linear regression (Fig. 9.35):

This yields a coefficient of 71.9 and an intercept of 70.9 from which $k_1 = 23.6$ and $k_2 = 95.5$ can be found using Eq. (9.5).

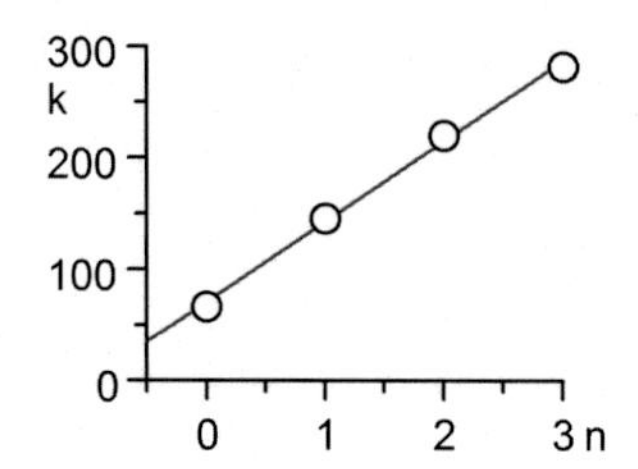

FIGURE 9.35 Linear regression of dwell times.

Not all dwell-time histograms can be fitted by a single exponential. In Fig. 9.36, activity of a single channel was recorded. The closed time histogram (n = 0) is poorly fitted by a single exponential. The plot of the residue-of-fit underneath the dwell-time histogram shows a triphasic time course that disappears if the histogram is fitted with two exponentials (n = 0, middle graphs). A nonrandom distribution of the data points around the fit as in the bottom left figure is an indication that the number of degrees of freedom of the fitting function is too low. The open time histogram (n = 1, rightmost graph) is well fitted with a single exponential. The Durbin–Watson method is an often used statistical test for the goodness of fit using this approach.

The fact that two exponentials are required to fit the closed time distribution indicates that two different pathways lead to the open state. The dwell-time histogram is actually the sum of two distributions: one describing the transition of C_1 to O and the other describing the

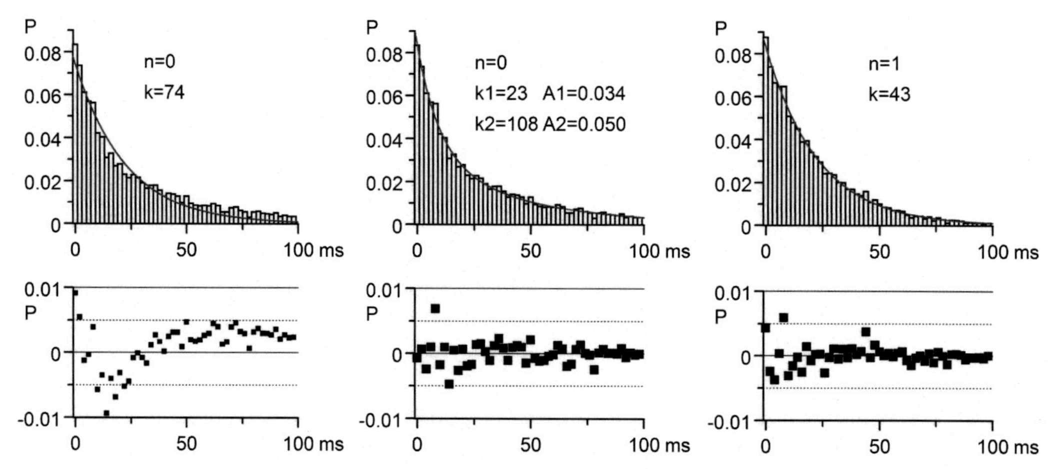

FIGURE 9.36 The residue-of-fit helps to decide whether a fit to data is adequate. It is obtained by plotting the differences between data and fit (bottom graphs). The closed time distribution was fitted by a single exponential (leftmost histogram) or two exponentials (middle histogram). The bump in the leftmost residues plot disappears if the histogram is fitted with two exponentials: the residues are now randomly distributed around the time axis. The open time distribution (rightmost histogram) is fitted well by a single exponential function.

transition of C_2 to O. This observation allows us to propose the following model to explain the data:

$$C_1 \underset{k3}{\overset{k1 = 23}{\rightleftharpoons}} O \underset{k2 = 108}{\overset{k4}{\rightleftharpoons}} C_2$$

Of k_3 and k_4 we know that $k_3 + k_4 = 43$. In equilibrium, net fluxes are 0:

$$k_1 \cdot C_1 = k_3 \cdot O$$

$$k_2 \cdot C_2 = k_4 \cdot O$$

hence, $\frac{k_3}{k_4} = \frac{k_1 \cdot C_1}{k_2 \cdot C_2}$ where $\frac{k_1 \cdot C_1}{k_2 \cdot C_2}$ corresponds to $\frac{A_1}{A_2}$, the ratio of the amplitudes of the two exponentials in the closed time distribution. With two equations, two unknowns can be solved, giving $k_3 = 17$ and $k_4 = 26$.

Unfortunately, our model ($C_1 \leftrightarrow O \leftrightarrow C_2$) is not the only one that explains a biexponential closed time distribution. Consider the following chain:

$$C_1 \leftrightarrow C_2 \leftrightarrow O$$

Here the open time distribution is monoexponential, while the open state may be reached directly from C_2 or indirectly from C_1. Without further information it is impossible to decide in favor of either of the two models.

If the channel is a ligand- or voltage-gated, we might force all the channels in state C_1 by removing the ligand or by hyperpolarization, respectively. Upon return to ligand/polarization, we measure the time lapse between the onset of the stimulus and the first channel opening. Doing so for many (1000 or more) stimulus presentations results in the so-called first latency distribution. If the first model is correct, the first latency distribution will be monoexponential with rate constant k_1; if the second model is correct, the first latency distribution will be biexponential. In that case, the chain, $C_1 \leftrightarrow C_2 \leftrightarrow O$, causes a delay in the channel opening and therefore the maximum probability of channel opening does not occur at $t = 0$, but at a later time point as in Fig. 9.37. The smooth line depicts the biexponential fit.

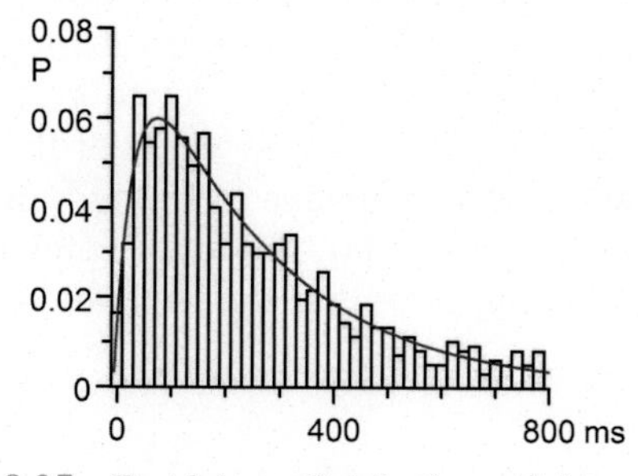

FIGURE 9.37 First latency distribution with biexponential fit.

CALCULATING DWELL-TIME HISTOGRAMS FROM MARKOV CHAINS

Above we were able to deduce some of the channel properties from inspection of the dwell-time histograms and exponential fits made thereof. The number of exponentials necessary to fit the dwell-time distributions and the shape of these distributions gave some clues about possible models describing the channel behavior. In this section, we will do the inverse: we postulate a model and then calculate the dwell-time histograms and the macroscopic current.

As we have seen above, a model of channel activity (e.g., $C_1 \leftrightarrow O \leftrightarrow C_2$) may consist of a number of states (e.g., C_1, O, C_2) interconnected by arrows depicting exponential decay from one state to another. In the section discussing spike intervals and Poisson processes and in the Appendix, it is shown that the rate constants leading from one state to another are most conveniently arranged in a square transition matrix and that such a matrix represents a set of linear differential equations. We will take up the arguments of that section, using them to calculate dwell-time distributions.

The First Latency Distribution

The first latency describes the time it takes for a channel to open for the first time. Suppose our model is

$$C_1 \underset{\mu 1}{\rightarrow} C_2 \underset{\mu 2}{\rightarrow} \ldots \underset{\mu(n-1)}{\rightarrow} C_n \qquad (9.6)$$

with all channels in state C_1 at $t = 0$ and transition rate constants μ_i.

Apart from the symbols (C for Closed and O for Open rather than S for State), this model is exactly the same as the one that we developed to calculate the waiting time distribution between successive spikes in a Poisson-distributed spike train (Eq. 9.4). Therefore, the math is the same and the result (for three closed states leading to the open state) is the same (see Eq. A.9):

$$Fl(t) = \mu_1\mu_2\mu_3\left(\frac{\exp(-\mu_1 t)}{((\mu_2-\mu_1)(\mu_3-\mu_1)} - \frac{\exp(-\mu_2 t)}{((\mu_2-\mu_1)(\mu_3-\mu_2)} + \frac{\exp(-\mu_3 t)}{((\mu_3-\mu_2)(\mu_3-\mu_1)}\right)$$

Of course, this result does not give the general solution to the first latency problem. For one the channel is not necessarily in C_1 at $t = 0$. Sometimes it might be in C_2 or in C_3 (let us restrict ourselves to $N = 3$ for the time being). In order to get the first latency distribution for arbitrary occupation of states at $t = 0$, the equivalents of the equation above have to be calculated for the cases that the channel is always in state C_2 or always in C_3 at $t = 0$. This will give us three first latency distributions Fl_1, Fl_2, and Fl_3 for the channel being in C_1, C_2, or C_3 at $t = 0$, respectively. Further suppose that the probability to be in state C_1 at $t = 0$ is

p_1, to be in state C_2 is p_2, and to be in state C_3 is p_3. The weighted sum of these distributions gives the more general result:

$$Fl(t) = p_1 \cdot Fl_1(t) + p_2 \cdot Fl_2(t) + p_3 \cdot Fl_3(t)$$

The quantitatively identical result would have been obtained if starting from:

$$c_1 = \begin{pmatrix} (\mu_3 - \mu_1)(\mu_2 - \mu_1) \\ \mu_1(\mu_3 - \mu_1) \\ \mu_1\mu_2 \end{pmatrix} \quad c_2 = \begin{pmatrix} 0 \\ \mu_3 - \mu_2 \\ \mu_2 \end{pmatrix} \quad c_3 = \begin{pmatrix} 0 \\ 0 \\ 1 \end{pmatrix}$$

and $p = z_1 c_1 e^{-\mu 1 t} + z_2 c_2 e^{-\mu 2 t} + z_3 c_3 e^{-\mu 3 t}$.

We would have chosen the vector p_{t0} as follows:

$$p_{t0} = \begin{pmatrix} p_1 \\ p_2 \\ p_3 \end{pmatrix} = z_1 c_1 + z_2 c_2 + z_3 c_3$$

and would have proceeded similarly from thereon.

It may seem that models containing more than one open state pose a second complication. However, the presence of multiple open states does not change anything fundamental to the math involved, as the only thing we are interested in is the total time spent in the closed states and we do not care to what open state the system exits. For example, the models having transition rate constants μ_1 and μ_2

$$C_1 \underset{\mu 1}{\rightarrow} O_1 \leftrightarrow O_2 \underset{\mu 2}{\leftarrow} C_2 \quad \text{and} \quad C_1 \underset{\mu 1}{\rightarrow} O \underset{\mu 2}{\leftarrow} C_2$$

evidently have identical first latency distributions, and the models

$$O_1 \underset{\mu 1}{\leftarrow} C \underset{\mu 2}{\rightarrow} O_2 \quad \text{and} \quad C \underset{\mu 1 + \mu 2}{\rightarrow} O$$

also have identical first latency distributions.

Now that we know what the ingredients are, we can formulate a general approach to calculate any first latency distribution. First, create the matrix A representing a set of differential equations (as in Eq. A.4, $X' = AX$) from the Markov transition matrix.

$$A = \begin{bmatrix} -\Sigma c_1 & k_{21} & k_{31} & k_{41} & k_{51} \\ k_{12} & -\Sigma c_2 & k_{32} & k_{42} & k_{52} \\ k_{13} & k_{23} & -\Sigma c_3 & k_{43} & k_{53} \\ k_{14} & k_{24} & k_{34} & -\Sigma c_4 & k_{54} \\ k_{15} & k_{25} & k_{35} & k_{45} & -\Sigma c_5 \end{bmatrix}$$

where k_{ij} represents the rate constants from state i to state j, and Σc_i, the sum of the rate constants in the ith column, hence the sum of all rate constants leading away from state i. Because we are only interested in the time spent in the closed states, part of the matrix is superfluous. It is therefore a good idea to reorganize the matrix such that transitions between closed states appear in the upper left corner. This is most easily done by interchanging rows and columns associated with the open and closed states. Suppose that in the above matrix columns 1, 2, and 5 correspond to closed states and columns 3 and 4 to open states. It suffices to switch columns 3 and 5 and rows 3 and 5 to obtain:

$$A = \begin{bmatrix} -\Sigma c_1 & k_{21} & k_{51} & k_{41} & k_{31} \\ k_{12} & -\Sigma c_2 & k_{52} & k_{42} & k_{32} \\ k_{15} & k_{25} & -\Sigma c_5 & k_{45} & k_{35} \\ k_{14} & k_{24} & k_{54} & -\Sigma c_4 & k_{34} \\ k_{13} & k_{23} & k_{53} & k_{43} & -\Sigma c_3 \end{bmatrix}$$

Now the upper 3×3 block of A represents transitions between closed states, giving the reduced transition matrix R:

$$R = \begin{bmatrix} -\Sigma c_1 & k_{21} & k_{51} \\ k_{12} & -\Sigma c_2 & k_{52} \\ k_{15} & k_{25} & -\Sigma c_5 \end{bmatrix} \tag{9.7}$$

Then, by the methods discussed in the section about Markov chains (see Appendix), we have a computer program calculate the eigenvalues and then solve for each eigenvalue λ: $|R - \lambda I| = 0$ to obtain the eigenvectors.

The only thing that still misses in order to be able to calculate the first latency distribution is the probability of occupancy of the closed states at $t = 0$:

$$P_{t0} = \begin{pmatrix} P_1 \\ P_2 \\ P_3 \end{pmatrix}$$

If the model is used to compare it with experimental results, then it is a good idea to design the experiment such that the probability to be in a specific closed state initially, equals 1 (e.g., by removing all ligand or by membrane hyperpolarization).

The case in which the system is in equilibrium will be discussed in the following section.

The Closed Time Distribution

The closed time distribution describes the time it takes for a channel to open given that it is closed. The problem of obtaining the closed time distribution much resembles the problem of obtaining the first latency distribution, except that we are interested not only in the first opening but also in subsequent openings. This suggests that we first take the first latency distribution, wait until the next closing, again take the "first" latency distribution and sum the two distributions, wait until the third closing, etc., until we reach the end of a period of length t for which we wish to calculate the closed time distribution. As we do not really follow individual closing events, our approach is probabilistic:

At every instant, the probability that a channel goes from an open state to a particular closed state is proportional to the probability to be in one of the open states multiplied by the rate constants leading from those open states to the particular closed state. So, if we want to know the average probability to enter a particular closed state during a stretch of t seconds following a situation that may be nonequilibrium, we have to calculate the integral of the evolution of each open state during t seconds and multiply these integrals by the rate constants leading to the particular closed state. This will give us (again for the model with three closed states):

$$p = \begin{pmatrix} p_1 \\ p_2 \\ p_3 \end{pmatrix}$$

where the vector p describes the average probability to enter each of the closed states during a period of t seconds after a (non)equilibrium condition. Calculating the first latency distribution with the vector p as initial condition gives the closed time distribution.

If the system is in equilibrium, the first latency distribution and the closed time distribution are identical and the calculations to obtain the vector p are simpler. In order to get the vector p, we have to solve three equations in three unknowns (p_1, p_2, and p_3). From Eq. (9.7), we have the three differential equations:

$$R = \left.\begin{matrix} -\Sigma c_1 & k_{21} & k_{51} \\ k_{12} & -\Sigma c_2 & k_{52} \\ k_{15} & k_{25} & -\Sigma c_5 \end{matrix}\right| \begin{matrix} 0 \\ 0 \\ 0 \end{matrix}$$

which should be read as follows (remember that $\Sigma c_1 = k_{12} + k_{15}$):

$$-p_1\Sigma c_1 + p_2 k_{21} + p_3 k_{51} = 0 \quad \text{or} \quad -p_1(k_{12} + k_{15}) + p_2 k_{21} + p_3 k_{51} = 0$$

$$-p_2\Sigma c_2 + p_1 k_{12} + p_1 k_{52} = 0 \quad \text{or} \quad -p_2(k_{21} + k_{25}) + p_1 k_{12} + p_3 k_{52} = 0$$

$$-p_3\Sigma c_3 + p_1 k_{15} + p_2 k_{25} = 0 \quad \text{or} \quad -p_3(k_{51} + k_{52}) + p_1 k_{15} + p_2 k_{25} = 0$$

As it should in equilibrium, the number of forward and backward transitions to a particular state is identical.

Unfortunately, one of the three equations is redundant; the matrix, R, is said to be singular. Therefore, a new one must replace one of the equations. We might impose that the sum of the probabilities to be in any of the closed states is unity:

$$p_1 + p_2 + p_3 = 1$$

The system of equations, with M the modified matrix R, can thus be modified into:

$$M = \left.\begin{matrix} -\Sigma c_1 & k_{21} & k_{51} \\ k_{12} & -\Sigma c_2 & k_{52} \\ 1 & 1 & 1 \end{matrix}\right| \begin{matrix} 0 \\ 0 \\ 1 \end{matrix}$$

This set is easily solved by hand for the case of only three equations but becomes much more difficult with increasing number of closed states. Fortunately, computer routines carrying out matrix inversion can do the job for us (see Watkins[6]).

The Open Time Distribution

The open time distribution is calculated in much the same way as the closed time distribution. The only thing that changes is that we swap columns and rows of the transition matrix such that all transitions between open states (rather than closed states) end up in the upper left corner of the matrix.

The Macroscopic Current

With all the knowledge we have now, the calculation of the macroscopic current associated with a given Markov chain is easy. First we set up a transition matrix, A, just like we have done in the first latency paragraph. Next we calculate the eigenvalues and eigenvectors. In the last step, we have to calculate the scaling factors for each of the eigenvectors from the initial conditions.

Pseudocode for routines to calculate the macroscopic current and dwell-time distributions can be found in Appendix.

Example: Simulation of the Hodgkin and Huxley Voltage-Gated Sodium Channel

As we have seen previously, the voltage-gated sodium channel according to Hodgkin and Huxley can be described by the action of four independent gates. Their mechanical model (Fig. 9.18) can be translated into the reaction scheme as shown in Fig. 9.38.

With $\alpha = 50\ s^{-1}$; $\beta = 12{,}000\ s^{-1}$; $\gamma = 400\ s^{-1}$; $\delta = 7\ s^{-1}$, which are the rate constants corresponding approximately to a membrane depolarization to -10 mV, the following transition matrix can be made (0 entries are left blank for clarity):

$$\begin{bmatrix} . & 50 & . & . & . & . & . & 7 \\ 36000 & . & 100 & . & . & . & 7 & . \\ . & 24000 & . & 150 & . & 7 & . & . \\ . & . & 12000 & . & 7 & . & . & . \\ . & . & . & 400 & . & 12000 & . & . \\ . & . & 400 & . & 150 & . & 24000 & . \\ . & 400 & . & . & . & 100 & . & 36000 \\ 400 & . & . & . & . & . & 50 & . \end{bmatrix}$$

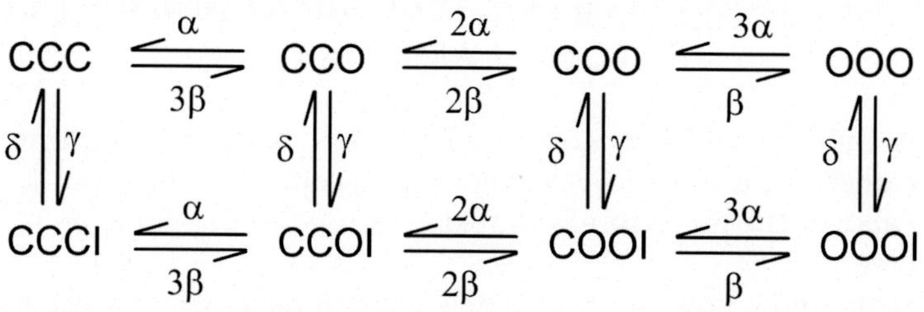

$$
\begin{array}{ccccccc}
\text{CCC} & \underset{3\beta}{\overset{\alpha}{\rightleftharpoons}} & \text{CCO} & \underset{2\beta}{\overset{2\alpha}{\rightleftharpoons}} & \text{COO} & \underset{\beta}{\overset{3\alpha}{\rightleftharpoons}} & \text{OOO} \\
\delta \upharpoonleft\downharpoonright \gamma & & \delta \upharpoonleft\downharpoonright \gamma & & \delta \upharpoonleft\downharpoonright \gamma & & \delta \upharpoonleft\downharpoonright \gamma \\
\text{CCCI} & \underset{3\beta}{\overset{\alpha}{\rightleftharpoons}} & \text{CCOI} & \underset{2\beta}{\overset{2\alpha}{\rightleftharpoons}} & \text{COOI} & \underset{\beta}{\overset{3\alpha}{\rightleftharpoons}} & \text{OOOI}
\end{array}
$$

FIGURE 9.38 Translation of the Hodgkin and Huxley model of the voltage-dependent sodium channel into a Markov reaction scheme. Each activation gate opens with a rate constant α and closes with a rate constant β. The forward and backward rate constants for the inactivation gate are γ and δ, respectively. CCC means all three activation gates are closed and the inactivation gate is open. CCOI means two activation gates closed, one open and the inactivation gate closed. Only the state OOO is conductive.

Then, applying the routines described above, the results shown in Fig. 9.39 are obtained.

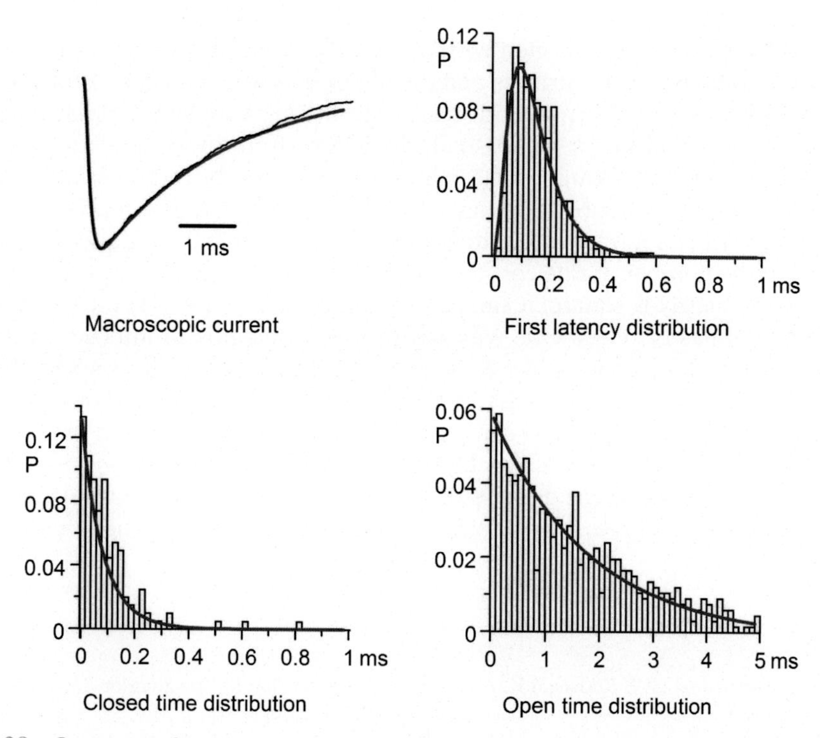

FIGURE 9.39 Comparison between dwell-time distributions obtained with the Markov chain approach and Monte-Carlo simulation of the Hodgkin and Huxley voltage-dependent sodium channel. Bar dwell-time histograms were obtained with Monte-Carlo simulation, and smooth red lines were calculated by eigen-decomposition.

ESTIMATING MODEL PARAMETERS FROM WHOLE-CELL AND PATCH-CLAMP DATA

It is feasible to estimate rate constants in a transition matrix starting from experimental data with the mathematical tools to calculate macroscopic currents and dwell-time histograms described above. The procedure is the same as used to fit a function to an experimentally obtained curve.

First, one needs to know how many channels were in the patch because the shape of the dwell-time histograms depends on it (see Fig. 9.34). It is best, because the most straightforward, to deal with just one channel. Second, one needs to propose a model. Some idea of the complexity of the kinetic scheme underlying the observed data may be deduced from the number of exponentials to fit the dwell-time histograms. For example, a model with two states has two eigenvalues (a model with N states has, in general, N eigenvalues). One of the eigenvalues is 0 (zero) corresponding to a stationary or DC level, and the other is an exponential relaxation time constant. This two-state model cannot be correct if one needs to fit more than one exponential to a curve. Hence, if a current or histogram needs m exponentials, then the model requires at least $m + 1$ states. Third, the rate constants to estimate are identified (many entries were left blank in the H&H transition matrix above). Just as in the H&H model, the rate constants to estimate follow from the arrows that point between states: not all states in the model are connected with one another. Fourth, one makes a guess of what the initial values could be. Fifth, currents and dwell times are calculated, and the mean square errors (or another measure of error) with respect to the experimental data are determined. This is done for the initial guesses and small deviations from those guesses. Like in curve fitting, parameters are then sought that minimize the error by nonlinear regression approaches such as the Levenberg–Marquardt and the simplex iteration routines.

The result of the fit of a model matrix does not necessarily give the unique solution. This is especially true if the number of rate constants to estimate is large with respect to the number of states. Since the matrix is square, it can have as much as $N \times (N - 1)$ rate constants to estimate if the model has N states. One way out of this problem is to impose constraints. To take the H&H transition matrix above as an example again, one could impose that several entries need remain identical (all entries "400" should always vary simultaneously).

This method is also very useful to compare different models that could explain the data. These different models do not necessarily have the same number of rate constants to estimate, and they have therefore different degrees of freedom. Statistical tests that take into account differences between the degrees of freedom, such as Fisher's F-test, can help to decide which model is best.

References

1. Peters RC, van Steenderen GW, Kotrschal K. A chemoreceptive function for the anterior dorsal fin in rocklings (Gaidropsarus and Ciliata: Teleostei: Gadidae): electrophysiological evidence. *J Mar Biol Assoc UK* 1987;**67**(4): 819–23.
2. Teunis PF, Bretschneider F, Bedaux JJ, Peters RC. Synaptic noise in spike trains of normal and denervated electroreceptor organs. *Neuroscience* 1991;**41**(2–3):809–16.

3. Hodgkin AL, Huxley AF. A quantitative description of membrane current and its application to conduction and excitation in nerve. *J Physiol* 1952;**117**(4):500–44.
4. Neher E, Stevens CF. Conductance fluctuations and ionic pores in membranes. *Annu Rev Biophys Bioeng* 1977;**6**:345–81.
5. DeFelice LJ. Fluctuation analysis in neurobiology. *Int Rev Neurobiol* 1977;**20**:169–208.
6. Watkins DS. *Fundamentals of matrix computations*. 2nd ed. New York: Wiley Interscience; 2002.

CHAPTER

10

Microscopy and Optical Methods in Electrophysiology

OUTLINE

A microscope is usually necessary for electrophysiological recording at the cellular level. It enlarges the specimen sufficiently to be able to place electrodes under visual control. In addition, microscopic techniques may be used to identify specific types of cells, often by immunological markers that make these cells or organelles fluoresce. Basically, the principle of the microscope can be understood by geometrical optics: A small object is illuminated by an artificial light source and subsequently enlarged by the appropriate lenses (or lens systems) to yield a useful image onto either the retina of the experimenter or on a recording device (video cameras and digital cameras). The first microscope made by van Leeuwenhoek

Introduction to Electrophysiological Methods and Instrumentation, Second Edition
https://doi.org/10.1016/B978-0-12-814210-3.00010-4

used a single, very strong positive lens, but all later microscopes use at least two stages: from object to intermediate image and from there to the final image. This is called a compound microscope and allows a larger magnification and a better optical resolution. For any kind of application, it is important to understand the principles of the techniques used, their possibilities, and their limits. Before coming to image formation by a microscope, we would first like to brush up our knowledge of the physical phenomena that are required for its understanding. As this is a book on electrophysiological methods and not on optics, the following sections will be succinct and reduced to the necessary minimum.

In the section on refraction, it will be demonstrated that a lens can be considered as an object that transforms a plane light wave, i.e., a wave whose wave fronts are parallel planes, into a spherical wave, and vice versa. The center of the sphere is the focal point of the lens. In the section on diffraction, it will then be shown that a lens creates a frequency spectrum of an object in the (back) focal plane if the object is illuminated by a plane wave. This implies that image manipulations such as filtering can be carried out in the focal plane of the lens, e.g., by suppressing or accentuating certain spectral components. It will be argued that because the focal plane is not infinite in practical microscopes, their spatial resolutions are necessary limited.

The second part of this chapter is devoted to optical techniques that make use of fluorescence in one way or another, either because the biological tissue is inherently fluorescent or because fluorescent molecules are introduced by the experimenter. We look into several microscopic techniques that allow for accurate optical sectioning such as confocal and multiphoton microscopy. Finally, the use of fluorescent probes to record membrane potentials from excitable cells and the optical excitation of these cells as alternatives to the classical electrical methods will be evaluated.

REFRACTION

The speed of light in transparent materials depends on the composition of these materials. It is slowed down in materials of higher optical density, i.e., materials with higher refractive index. The refractive index, n, is defined as the ratio of the speed of light in vacuum over that in the medium. Given the refractive indices, n_1 and n_2, the angle of refraction at an interface between optic media can be calculated by Snell's law:

$$\frac{\sin(\theta_1)}{\sin(\theta_2)} = \frac{n_2}{n_1} \tag{10.1}$$

where θ_1 and θ_2 are the angles that the incident wave and refracted wave are making with the normal to the surface between the media (Fig. 10.1A). The blue lines in the figure represent positions in the plane wave with equal phase. It is called a plane wave because the wave fronts form a plane, drawn as a straight line in the 2D representation in Fig. 10.1. As soon as the wave hits the surface, it is retarded with respect to those positions in the wave front that are still in the low-density medium, but phase coherence is not lost. It is said that all points in a wave front have followed paths of equal "optical length," where optical length (p) means the product of physical distance (d) and refractive index or $p = n \cdot d$.

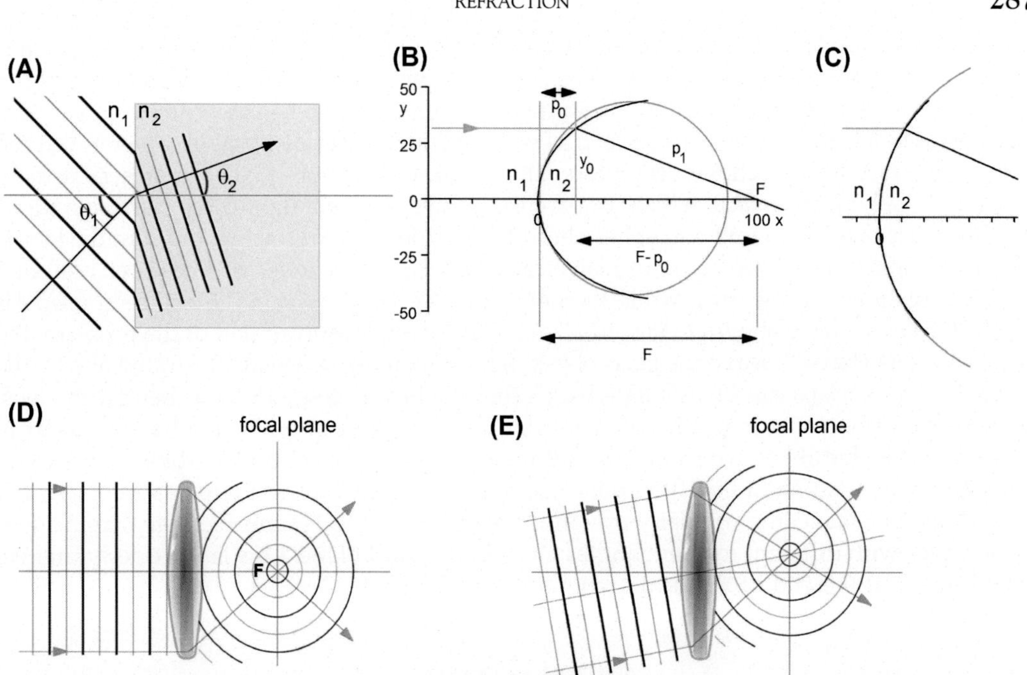

FIGURE 10.1 (A) A plane wave is refracted at the interface of media with different refractive indices. The angle of diffraction is calculated by Eq. (10.1). (B) A plane wave parallel to the x-axis travels through air and is then diffracted by a glass ovoid such that all paths of equal optical length converge in the focal point F (it is half a lens) ($n_1 = 1$, $n_2 = 1.5$). The blue line depicts a perfect circle. (C) The overlap between ovoid and circle improves with augmented n_2 ($n_2 = 3$). The dimensions of x and y axes are arbitrary. (D) An ideal lens converts a plane wave into a spherical wave with its center in the lens' focal point if the plane wave runs along the optical axis or (E) at a point on the focal plane if it comes in at an angle with the optical axis.

Everyone knows that an important application of a lens is to focus sunlight onto a small spot. The Sun is so distant that the light wave reaching earth can be considered a plane wave. Hence, a lens converts a plane wave into a single spot or focal point. The question we would like to address now is: "What needs the shape of a glass object be to do the job"? Suppose the light beams travel in air with a refractive index $n_1 = 1$ and then from the surface through the glass with $n_2 = 1.5$ to the focal point. For this to work, the optical paths of all beams arriving at the focal point need to be identical. The problem is shown in Fig. 10.1B. A plane wave impacts the glass object parallel to the x-axis. The plane at $x = 0$ is taken as the reference. The path length of the beam that runs along the x-axis (at $y = 0$) is $F \times n_2$. The path length of a beam coming in somewhat higher, running at $y = y_0$, is $p_0 \times n_1 + p_1 \times n_2$. As said, the optical paths have to be equal and hence $p_0 \times n_1 + p_1 \times n_2 = F \times n_2$. Therefore, if we choose p_0, p_1 can be calculated by:

$$p_1 = \frac{F \cdot n_2 - p_0 \cdot n_1}{n_2} \tag{10.2}$$

As p_0 and p_1 are known, y_0 can be easily calculated from the triangle with sides y_0, p_1, and F-p_0. This gives rise to an ovoid surface as shown in the figure, which is also known as the Cartesian oval after René Descartes who studied the problem first. Lenses with ovoid surfaces are difficult to fabricate. They can be cast, as is done with certain disposable cameras, but the surface contains imperfections. Therefore, the glass surface for quality optics is usually ground to a spherical surface, which, in practice, comes close enough to the Cartesian ideal. The difference between ovoid and sphere becomes smaller if the refractive index of the lens is high (to demonstrate this point, an unrealistically high refractive index of 3 has been assumed in the calculations for Fig. 10.1, while the normal index for glass, $n = 1.5$, has been used for Fig 10.1B). The surfaces shown in Fig. 10.1B and C have only resolved half of the problem; the light beams also have to leave the glass object. By a similar approach as illustrated above, the second, "exit," surface can be calculated such that the beams come to focus behind the lens. An important point that logically follows from the constraint of all beams having identical optical paths is that the wave fronts around the focal point are concentric spheres, or circles in the 2D representation of Fig. 10.1. Hence, an ideal lens may be considered as an object that converts a plane wave in a circular wave and vice versa (Fig. 10.1D and E). Note that incident plane waves with different incident angles focus in the focal plane. This is a property that we come back to at the end of the next section on diffraction.

DIFFRACTION

Diffraction is prominent when light passes through a small slit, pinhole, or object. According to Huygens' principle, every point in a propagating light wave is a source of circular waves. Many of these circular waves may interfere with one another to form a plane wave. When a plane wave encounters a small slit in a wall, the slit becomes the only source of a circular wave behind it (Fig. 10.2A). If the wall contains two slits, two interfering circular waves are generated that enhance and extinguish each other forming an intricate pattern (Fig. 10.2B). The waves enhance each other at points for which the distances to the two slits differ by a multiple of the wavelength. Note that the points in Fig. 10.2B for which this is valid

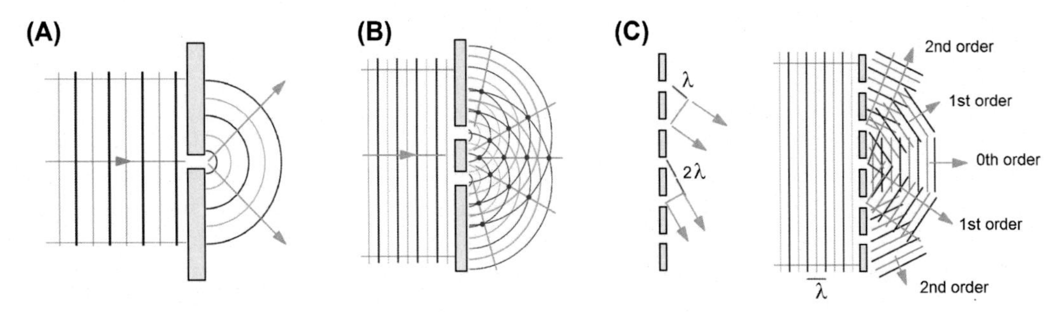

FIGURE 10.2 Diffraction of a plane wave by a single slit (A), two slits (B), and a grating (C). Circular waves emanating from each slit interfere constructively at points for which the distances to two adjacent slits are multiples of the wavelengths (nλ). Points of equal phase are shown in shades of blue. Interference is destructive at points of opposite phase, i.e., at distances of $(n+{}^{1}/_{2})\lambda$; n may be 0 (0th order wave in (C)).

are located on straight lines. A grating is created by largely increasing the number of slits. The multiple circular waves interfere to form plane waves that are sent in directions for which the differences in path lengths between adjacent slits are multiples of the wavelengths (Fig. 10.2C). This gives rise to a 0th order, undiffracted, plane wave that travels along the same axis as the incoming wave and diffracted waves of 1, 2, … nth order depending on the path length differences.

We have seen in Fig. 10.1 that a lens transforms a plane wave into a spherical wave. For this reason a grating produces a series of dots in the focal plane of a lens, in general, and the back focal plane of a microscope objective, in particular (Fig. 10.3A). In this figure, the object

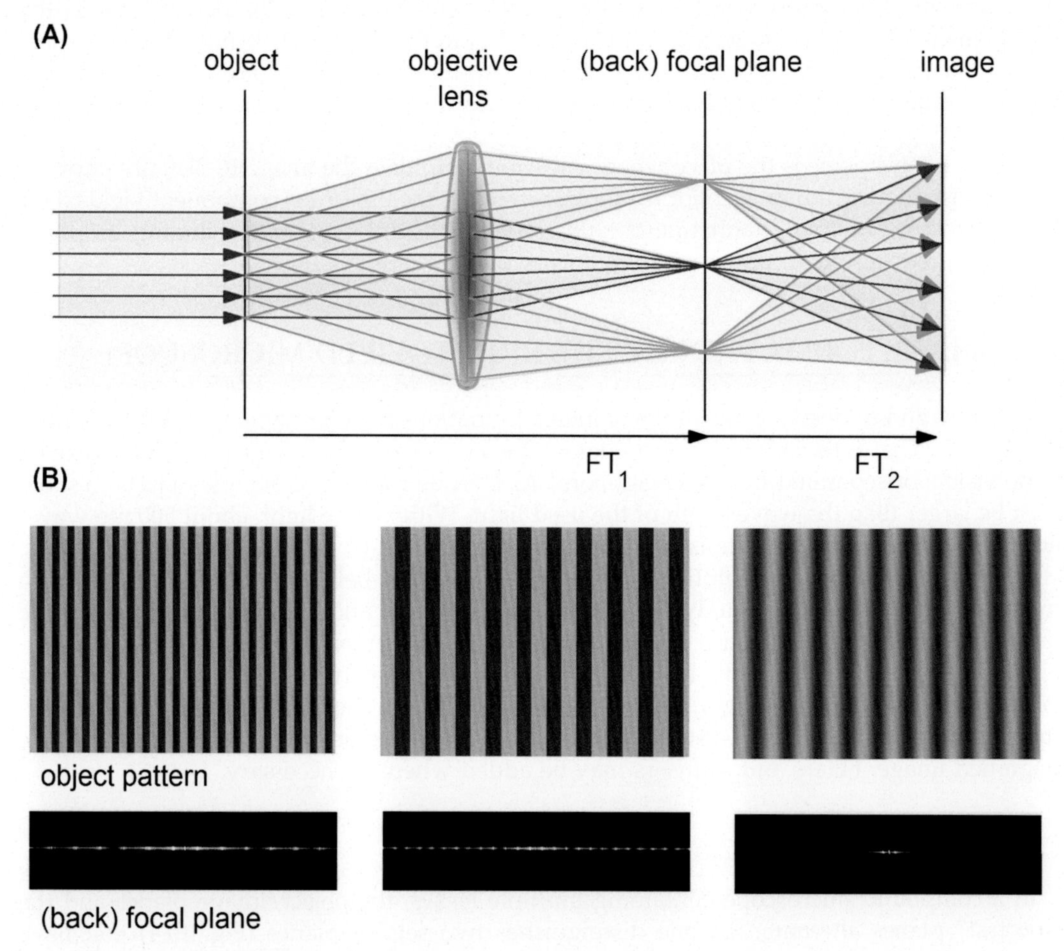

FIGURE 10.3 Single-lens Fourier transformation. (A) A diffractive object sends plane waves in various directions, which are focused by an objective lens onto the (back) focal plane at different locations depending on spatial frequency. The image is reconstituted on the image plane. The process involves a forward, FT1, and a backward, FT2, Fourier transformation. (B) Three gratings and their back focal spectra. Left and middle are rectangular contrast profiles, while the grating on the right is sinusoidal-modulated profile, giving the zeroth-order dot and two first-order dots of -f and f, where f stands for the spatial frequency of the grid.

is the grating, or any object that diffracts light. The diffracted plane waves focus in the back focal plane and produce an image in the "image" plane. It can be seen in Fig. 10.3B that if the spatial frequency of the grating diminishes (middle panel) with respect to the grating to the left of Fig. 10.3B, the dots are more closely spaced. Actually, the images in Fig. 10.3B were not obtained with gratings but with dark and light bands on photographic film, but the principle remains the same. The advantage of this approach is that the absorbance can be modulated at will. If the light and dark bands are modulated in a sinusoidal fashion rather than a rectangular fashion, only three dots appear in the back focal plane (right panel in Fig. 10.3B). Like for the rectangular modulated "gratings," the dots become more widely spaced if the spatial frequency of the sine on the film increases (not shown). Hence the distance between a dot and the middle dot (0th order) is a measure of spatial frequency. In fact, the pattern seen in the focal plane is the Fourier transform of the object (FT_1 in the figure). A second, inverse, transformation creates the magnified image. A grating produces multiple equidistant dots because the rectangular form contains harmonic frequencies that we have already encountered in Chapter 7 (Fig. 7.7D and E).

The back focal plane is the place where one can manipulate the image to visually improve or otherwise modify the image. For example, removing the 0th order component yields dark field microscopy, while attenuating the 0th order by about 70% and changing its phase by −90 degrees results in phase-contrast microscopy.

IMAGE FORMATION IN THE BRIGHT-FIELD MICROSCOPE

To assess and understand the limits of image formation, some principles of physical optics are needed: What is the resolution of a microscope, i.e., what is the smallest object that can be made visible or separated from its neighbors? As a crude rule of thumb, the objects to show must be larger than the wavelength of the used light. With visible light, about 500 nm wavelength, the order of magnitude is half a micrometer.

In practice, the limit of resolution is somewhat lower (i.e., better) but depends very much on the proper equipment (mainly the microscope objective) and, equally important, on the optimal adjustment of the entire optical "train" from light source to final image.

This optical train consists of a lamp (LEDs, a tungsten filament, or a gas discharge tube), a collector lens to bundle the light, a condenser to concentrate the light further onto the specimen, then the objective lens (system), and finally an eyepiece or projective lens to create the magnified image. Filters and diffusers may be added wherever necessary.

The Optical Train of a Microscope

In a compound microscope containing multiple lenses, the object/image planes and the "spectral" planes alternate. So, one distinguishes two sets of planes (Fig. 10.4) that have different content. The first set, called "orthoscopic" planes, consists of the collector lens (or the field iris diaphragm close to it), then the object plane, the primary image (in the upper part of the tube), and the final image on retina or screen.

All planes in this set are called "conjugate." Obviously, these all deal with the main purpose of the microscope, i.e., making an enlarged image of a small object. Different positions in

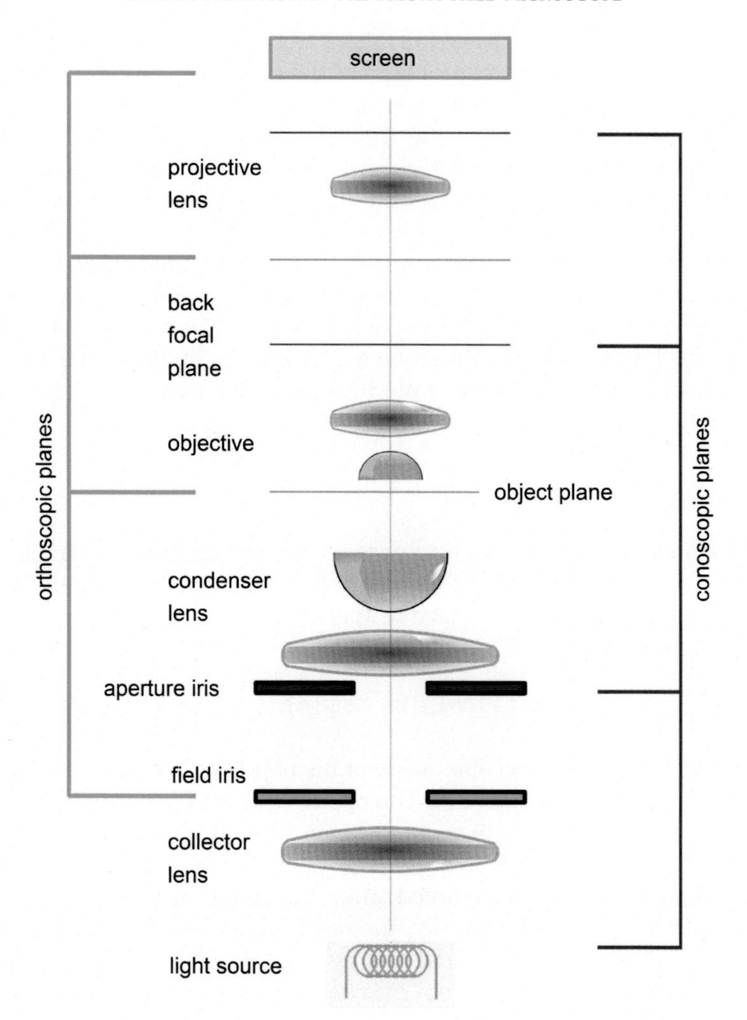

FIGURE 10.4 Schematic drawing (not to scale) of the two sets of planes in the optical train of the microscope.

the object plane are mapped onto corresponding positions in the image planes, whereas the size of the field iris determines how large a part of the object is illuminated. The latter is called the illuminated "field." Note that this also means that dust on or near the collector lens is projected onto the object and so will show up in the final image.

The second set of planes, interspersed between the former ones, is called "conoscopic" because the distribution of light here is related to the angle of incidence of the light beam, which usually has a conical shape. Here, the illuminated area, i.e., the size of the iris, is called the "aperture." The aperture iris is used to modulate image contrast and depth of field as does the diaphragm on a photo camera. Thus, a pinhole in the object plane means a small illuminated object or field and is imaged as a small (be it magnified) spot on the screen. To the contrary, a pinhole aperture iris may illuminate the entire object plane, and hence the entire screen, but with a small angle, i.e., an almost parallel bundle and therefore almost a plane wave. The back focal plane of the objective is the most important conoscopic plane for the resolution of the microscope (the sharpness of the image). This plane lies in or near the objective mount inside the microscope tube and can be observed by removing an eyepiece and looking directly down the tube. Under the usual circumstances of an (almost) fully opened condenser iris and an extended object, the light distribution in the back focal plane is filling the aperture almost uniformly. In phase-contrast microscopes, one simultaneously sees the ring-shaped aperture iris and the ring-shaped attenuation filter present in the objective focal plane. The two rings should coincide for proper phase-contrast and can usually be made to do so by turning screws on the aperture iris holder. A Bertrand lens may be inserted into the microscope tube to enlarge the image of the aperture iris. The lens was named after the mineralogist Emile Bertrand who used it to examine conoscopic images of crystals.

Optical Resolution

Although the conoscopic planes are rarely inspected during normal microscopy, the light distribution in the conoscopic planes has important consequences for image formation. The relationship between the microscope objective aperture and the half-angle of the transmitted light cone is determined by:

$$NA_{obj} = n\sin(\alpha) \tag{10.3}$$

where n is the refractive index at the object side of the objective lens, and α is the half-angle of the cone of light admitted by the objective. The objective aperture is usually expressed as this "numerical aperture" (NA), rather than in distance units (e.g., mm), because the angular width of the entering light cone determines the resolving power of the microscope. The NA of a microscope objective is engraved after the magnification factor. Examples are "40 × 0.65" and "100 × 1.30." Traditionally, the resolving power is given in the form of the "Rayleigh criterion." This measure was defined by Lord Rayleigh at the end of 19th century from the shape of diffraction patterns from two nearby points. The image from an infinitely small object, i.e., a theoretically small point in the object plane, is spread out somewhat in the image plane because the higher harmonics in the spectrum are cut off. It consists of a central light peak, surrounded by a series of dark and light rings of quickly diminishing intensity. This is called an Airy disc. A cross section of this intensity pattern is given in Fig. 10.5A. Rayleigh contended that the limit of resolution is reached if the first intensity minimum of one pattern coincides with the peak of the second. In this case, the intensity distributions merge into one larger disc, with a dip in the center of about 25% (Fig. 10.5B). For direct visual microscopy, this is indeed a good measure, since the human eye is not particularly sensitive to intensity gradients, so that the dip will be hardly visible, if at all. It must be kept in mind, however, that the Rayleigh criterion, although useful, is not the fundamental limit of resolving power. Especially since the advent of video techniques, small contrasts may be boosted by one or two orders of magnitude so that even the very small central dip in Fig. 10.5C may be made visible. The fundamental limit, called Sparrow limit, seems to be reached when there is no dip at all, that is, when the combined diffraction image of the

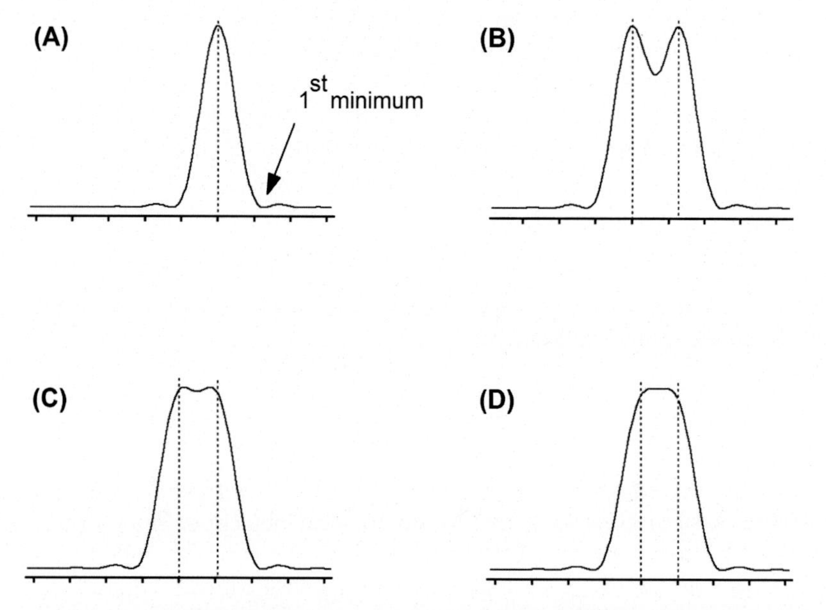

FIGURE 10.5 Cross sections of diffraction discs of one illuminated point (Airy disc) and of two close points. (A) Single Airy disc. (B) Two points at a distance according to the Rayleigh criterion (1.22 normalized units). (C) Two points closer together (about 1.04 units). (D) Two points at the Sparrow limit (one unit). The horizontal axis is subdivided in normalized distance units, i.e., one unit corresponds to the Sparrow limit. The vertical axis reads relative intensity in arbitrary units. The vertical, *dashed lines* indicate the geometrical positions of the points.

two points has only a single, central, intensity maximum (Fig. 10.5D). Recent research suggests that even this limit may be surpassed because the Airy disc of two closely spaced points is slightly wider than a single-point Airy disc so that in principle the difference may be detected.

For normal (digital or analog) video microscopy, the Sparrow limit is still the best measure of resolving power. For self-luminous objects, such as fluorescent ones, the Sparrow distance, d_S, is dependent on wavelength and objective aperture according to

$$d_s = \frac{\lambda}{2NA} \quad (10.4)$$

where λ is the wavelength. Since microscope objectives may have an NA of up to about 1.4, the Sparrow limit for green light ($\lambda \approx 550$ nm) is about 0.2 μm.

For objects in bright-field microscopy, the sparrow distance is dependent on both objective and illumination aperture:

$$d_s = \frac{\lambda}{NA_{obj} + NA_{ill}} \quad (10.5)$$

This means that the resolving power at full illumination aperture is twice the resolving power that occurs with pinhole ($NA_{ill} \approx 0$) illumination. This demonstrates that it is important to keep the condenser iris as wide open as possible for the observation of details. In practice, there is a trade-off between the deterioration of the image by stray light that may occur at full condenser NA and the resolving power, which is the reason for the rule of thumb to use a slightly lower condenser NA (about 90% of NA_{obj}). If the low contrast due to stray light can be compensated by video enhancement, the condenser iris may be opened fully to get the best resolution.

Optical Contrast

Contrast is usually defined as the difference of the highest and the lowest light intensities attainable in the image plane divided by their sum:

$$C = \frac{I_{max} - I_{min}}{I_{max} + I_{min}}$$

Contrast as a function of spatial frequency is called the "contrast transfer function" (CTF) of the microscope system consisting of illuminator and objective. Since optical objects are complex entities, the complete transfer function of the microscope, called "optical transfer function" (OTF), consists of the CTF together with a phase transfer function. However, the latter part may often be omitted because image detectors generally are not phase sensitive.

Near the resolution limit, smaller and smaller objects do not disappear abruptly. Rather, the contrast diminishes gradually with increasing spatial frequency. This is illustrated in Fig. 10.6.

With central pinhole illumination (field iris closed), the zero-order is centered in the objective aperture, whereas the diffracted orders are shifted outward by an amount dependent on the spatial frequency of the object. At the highest visible spatial frequency,

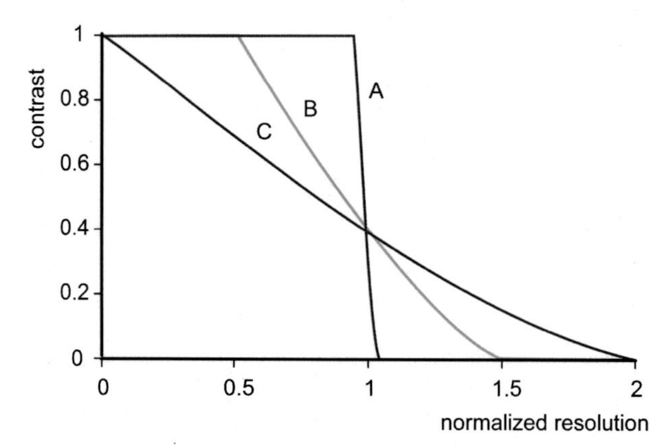

FIGURE 10.6 Contrast transfer functions of the microscope objective at three illumination aperture sizes: A, pinhole illumination (condenser diaphragm virtually closed); B, larger illumination aperture; and C, full objective aperture illuminated.

the first-order dots are at the edge of the aperture. The normalized resolution at this limit is chosen as unity, where

$$\text{normalised resolution} = \frac{\lambda}{\text{NA}}$$

Spatial frequencies higher than this limit (see Eq. 10.5) are diffracted at still larger angles, falling outside the objective, and thus contrast collapses to zero. With a completely opened field iris, resolution is increased at the expense of contrast (curve C in Fig. 10.6).

Microscope Tube Length

A microscope objective lens is a delicate combination of elements made from different types of glass. This is needed to reduce lens errors like chromatic aberration and astigmatism to a minimum, at the same time taking care that the resolution remains close to the theoretical limit just described. Formerly, the design was made by traditional techniques and improved by trial-and-error. Today, the properties of microscope objectives are largely calculated and optimized by computer programs. However, even with optimized designs, objective lenses yield their best imaging properties only under certain conditions, an important one being the tube length.

For most of the history of the microscope, the tube length was standardized at 160 or 170 mm. More recently, this fixed tube length was becoming a problem because of the need to add components like filters, epi-illuminators, beam splitters, and the like. So the best research microscopes today use a different lens design, called "infinity-corrected." The objectives are designed to cast their (intermediate) images in infinity. In other words, the outgoing image-forming rays are parallel. This gives freedom to use an arbitrary tube length. The intermediate image is finally projected at a finite distance by an accessory lens higher up in the tube.

Media Between Object and Objective Lens

A second condition necessary to obtain optimal image formation is the medium between the object (the specimen) and the front lens of the objective.

For low-magnification lenses, this is mostly air, but usually the specimen is covered by a thin coverslip that is assumed by microscope manufacturers to be $170 \pm 10\ \mu m$ thick and having a refractive index of 1.5 (see Fig. 10.7A). Using coverslips with other specs deteriorates image quality. For higher NA lenses, a different medium than air is necessary to get the high-angle rays into the objective lens. For stationary specimens, immersion oil is used, a type of inert, highly qualified oil having a refractive index of about 1.5, close to that of most glass types. These objectives perform very badly if the oil is omitted.

However, in electrophysiological experiments, the specimen is alive and can be anything from a single cell to an entire organism. In this case, since one needs to access the specimen with electrodes, a cover glass is not an option. Two solutions exist: water-immersion objectives and inverted microscopes.

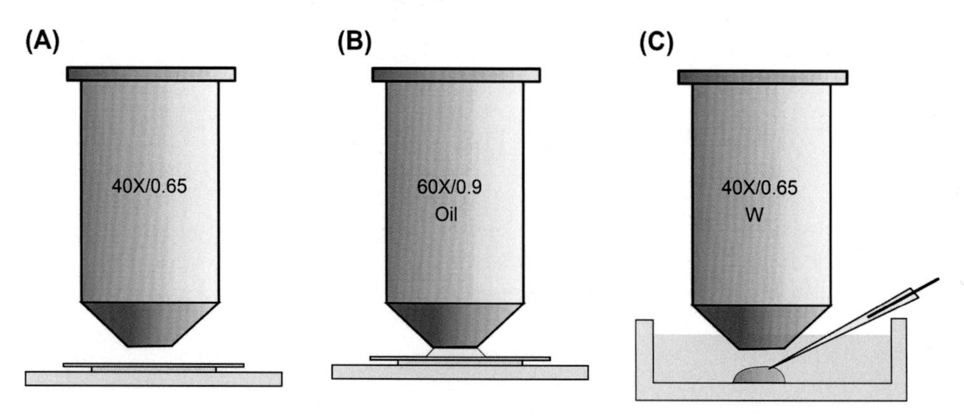

FIGURE 10.7 Media between condenser and objective lens. (A) A specimen slide with a flat object and a coverslip. (B) Idem but with an oil-immersion objective. (C) An extended specimen, impaled with a micropipette, and a water-immersion objective.

Water-immersion objectives are designed to yield optimal imaging having water between the object and the front lens. These are labeled with "W" (see Fig. 10.7C). Some manufacturers label the lens "WI" if a coverslip is supposed to be used with the objective.

An inverted microscope has its objective lenses (dry or oil immersion) below the microscope stage, and the illuminator and condenser above it. This means folding the optical path by mirrors, resulting in a very different construction with infinity-corrected design. In this case, the specimen holder must consist of optical-grade glass and for high NAs rather thin at that.

Achromatic and Apochromatic Objective Lenses

Last but not least: microscope objectives come in sorts and varieties, having properties and prices for different budgets and applications, from the simplest study microscope to the advanced confocal and fluorescence systems. The simplest type of objective, sufficient for most schools and courses, is called "achromat." Here, the chromatic aberration is corrected for two points in approximately red and green regions of the visual spectrum. Apochromats are corrected for blue in addition and are therefore more expensive. The prefix "plan" is added if the lens is moreover corrected for image field curvature (hence, plan-achromat and plan-apochromat). These lenses have a wider and flatter image field. Finally, lenses for specific tasks are developed, e.g., for fluorescence, ultraviolet, or infrared applications.

FLUORESCENCE MICROSCOPY

Fluorescence is the process in which a fluorescent molecule emits light almost immediately after being excited by light of a somewhat shorter wavelength (explained later; Fig. 10.9A). The molecules in question may be, for example, tags to antibodies or dyes that report intracellular calcium concentration or (plasma) membrane potential. Epifluorescence microscopy is a very popular configuration (Fig. 10.8A) used in electrophysiology to monitor and record

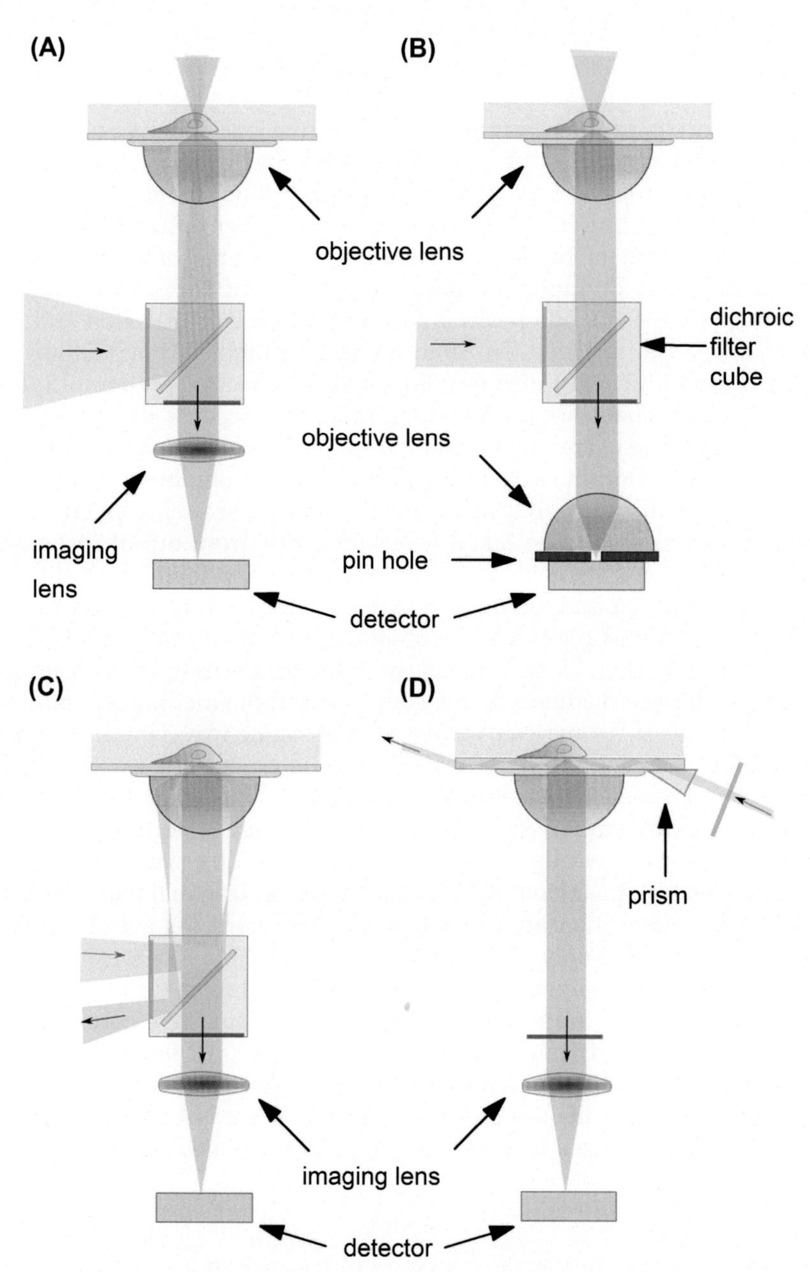

FIGURE 10.8 (A) Epifluorescence, (B) confocal microscopy, and (C and D) TIRF microscopy. The excitation beams are in green and the emitted fluorescence is shown in red. The preparation is separated from the objective lens by a glass coverslip and immersion oil.

images of fluorescent probes. A beam of light is sent to the preparation via the objective lens of the microscope. This beam converges onto the preparation, illuminating the entire field of view of the objective.

Emitted fluorescence is captured by the same objective lens and focused on an imaging device such as a CCD camera. A dichroic filter/mirror and usually two additional bandpass interference filters, together making up a dichroic filter cube, are used to separate the excitation wavelengths from those emitted. A dichroic filter reflects light of short wavelengths and transmits light of higher wavelengths. The source of light is usually a gas discharge lamp, but LED illumination gains terrain in epifluorescence microscopes. In confocal microscopy, the excitation beam is focused onto the preparation and the emitted light focused by a second objective lens onto a narrow pinhole. This hole has a diameter of 0.5–1 time the size of an Airy disc (see Eq. 10.4). A detector (photomultiplier or CCD camera) behind the hole quantifies the incoming radiation. Because the confocal configuration results in the recording of emission coming from only a small voxel, the excitation beam needs to scan the preparation in two or three dimensions to obtain an image. The various ways to achieve this will not be discussed here. Confocal scanning microscopy has the advantage that crisp images can be taken that suffer little from out of focus background fluorescence. Disadvantages are increased bleaching of fluorophores due to the higher illumination intensities required and low temporal resolution. High temporal resolution while maintaining low background noise may be obtained with total internal reflection fluorescence (TIRF) microscopy (Fig. 10.8C and D). In TIRF, an exciting laser beam is reflected at the glass/water interface made up by the coverslip and the medium containing the cells to record from. A so-called "evanescent" wave with the same wavelength as the laser beam extends from the interface for about 100 nm into the liquid thereby exciting fluorophores. The emitted light is focused on a sensitive camera (e.g., an electron-multiplying CCD). Globally, two methods are employed to deliver the excitation laser light. The first consists of delivering the light by a specially designed TIRF high NA objective. The laser light beam leaves the objective lens at an angle that needs to be greater or equal than the angle of total reflection at the glass/water interface (~60 degrees). Therefore, the light beam has to enter the objective at its border and only at its border (Fig. 10.8C). Most of the excitation light re-enters the objective after reflection and needs to be separated from the emitted fluorescence light by a dichroic filter cube. Scattering of light within the objective may cause stray light that enters the coverslip at any angle, thus causing background fluorescence. The second method prevents this to happen by delivering the laser light via a prism into the coverslip (Fig. 10.8D). It also allows for the use of objectives with lower NA. It should be noted here that TIRF can only make images of structures close to the glass/water interface and that deeper structures remain out of reach.

MULTIPHOTON MICROSCOPY

The study and visualization of cells in deeper layers of (live) tissue is a major challenge, especially if the integrity of overlaying structures needs to be maintained. Multiphoton microscopy has several properties that face this challenge such as reduced light scattering and foremost: excellent optical slicing.

Several phenomena in which two or more photons interact are of interest to microscopy and in extension to the practice of electrophysiology. These phenomena only take place in particular media and if light intensity is very high. In biphoton fluorescence microscopy, two photons are absorbed simultaneously to excite a fluorescent molecule. The molecule then emits a single photon having an energy slightly less than the combined energies of the two absorbed photons. The Jablonski diagram corresponding to this situation is shown in Fig. 10.9B.

Jablonski diagrams schematically show the electronic energy levels of a given molecule. The resting ground state is S_0 and excited states are labeled S_n. At room temperature, each state S is subdivided by vibrational states depicted in Fig. 10.9 by thin horizontal lines 0, 1, 2 .... In regular fluorescence, a single photon is absorbed (the blue arrow in Fig. 10.9A) and a single photon is emitted (the green arrow). The energy of the emitted photon, depicted in Jablonski diagrams by arrow length, is always less than the photon absorbed. The difference in energy between absorbed and emitted photons determines the Stokes shift named after the physicist who described it. In biphoton fluorescence, two photons are absorbed (almost) simultaneously creating a "virtual" state in between S_0 and S_1. A virtual state is a quantum state that possibly only exists in the imagination of theoretical physicists but which has its use in Jablonski diagrams. As for regular fluorescence, some energy is lost to vibrational relaxation.

Second harmonic generation (SHG) is also a biphoton process, but that does not involve absorption of light by molecules even though the light interacts with those molecules. It is a form of microscopy that shares some properties with biphoton absorption microscopy just discussed in that it allows for fine optical sectioning and crisp images of fluorescence.

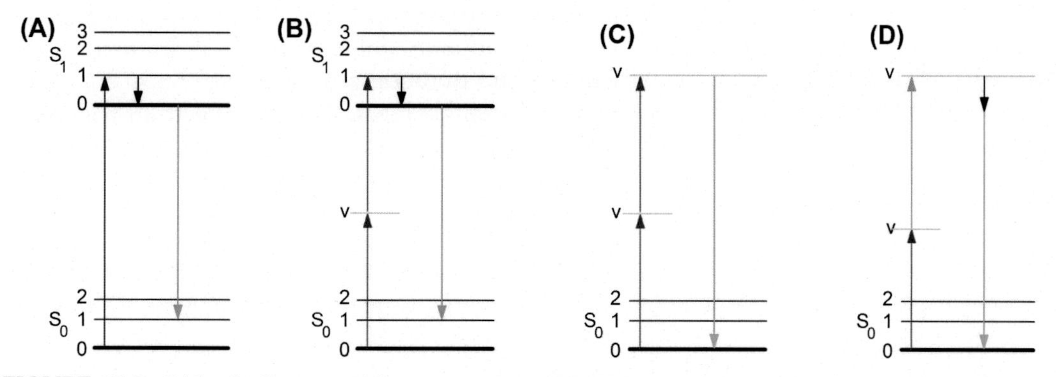

FIGURE 10.9 Jablonski diagrams of fluorescence (A and B) and second harmonic generation (C and D). (A) In classical fluorescence, an incoming photon excites the molecule from the ground state to S_1. It then rapidly loses vibrational energy, after which it emits a photon to the ground state. (B) Simultaneous absorption of two photons of half the energy required to reach an excited state may excite a molecule giving rise to emission of a single photon with an energy somewhat less than the sum of the two absorbed photons. (C) Second harmonic generation occurs if light interacts with a medium consisting of molecules with asymmetrical chemical structure. In such a medium two photons combine to yield a single photon of double the energy. The medium becomes electrically polarized during the process. (D) If light pulses are in the femtosecond range, medium polarization fluctuates rapidly, giving rise to Terahertz radiation (the black vector in (D)). Intermediate states that have no correspondence to the molecules electronic states are "virtual" (v).

SHG has some advantages over the other optical methods that are worth mentioning. Neatly aligned molecules, locally forming an electrically polarized region such as tubulin and collagen, are biological relevant molecules that constitute a medium in which SHG may occur. Bipolar molecules such as several voltage-sensitive (styryl) dyes insert into the plasma membrane in a regular fashion and thus also form a polarized region suited for SHG. Solutions in which molecules are randomly oriented do not generate second harmonics. In SHG, two photons with the same phase relation combine to form a single photon with double the energy and hence, half the wavelength. Third harmonic generation, the process that adds three photons (THG), also exists. In either case, the newly formed high-energy photon propagates in more or less the same direction as the incoming photons. The degree of scattering in other directions than the forward direction depends on the size and orientation of the interacting molecular structure. This is a distinguishing property of SHG and THG with respect to (biphoton) fluorescence in which the emitted photon is scattered almost evenly in all directions. SHG is of interest if used with voltage-sensitive dyes. Because SHG only takes place on ordered interfaces, the method is completely insensitive to the presence of dye molecules in the cytosol or the external medium, where their orientation is random. This considerably increases the membrane/cytosol contrast in fluorescence as compared to fluorescence by single-photon excitation.

The probability that two or more photons arrive simultaneously on a molecule increases to a great extent with light intensity. For this reason, focused laser illumination is used for biphoton and SHG. This way of illumination has as a consequence that light intensity may be made just high enough to produce emission in only a small region around the focal point (typically in the order of a cubic micrometer). Hence, emission from deeper layers, as occurs in classical single-photon fluorescence microscopy, is absent. This allows for fine optical slicing of the object (Fig. 10.10).

About 1 W of light energy per cubic μm is necessary to produce the optical nonlinear events mentioned. As so much energy input per volume, if maintained, would rapidly and irremediably damage tissue, pulses of femtosecond duration are used.

This unfortunately spreads the spectrum of the single line laser light. Spectrum widening may extend into the single-photon absorption region of fluorescent target molecules causing fluorescent emission from out of focus regions. If multiple illumination is used to excite multiple probes, the problem may become more severe.

Heisenberg's uncertainty principle is at the base of the spectrum broadening when very short laser pulses are being used. According to this principle, the precision with which certain pairs of properties can be known is limited. The energy-time uncertainty equation says:

$$\Delta E \Delta t \geq \frac{h}{4\pi}$$

where ΔE is the standard deviation in the energy; Δt, the standard deviation in the time domain; and h, the Planck constant. According to the Planck–Einstein relation, the energy (E) of a photon with frequency (f) is:

$$E = hf$$

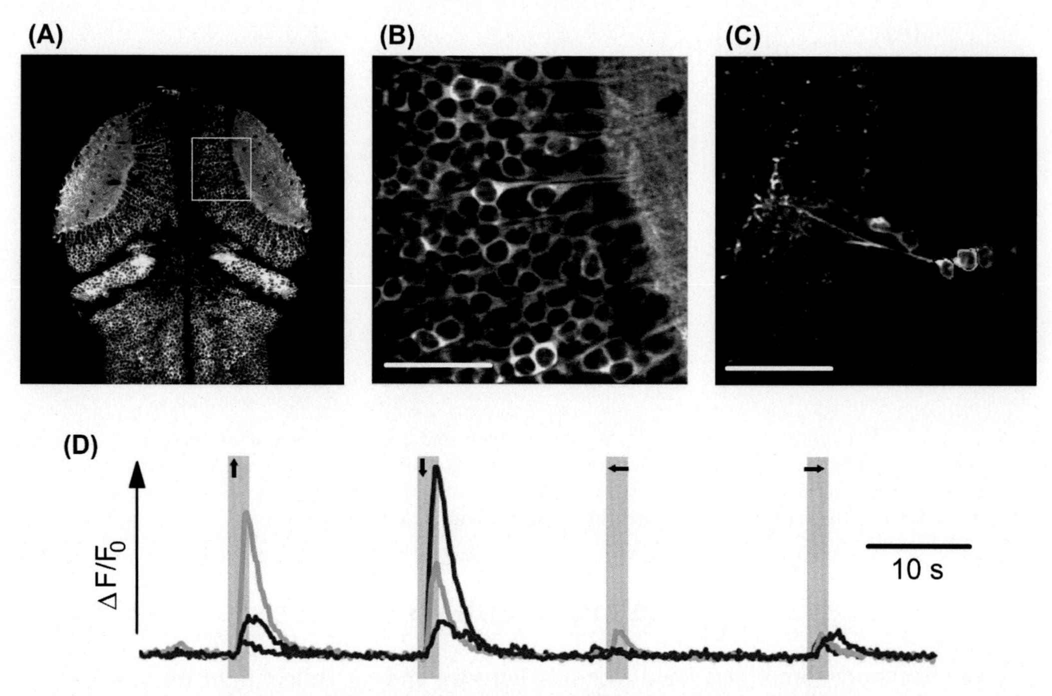

FIGURE 10.10 Two-photon imaging in the zebrafish brain. (A) A two-photon optical section through the brain of a fish expressing a pan-neuronal fluorescent marker. (B) Detail image showing subcellular resolution of neurons in the tectum. (C) Imaging of visually evoked, direction-selective calcium signals in tectal neuron somata. Tectal neurons transiently expressed the genetically encoded indicator GCaMP5G. Scale bars in (B) and (C): 25 μm. (D) Fluorescence fluctuation traces for the three neurons during presentation of dark bars that sweep across a screen below the fish more than 2 s in four different directions. First bar is in tail to head direction (note the upward arrow in the gray bar); average of 2 repeats. *Adapted from Renninger SL, Orger MB. Two-photon imaging of neural population activity in zebrafish. Methods 2013;62(3):255–67, permission Elsevier.*

Combining the two equations gives:

$$\Delta f \Delta t \geq \frac{1}{4\pi} \tag{10.6}$$

Hence, the spread of the probability density function of a photon in time is inversely proportional to the spread in its frequency spectrum. This is something easy to understand if one considers that Δf is related to Δt by its Fourier transform. Richard Feynman was the first to depict a photon as a pure sine wave modulated by a bell-shaped envelope (a Gaussian function that stands for the quantal probability density function, Fig. 10.11). As we have seen in Fig. 7.6H, the Fourier transform of a Gaussian-modulated sine wave is a Gaussian function. Suppose for the sake of the example that the color of the photon is in the visible range (center frequency 500 nm, corresponding to the color cyan in the example of Fig. 10.11). When the width of the Gaussian envelope of the photon comes close to zero (i.e., when the photon duration is a few femtoseconds or less), the spectrum widens so much that it spans a large part of the visible spectrum.

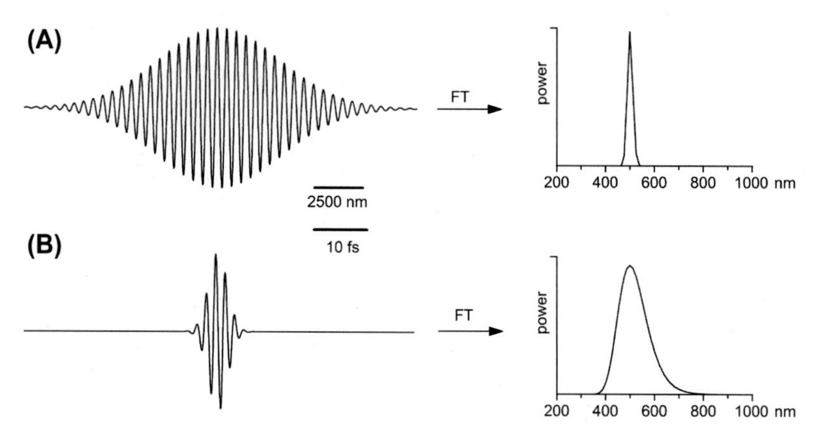

FIGURE 10.11 Two photons: one spread out wide in time and space and the other much more precisely localized. The wide photon (A) has a sharp peak in the frequency domain, therefore its energy is relatively precisely known, while the short photon (B) has a very wide spectrum, meaning that its energy is not well determined. Note that the frequencies on the axes in the left panels have been converted to wavelength.

OPTICAL FIBERS

Optical fibers are sometimes used to guide light to, and/or collect light from, structures deeper in tissue and also form an integral part of so-called optrodes, discussed later in this chapter.

Optical fibers consist of a core of glass or plastic with a high refractive index clad with a material of lower refractive index. When light enters the core at a grazing angle with the normal to the entry surface, it is reflected by the core-cladding interface and remains bouncing within the fiber until it leaves the fiber at its other extremity.

Total reflection at the interface occurs for $\sin(\theta) > n_1/n_2$, where n_1 is the refractive index of the cladding and n_2, the refractive index of the core. The refractive index of the external medium (air or water) is n_0. Suppose that light enters the fiber making an angle α with the normal at the surface of the fiber (Fig. 10.12).

According to Snell's law (Eq. 10.1): $n_0 \sin(\alpha) = n_2 \sin(\varphi)$, hence $\sin(\varphi) = n_0/n_2 \sin(\alpha)$.

It is easy to see that $\cos(\varphi) = \sin(\theta)$. With $\sin(\varphi)^2 + \cos(\varphi)^2 = 1$, this yields:

$$\sin(\propto) < \frac{\sqrt{(n_2^2 - n_1^2)}}{n_0}$$

The angle, α, is called the acceptance angle and $\sin(\alpha)$, the numerical aperture, NA, of the fiber.

After the light has traveled through the fiber, it leaves it making the same angle as the acceptance angle, causing the light intensity to diminish as a function of distance, d, from the exit point.

If r_0 is the radius of the fiber core, r_d is the radius of the cone of light at a distance d and x as drawn in the figure, then $x/r_0 = \text{cotg}(\alpha_0)$ and $r_d/r_0 = (x + d)/x$. The light intensity

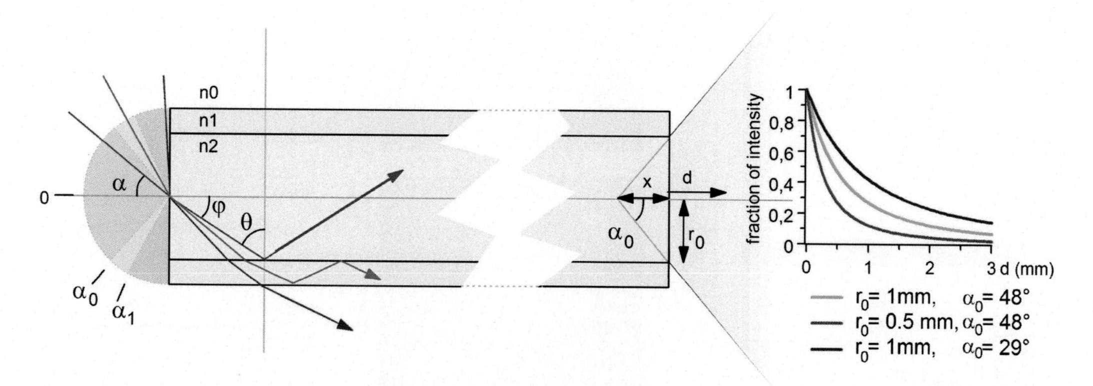

FIGURE 10.12 Light entering an optical fiber making an angle with the normal to the front surface of the fiber of α_0 is reflected internally against the interface between core (blue) and cladding (green). Light entering between α_0 and α_1 remains captured by the cladding layer. Light entering at an angle larger than α_1 traverses the fiber wall. The figure is based on the refractive index of water, $n_0 = 1.33$, and common values for $n_1 = 1.5$, $n_2 = 1.8$. In that case $\alpha_0 = 48$ degrees and $\alpha_1 = 66$ degrees. The graph on the right shows the decrease in light flux as a function of distance from the exit point of the fiber (d) for three conditions.

diminishes proportional to the square of the radius. Combining the equations then gives the decrease of light intensity as a function of distance d:

$$\left(\frac{r_0}{r_d}\right)^2 = \left(\frac{\text{cotg}(\propto_0)}{\frac{d}{r_0} + \text{cotg}(\propto_0)}\right)^2$$

Light penetrates farther if the angle of the exit cone is reduced, e.g., by increasing the refractive index of the cladding, while reducing fiber diameter has the opposite effect (Fig. 10.12, graph on the right).

OPTICAL SINGLE-CHANNEL RECORDING

Calcium currents flowing through plasma membrane ion channels can be detected with fluorescent dyes. These dyes are usually structurally based on the Ca^{2+} chelator BAPTA. First, cells are grown on dishes with thin glass bottoms, the thickness of a coverslip, such that a high-magnification oil-immersion microscope objective can focus onto the cells. These are then voltage-clamped in whole-cell mode using a pipette containing the fluorescent dye. The influx of Ca^{2+} through Ca^{2+}-permeable ion channels can subsequently be estimated by monitoring fluorescence. The resolution of a normal microscope allows for the recording of calcium variations corresponding to whole-cell currents when in patch-clamp but is not enough to resolve single-channel events. To augment resolution, a scanning confocal microscope can be used. Variations in intracellular Ca^{2+} may be observed using the calcium-sensitive dye Fluo-4. An example is shown in Fig. 10.13. The dye has an absorption maximum

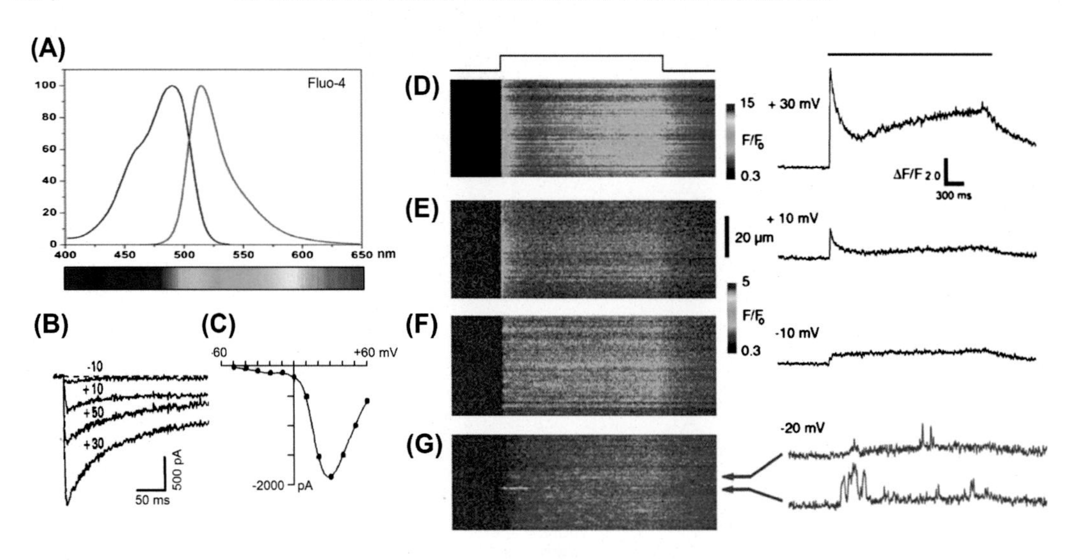

FIGURE 10.13 (A) Absorption and emission spectra of Fluo-4. (B) N-type Ca^{2+} currents elicited by voltage steps from −60 mV to the voltages indicated. (C) The size of the peak current as a function of step voltage (I/V curve). (D–G) The membrane potential of a Xenopus oocyte was stepped from −60 mV to the voltages indicated and the fluorescence more than a 50 μm scan line (shown vertically) was recorded (time is running horizontally). The ratio of fluorescence during the step over the fluorescence at −60 mV is shown in false colors. Note that the scale in (D) is different from the others. Mean fluorescence averaged over the scan line is shown to the right. *(B, C) Adapted from Fujita Y, Mynlieff M, Dirksen RT, et al. Primary structure and functional expression of the omega-conotoxin-sensitive N-type calcium channel from rabbit brain. Neuron 1993;10(4):585–98, permission Elsevier; (D–G) Adapted from Demuro A, Parker I. Optical single-channel recording: imaging Ca2+ flux through individual N-type voltage-gated channels expressed in Xenopus oocytes. Cell Calcium 2003;34(6):499–509, permission Elsevier.*

at 490 nm and an emission maximum close to 510 nm. It was injected into oocytes heterologously expressing the subunits forming the N-type Ca^{2+} channel. The voltage and time dependence of N-type Ca^{2+} currents is shown in Fig. 10.13B and C. These currents are maximally activated at +30 mV and inactivate slowly.

The increase in fluorescence during depolarizing steps due to the entry of calcium into the oocyte is shown in the right panel of Fig. 10.13D–G. At the beginning of the step, fluorescence rises sharply and comes down again, just as the N-type currents do in Fig. 10.13B. However, fluorescence is seen to rise again at the end of a depolarizing pulse to +30 mV (Fig. 10.13D). This is most likely due to calcium accumulation below the plasma membrane as a result of sustained Ca^{2+} influx. Most interestingly, events corresponding to Ca^{2+} entry through single channels can be observed at step potentials where the probability of channel opening is low enough to separate events, e.g., −20 mV (G).[3] The excellent spatial resolution of about 0.5 μm is determined by confocal microscopy. The temporal resolution is of course determined by the scan frequency (~100 Hz in the example of Fig. 10.13); the more points sampled, the less time resolution will be. It is clear that the method is not suited for the accurate measurement of rapid channel kinetics. Its strength lies in its noninvasive nature. Ion channel responses depending on intracellular signaling pathways can be recorded (almost) without perturbation of these pathways.

One might expect that temporal resolution would be much improved, if TIRF microscopy instead of scanning confocal microscopy were used to collect the images. This appears the case indeed, since calcium-induced fluctuations in Fluo-4 fluorescence due to influx through acetylcholine-operated ion channels can be recorded with a few milliseconds resolution using TIRF.[4]

OPSINS, LIGHT-SENSITIVE ACTUATORS

Unicellular flagellates direct their movement toward locations where their photosynthesis is optimal. They do this through the intermediate of light-sensitive ion channels (channel-rhodopsins). These channels are members of a large group of related proteins that are light-sensitive, the opsins. Opsins give our retinas their sensitivity to light. Rhodopsin is present in human rods, while L-, M-, and S-opsins are found in the three types of cone. In contrast to those in the flagellates, human opsins are G-protein-coupled receptors rather than channels. Common features of these receptors and channels are that they consist of seven transmembrane domains and incorporate a molecule of retinal, which induces a conformational change of the protein upon absorption of a photon. Opsins have found their way to a discipline coined optogenetics with interesting applications in electrophysiology. Two channel-rhodopsins, one a proton channel, ChR1, and the other a cation channel, ChR2, have been cloned from the flagellate alga, *Chlamydomonas reinhardtii*. They can be expressed in vertebrate cells where they insert correctly in the plasma membrane. If expressed in neurons, a flash of blue light (450–500 nm) may provoke depolarization and firing (Fig. 10.14). Hence, cells can be stimulated using light pulses instead of electric pulses delivered by an electrode. This is particularly handy, when the cells in question are contractile like cardiac muscle cells.

Proteins with much structural homology to the channel-rhodopsins are synthesized by halobacteria. The light-sensitive element of these proteins is also retinal. These halorhodopsin proteins transport chloride into the cell. One of those is the halorhodopsin, NpHR, from *Natronomonas*, which is activated by yellow light. Other bacteria, such as the proteobacteria, use a light-sensitive pump to extrude protons. In both cases, expression of ion transporters in mammalian cells can be used to cause light-induced hyperpolarization and hence, inhibition of spike activity. Since the first utilization of light-driven excitation of neurons by ChR2 in 2005, wild-type opsins have been engineered to counteract certain inconveniences and to meet specific needs. To give just two examples: Addition of an endoplasmic reticulum (ER) export sequence in NpHR prevents its accumulation in the ER. A point mutation in ChR2 augments the speed of channel closure such that rapid stimulation of more than 50 Hz becomes possible. Many other amendments have been made since.

What are the benefits of opsins and what are their drawbacks?

Electrical stimulation with metal or glass electrodes is known to cause stimulation artifacts that may be large and long enough to mask (rapid) responses. Photonic stimulation alleviates this problem. However, light may induce a current in metal recording electrodes, a

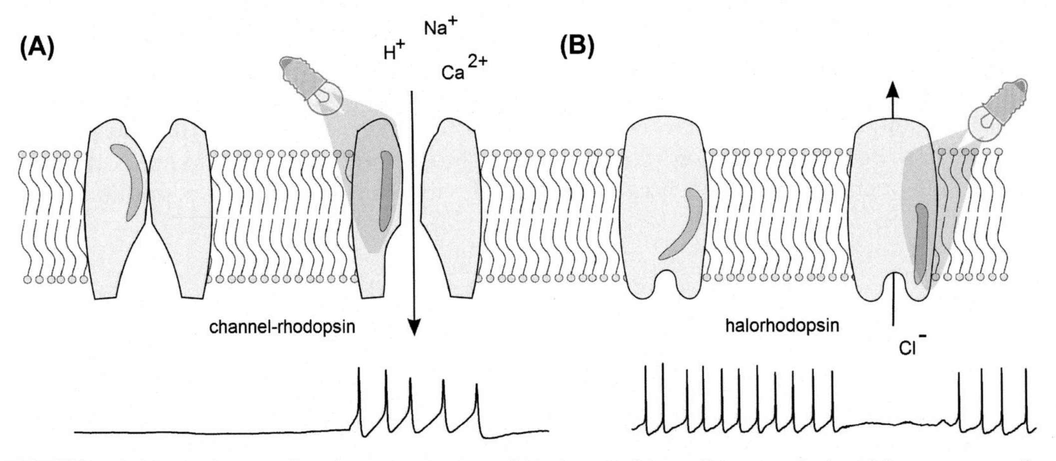

FIGURE 10.14 Light provokes the isomerization of 11-cis retinal into all-trans retinal which causes a conformational change of the rhodopsin protein. (A) Excitable cells expressing channel-rhodopsin can be stimulated by brief light pulses to depolarize and fire action potentials. (B) Inhibition of ongoing electrical activity is achieved by illuminating halorhodopsin-expressing cells.

phenomenon known as the photoelectric effect. This possible source of artifact is absent when using carbon or glass recording electrodes.

Gene transfer can be carried out with viral vectors. Injection of viruses in tissue such as the brain allows for the transfer of conditionally expressed genes or genes under the control of chosen promoters/enhancers such that specific populations of neurons may be (in-)activated by light pulses. For those who do not want to waste time with biomolecular techniques, many transgenic mouse strains are commercially available from Jackson Laboratories and other companies. Using transgenic animals has the additional benefit of a more homogenous expression of the transgene among cells. Infection or transfection methods show much more variability between cells, some of them largely overexpressing the gene, while others express next to nothing. If expression of the transgene is under the control of a specific promotor, then it is possible to stimulate a known cell type without stimulating its neighbor, even if the two cell types are close together. Excitation or inhibition of specific cell types also allows to estimate the role of these cells in large neuronal circuits and in the behavior of freely moving animals.

VOLTAGE-SENSITIVE FLUORESCENT PROBES

In experimental situations where electrode-based techniques might not offer enough information on the distribution of electrical responses, all-optical methods could be envisioned by combining expression of opsins with voltage-sensitive dyes. For example, the blue-sensitive ChR2 could be used to stimulate cells in conjunction with the yellow-sensitive NpHR to inhibit cells where appropriate and the infrared voltage-sensitive dye, rh-155, to record membrane potential.

Voltage-Sensitive Amphiphilic Dyes

Unfortunately, a major problem with rh-155 and similar fast-response fluorescent dyes, such as rh-1691, rh-1838, and di-4-ANEPPS, is their small change in fluorescence as a function of membrane potential. It ranges between 1% and 10% per 100 mV. The signal-to-noise ratio (S/N), somewhere between 1 and 5, leaves to be desired as well. Internalization of the dye and its occasional accumulation in intracellular organelles is a cause of low S/N. Even so, given the current interest in these optical methods, performance of new fast voltage-sensitive dyes is likely to improve in the future.

A usually minor problem is photo-bleaching of the dyes: The dyes are destroyed by the excitation light and then stop emitting light. Hence, the optical signal tends to diminish during an experiment. The problem can be partly solved if the rate of bleaching is known. One can, for example, multiply the fluorescence signal by a factor that depends on the number of images or scans taken. Another possibility is to reduce excitation intensity. The price to pay is reduced S/N ratio. The leak of dye from the plasma membrane also contributes to a loss of signal and reduced S/N during an experiment.

Voltage-Sensitive Protein Probes

An alternative to voltage-sensitive fluorescent dyes are voltage-sensitive fluorescent proteins having the convenience that they may be conditionally expressed in target tissue. Green fluorescent protein (GFP) from the jellyfish, *Aequorea victoria*, when inserted into a voltage-sensitive channel protein, can yield a voltage-sensitive fluorescent probe. The first fluorescent fusion protein was developed at the end of the last century and consisted of GFP fused with the Shaker potassium channel. The construct had been made nonpermeant for ions by a point mutation in the channel pore.[5] The Shaker channel does not have very fast on and off kinetics, and the fluorescent responses of the construct are correspondingly slow: actually too slow to resolve spike activity. Since then, many voltage-dependent channels with different kinetics have been fused with variants of GFP that may also have different colors. In order to create a rapid voltage sensor, the voltage-dependent Na^+ channel is a good candidate because it has rapid kinetics of activation and deactivation. Indeed, insertion of GFP between the second and third repeats of six transmembrane domains of this Na^+ channel makes for a rapid fluorescent sensor.[6] Remarkably, the voltage-dependencies of fluorescence of channel-GFP constructs correspond to those of the channel's gating charge displacements. In contrast to transmembrane currents, these charge displacements show kinetics which do not inactivate. This means for the Shaker construct that steady fluorescence changes can be observed between −60 and −10 mV. The range is between −80 and −20 mV for the Na^+ construct. Hence, the voltage range and kinetics are determined by the channel's gating properties. Constructs based on the voltage-sensing domain of voltage-sensitive phosphatase may have a broader voltage range, i.e., between −70 and +50 mV.

The percent of change in fluorescence and S/N ratio of the protein probes are usually small and similar to the voltage-sensitive dyes. A considerable advantage over the dyes is that the protein probes can be expressed conditionally so as to target specific populations of cells. Transgenic mice expressing fluorescent voltage sensors are commercially available.

If voltage-sensitive probes are to be used in live animals, artifacts due to heartbeat and respiration may contaminate the recordings. Light scattering will blur the image if emission sources are located at some depth in the tissue (e.g., as in brain slices or whole brain). Infrared biphoton microscopy is a good way to maintain good spatial resolution in those cases.

Illumination Warms Up Tissue

Illumination of light-sensitive actuators and/or sensors is a major challenge if they are localized in deeper structures. Actuators require a minimum of incoming light energy to effectively modulate membrane potential. This corresponds to roughly 1–5 mW/mm^2 for opsin channels and to 5–20 mW/mm^2 for the inhibiting transporters. Absorption of light by tissue depends on wavelength; blue light is approximately absorbed five times more by brain tissue than near-infrared. The absorption coefficient of green light (560 nm) of brain tissue is approximately 0.25/mm.[7] A cubic millimeter of brain tissue receiving a 10 ms light pulse of 5 mW absorbs $10e^{-3} \times 5e^{-3} \times 0.25 = 12.5$ μJ or 3 μcal. Because brain tissue is essentially water, the specific heat capacity is more or less the same as for water: 1 cal/g. Hence, the tissue within a 1 mm^3 volume may heat up $3 \times 1e^{-6}/1e^{-3} = 0.003$°C during a single pulse. It will be more of course if repetitive stimulation is carried out. Stimulation with 10 ms, 9 mW blue LED light pulses at 20 Hz causes a 0.1°C rise in temperature.[8] Taken together, these figures indicate that light-driven stimulation does not really overheat the tissue.

The situation becomes entirely different when light is used to sample fluorescence of voltage sensors continuously or with high frequency. Illumination of cardiac tissue with 5 mW/mm^2 with green (532 nm) light for 1 min raises the tissue temperature with more than 4°C.[9] With 660 nm, this is 1°C. Since S/N ratio of emission of voltage-sensitive dyes increases with increasing excitation energy, excitation energies above 1 mW/mm^2 are generally required to obtain exploitable signals. Often recording periods are short (0.1 s to a min) and interspaced with dark periods to recover from heating and prevent photodamage. As animal tissue absorbs infrared (IR) radiation much less than visible light, fluorescent probes can be excited by IR using two-photon microscopy. Radiation fluxes in the order of 10 MW/mm^2 to 1 GW/mm^2 are needed for the simultaneous absorption of two IR photons to occur at an appreciable rate. To keep the average light flux below a few mW/mm^2, femtosecond pulsed lasers are the most frequently used. This shortening of the light pulse, as we have seen, spreads the wavelength spectrum of the radiation, thereby possibly causing overlap with the excitation spectra of other fluorophores used in an experiment.

OPTRODES

Fluorescence imaging of cells with two-photon scanning microscopy can be carried out as deep in the tissue as 1 mm if using infrared radiation (~900 nm), which traverses tissue relatively well. Light in the visible spectrum, in particular blue light, does not penetrate tissue very deep (extinction coefficient for brain tissue is 0.37/mm at 480 nm[7]). Hence, other methods have been sought to shine light on deeper structures. Several more or less equivalent solutions have been experimented with and consist of delivering light by glass electrodes or optical fibers. The light sources are often lasers that output at a single wavelength. These lines can then be combined for multimodal excitation and recording. The benefit of single wavelength excitation as opposed to broadband excitation using gas discharge lamps is that multiline dichroic filters can be implemented to reject or combine multiple light sources with little loss of collected fluorescence (Fig. 10.15A and B). The dichroic filter #2 in Fig. 10.15A reflects only at specific wavelengths and transmits most of the visible spectrum (Fig. 10.13B). The combination of a light-conducting element and a classical electrode forms an optrode. The example in Fig. 10.15C shows a setup in which opsins are stimulated by blue light and a red fluorescent sensor by green light. The fluorescence signal picked-up by the optrode is small and needs amplification by a photomultiplier tube. Of course, many other configurations can be envisioned such as three-wavelength fluoresce probes to name one.

Classical silicon-based light-emitting diodes (LEDs) could eventually be used as less expensive light sources for optrodes. Their emission bands are about 25 nm wide and possibly need trimming somewhat if used in multifluorescence applications. Organic LEDs

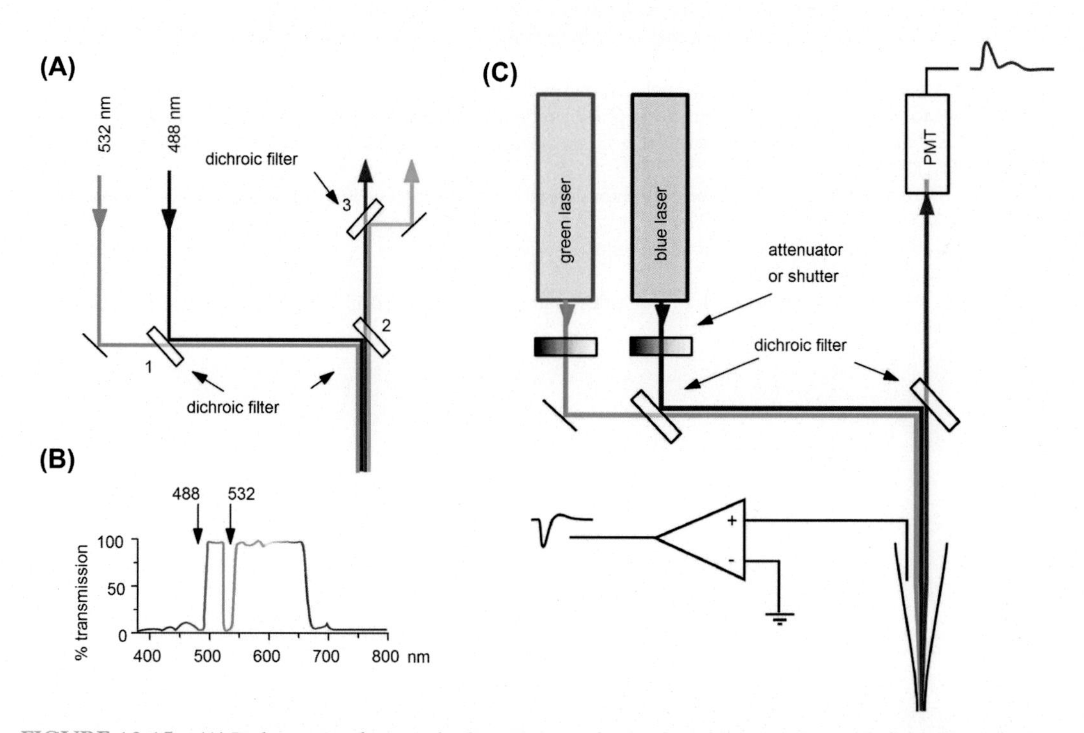

FIGURE 10.15 (A) Light coming from multiple sources can be combined using dichroic filters (1). Filter 2 reflects light at narrow bands and transmits the rest of the visible spectrum. Dichroic filter splits the red and green components of the emitted fluorescence. (B) The transmission characteristics of filter 2. (C) Optrode setup to record field potentials and emission of a fluorescent probe in response to a light pulse activating opsin actuators. Laser emission lines in the blue and green are combined into one beam and sent down to the tip of a glass or optic fiber electrode (optrode). Red fluorescence is collected by the same optrode and amplified by a photomultiplier tube (PMT) resulting in an electric signal. The optrode contains an electric conductor (wire and/or salt solution) to record extracellular potentials.

(OLEDs) have been proposed as light sources because they can be deposited in very small dimensions. Unfortunately, their emission spectra are very wide, and different colored OLEDs show therefore much overlap in their spectra such that they are of little use in multimodal optrodes. Even so, OLEDs can be used when only one light source is needed, e.g., as in optical stimulation of cells in conjunction with electrical recording.

In conclusion, optical recording methods are somewhat difficult to implement and suffer from a certain number of artifacts and drawbacks but can be very useful if entire neuronal circuits are being studied or if the roles of subpopulations of cells are to be determined. They may furnish results akin to those obtained with EEG or electrode arrays but with much higher spatial resolution.

References

1. Renninger SL, Orger MB. Two-photon imaging of neural population activity in zebrafish. *Methods* 2013;**62**(3):255–67.
2. Fujita Y, Mynlieff M, Dirksen RT, et al. Primary structure and functional expression of the omega-conotoxin-sensitive N-type calcium channel from rabbit brain. *Neuron* 1993;**10**(4):585–98.
3. Demuro A, Parker I. Optical single-channel recording: imaging Ca2+ flux through individual N-type voltage-gated channels expressed in Xenopus oocytes. *Cell Calcium* 2003;**34**(6):499–509.
4. Demuro A, Parker I. Imaging single-channel calcium microdomains. *Cell Calcium* 2006;**40**(5–6):413–22.
5. Siegel MS, Isacoff EY. A genetically encoded optical probe of membrane voltage. *Neuron* 1997;**19**(4):735–41.
6. Ataka K, Pieribone VA. A genetically targetable fluorescent probe of channel gating with rapid kinetics. *Biophys J* 2002;**82**(1 Pt 1):509–16.
7. Johansson JD. Spectroscopic method for determination of the absorption coefficient in brain tissue. *J Biomed Opt* 2010;**15**(5):057005.
8. Kim TI, McCall JG, Jung YH, et al. Injectable, cellular-scale optoelectronics with applications for wireless optogenetics. *Science* 2013;**340**(6129):211–6.
9. Kanaporis G, Martisiene I, Jurevicius J, et al. Optical mapping at increased illumination intensities. *J Biomed Opt* 2012;**17**(9). 96007–96001.

Appendix

SYMBOLS, ABBREVIATIONS, AND CODES

SYMBOLS

Quantity		Unit	
A	Area	m^2	
A	Ion activity	M	molar
C	Capacitance	F	farad
C	Ion concentration	M	molar
D	Distance, diameter	m	meter
E	Electromotive force	V	volt
F	Force	N	newton
F	Frequency	Hz	hertz
F	Rate	1/s or Ad	adrian
F	Ion activity coefficient	(Dimensionless)	
G	Gain	(Dimensionless, often in dB)	
g	Conductance	S	siemens
H	Magnetic inductance	T	tesla
I	Current	A	ampere
J	Current density	A/m^2	
L	Self-inductance	H	henry
l	Length	m	meter
P	Power	W	watt
Q	Charge	C	coulomb
R	Resistance	Ω	ohm
r	Radius or distance	m	meter
S	Ionic strength	M	molar
t	Time	s or sec	second

(*Continued*)

	Quantity	Unit	
U	Tension, voltage	V	volt
u	Ion mobility	$cm^2/(s \cdot V)$	
V	Voltage	V	volt
W	Work, energy	J	joule
X	Reactance	Ω	ohm
Z	Impedance	Ω	ohm
ε	Dielectric constant		
ρ	Resistivity	Ωm	ohm meter
μ	Magnetic permeability		
T	Time constant	s	
ω	Angular frequency	rad/s	

ABBREVIATIONS

AC	Alternating current
AD	Analog to digital
ADC	Analog-to-digital converter
AVO	Ampere, volt, ohm (meter)
BCD	Binary-coded decimal
BIOS	Basic input/output system
BW	Bandwidth
CMR(R)	Common-mode rejection (ratio)
CPU	Central processing unit
DA	Digital to analog
DAC	Digital-to-analog converter
DC	Direct current
DOS	Disk-operating system
DRAM	Dynamic random-access memory
DVM	Digital voltmeter
ECG	Electrocardiogram
EEG	Electroencephalogram
EMF	Electromotive force
ENG	Electroneurogram

ERG	Electroretinogram
FET	Field-effect transistor
GND	Ground
jFET	Junction field-effect transistor
LDR	Light-dependent resistor
LIX	Liquid ion exchanger
LJP	Liquid junction potential
MOSFET	Metal-oxide semiconductor field-effect transistor
OS	Operating system
PCB	Printed circuit board
RAM	Random-access memory
RFI	Radio frequency interference
ROC	Receiver operating characteristic
ROM	Read-only memory
RMS	Root mean square
SPST	Single-pole, single-throw (switch)
SPDT	Single-pole, double-throw
VCO	Voltage-controlled oscillator
XOR	Exclusive OR (function, circuit)

DECIMAL MULTIPLIERS

T	Tera	10^{12}
G	Giga	10^{9} [a]
M	Mega	10^{6} [a]
K	Kilo	1000 [a]
m	Milli	10^{-3}
μ	Micro	10^{-6}
n	Nano	10^{-9}
p	Pico	10^{-12}
f	Femto	10^{-15}
a	Atto	10^{-18}

[a] *Note that in computer jargon, "k" stands for 1024, "M" for 1024^2 (1,048,576), and "G" for 1024^3 (1,073,741,824).*

COLOR CODE FOR RESISTORS

Silver	10^{-2} or 10% tolerance	
Gold	10^{-1} or 5% tolerance	
Black	0	No zeros
Brown	1	0 or 1% tolerance
Red	2	00 or 2% tolerance
Orange	3	000
Yellow	4	0000
Green	5	10^5
Blue	6	10^6
Violet	7	10^7
Gray	8	
White	9	

Use:
For carbon resistors: 4 rings ddm t (2 digits, multiplier, and tolerance).
For metal film resistors: 5 rings dddm t.
Occasionally for capacitors: ddm signifying value in pF.

SYMBOLS FOR CIRCUIT DIAGRAMS

Fig. A.1.

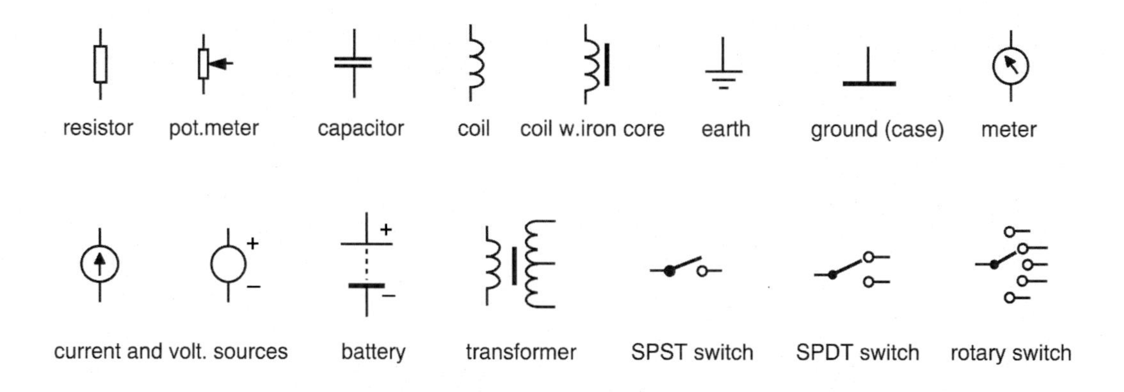

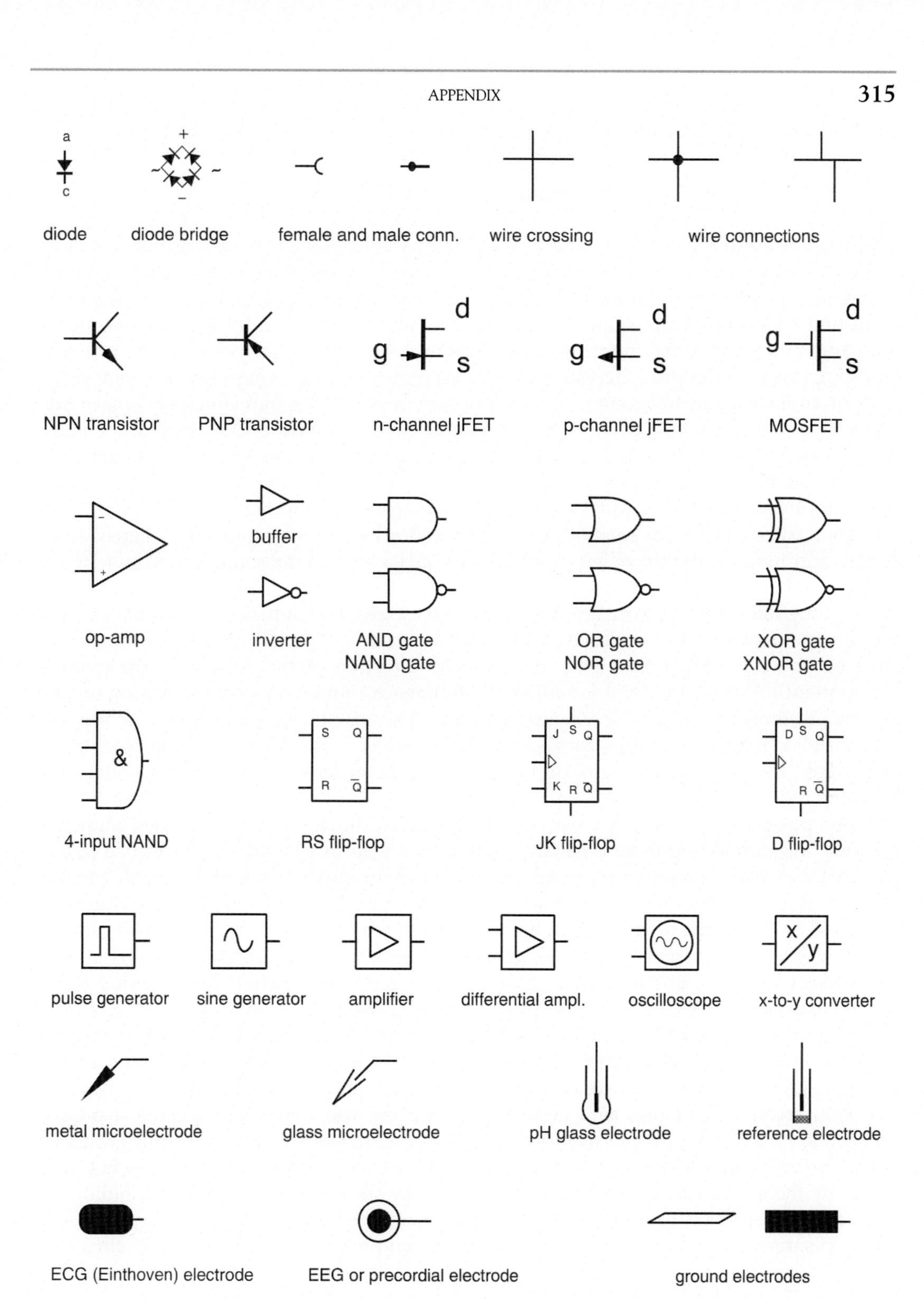

FIGURE A.1 Symbols for circuit diagrams.

ELECTRICAL SAFETY IN ELECTROPHYSIOLOGICAL SET-UPS

REGULAR INSTRUMENTS

Regular electronic instruments used for electrophysiological work are reasonably safe as to the hazard for electric shocks from the mains supply. In principle, any set-up is grounded, which means that all metal parts that can be touched are connected to earth. The earth or ground connection runs along the mains wires and is provided by the local electricity company, at least in most countries. The types of connectors differ per country: rim clip, third pin, round or flat pins, etc. All serve the same purpose, viz. connecting the joint metal parts of an instrument to ground. If a fault occurs in which the mains voltage comes into contact with such a part, the very low impedance of the ground cables limits the voltage that will build up to safe values. Usually, the corresponding mains fuse blows, signaling the fault condition.

Electrophysiologists face a dilemma though. The ground connection provided along with the mains wires suffices for safety but is often too burdened with interference signals to be used as a ground for the recording of weak electrical signals. Many sources contribute to the spurious voltages on the ground wires. In the first place, gas discharge lamps (fluorescent lighting, high-intensity microscope illuminators, etc.) produce inductive transients because they ignite and quench twice in each period of the mains frequency (twice because the voltage must be above a certain threshold, whereas the direction of the current in the lamp is unimportant). A second notorious source of interference signals is motors switching on and off, such as those in refrigerators, centrifuges, and thermostatic baths. Therefore, many labs have installed their own "clean" earth connection, usually in the form of an iron pipe driven deeply into the ground. Such grounding points can be ordered from companies installing lightning rods.

If one possesses such a clean extra earth connection, however, one is blessed with two grounds. Interconnecting them is certainly not a good idea! A ground loop is formed in that case that may encompass many square meters. This may cause substantial currents to flow, again deteriorating the properties of the grounding point as a reference for measurement circuits (Fig. A.2A). The dilemma then lies in choosing which ground to use. If one decides to use the clean measurement ground, all ground cables and connectors to the mains ground circuit must be cut. This violates the safety rules because any instrument connected to the mains outlet would lack its proper safety ground until the connection to the clean ground is made. Apart from the aspect of safety for the people that operate laboratory instruments, the safety of the instruments itself is at stake when one abandons the normal ground circuit. Failing to ground a digital instrument such as a computer may blow all its integrated circuits in a split second. Therefore, as a golden rule, "never" use computers without a ground connection. True, some people use their PCs at home, in rooms not fitted with grounded outlets, seemingly without nasty consequences. However, proper functioning is not warranted by the manufacturer, and damage to data may show up very irregularly, hiding the true cause. In addition to leakage currents from the power line, static electricity that may build up simply by walking across a nylon carpet can play havoc with electronic circuitry.

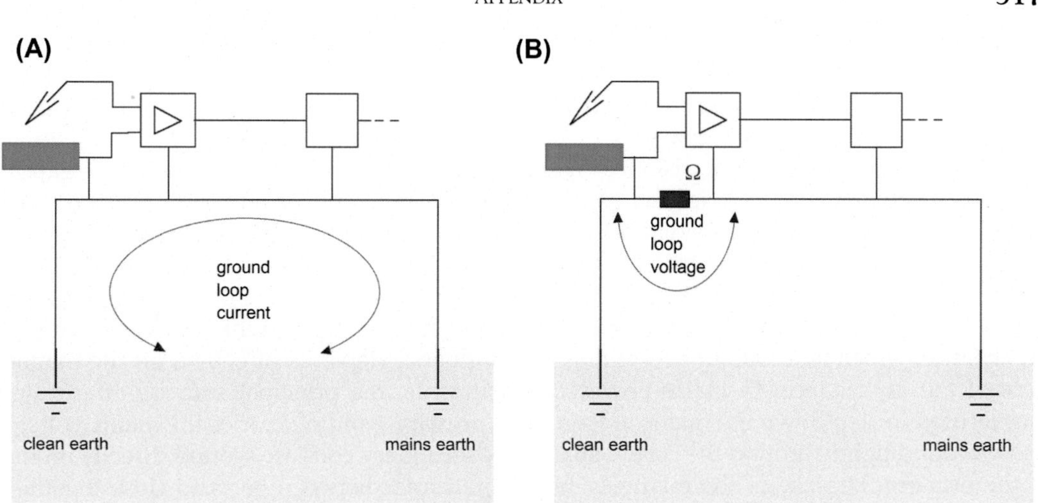

FIGURE A.2 Grounding methods: (A) conventional and (B) with ground loop resistor.

A compromise is possible, however, by interconnecting the two grounds through a small series resistance (Fig. A.2B). In this case, oscilloscopes and other instruments can be grounded safely to the mains ground, whereas a preamp or preparation grounding point is connected to the clean instrument ground. This may seem a contradiction, but remember that the spurious voltages across a ground circuit amount to no more than a few millivolts.

The reason these low voltages are harmful to sensitive measurements stems from the low impedance of the ground circuit. Interference signal of 5 mV across 10 mΩ of cable resistance causes a current of 0.5 A to flow through the entire ground circuit. If the ground loop resistance is increased to 10 Ω, which is a factor of 1000 higher, the current is reduced by the same amount. In this case, almost the entire interference voltage appears across this 10 Ω resistor. In the case depicted in Fig. A.2B, that voltage forms only a minor contribution to the amplified output voltage. The same principle of segregating ground circuits is employed in some hybrid (digital-and-analog) instruments, such as multimedia computers. Here, the ground circuit is split into a "digital ground" and an "analog ground" because the fast-switching digital circuits produce spurious transient currents to ground.

What about the inherent safety of the instruments involved? Laboratory instruments are (or should be) built according to international standards guarding the safety of mains operated gear. These standards are set up by international committees such as the IEC (International Electrotechnical Committee) and the ECMA (historically the European Computer Manufacturers Association) and published in a large series of numbered volumes. For lab instruments, the safety standards can be found, e.g., in ECMA 287 (safety of electronic equipment).

Like household apparatus, lab instruments employ double insulation in cables and in any internal connections that are exposed to wear or stress. Double insulation prevents one from touching a dangerously high voltage under a so-called single-fault condition. This means any situation in which the basic layer of insulation around a live wire is interrupted or damaged or any other single inadvertent contact with a mains voltage carrying conductor.

ECMA 287 specifies, among other things, the minimum distances required between conductors carrying high voltages. In general, any AC voltage of more than 30 V RMS (42.4 V peak) and any DC voltage more than 60 V is considered as a "hazardous live." Obviously, mains voltages belong to this category.

By specifying the minimum distances named "clearance," meaning the size of the gap in air, and "creeping," which is defined as the shortest distance across an insulating body, the risk of an electric shock is considered to be reduced to acceptable levels. However, even if these standards are met, an electric current may flow between circuits that are connected to the mains and the circuits that may be touched by humans. This current, called the leakage current, arises in principle in all parts that carry alternating current and have mutual capacitance, which amounts to everything. The influence depends of course on the magnitude of the capacitance. In mains-powered instruments, the principal source is the transformer used to step down the mains voltage. The primary winding carries the mains voltage and is wound tightly around the core. Usually, the secondary coils are wound directly on top of the primary. Because of the relatively high capacitance between parts so close together, both the core and the secondary windings carry a leakage current. Under circumstances of proper grounding, this current will flow to ground. However, in an ungrounded instrument, the leakage current may develop sizeable potentials on the metal parts that can be felt if touched.

Another important source of leakage current at the mains frequency stems from interference filters built into most electronic equipment. Some instruments, such as vacuum cleaners, refrigerators, and other frequently switching machines, generate short peak currents and/or voltages that may flow back into the power lines. In other instruments, especially digital ones, these peaks may cause a malfunction. To damp the peaks, most instruments are fitted with an interference filter, usually an LC low-pass filter, at the entry point of the mains/power cable. The most common type is shown schematically in Fig. A.3.

The values of the components differ a bit between brands but are in the order of magnitude given below:

C_1: about 10–100 nF,
C_2: about 2–5 nF,
L: about 0.5 mH.

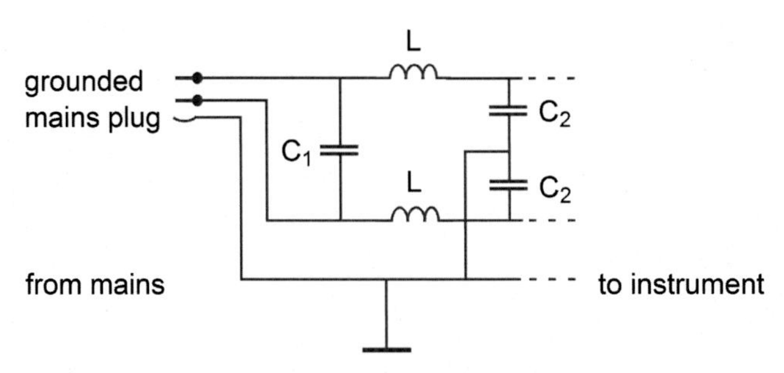

FIGURE A.3 Mains interference filter.

Note the capacitive voltage divider formed by the two C_2 capacitors. If the ground lead is omitted, the instrument case will float at half the mains voltage (i.e., 55 V AC in the United States, 115 V AC in Europe). If one touches the case of an instrument under such conditions, a current will flow through the circuit: mains—capacitor C_2—body—shoes—ground. The values of the capacitors involved (C_2) limit the current to a harmless value of about 0.36 mA (at 230 V, 50 Hz, C2 = 5 nF). However, if the (single!) insulation between the two plates of that capacitor would give way, one would be in direct contact with the full mains voltage and power.

MEDICAL INSTRUMENTS

For medical apparatus, i.e., instruments that are intended to be connected to a human subject, more stringent safety rules apply. Thus, one is allowed to connect intact animals or preparations to a regular set-up, but no human subjects. For use on human subjects, more strict rules apply as to shock hazard, leakage currents from the mains voltage, etc. These rules are laid down in IEC 60601 "Medical Electrical Equipment" and its later amendments. In short, IEC 60601 defines the apparatus and parts used and states rules as to the specifications to which medical electronic instruments must be built as well as the methods to test them. Examples of these standards are given below.

To reduce the risk of an electric shock, medical electronic instruments are built with a higher degree of electrical insulation between such components as the primary and secondary windings of transformers. Additional circuits serve to insulate the patient circuit from the rest of the mains-operated set-up or instrument. These additional safety measures explain at least part of the higher price one must pay for medical grade apparatus. Alas, painting an oscilloscope white does not turn it into a medical grade one!

The most reliable circuits to insulate an input circuit from mains-operated instruments are optical and radio frequency (RF) couplers or isolators. These consist of a transmitting part at the patient side and a receiving part at the mains-operated side. By this principle, an optocoupler consists of a light-emitting diode, the intensity of which is modulated by the signal (patient ECG), a transparent plastic layer that can withstand high voltages, and a photodiode or similar light detector to pick up the emitted light signal (see Fig. A.4A). An RF isolator (Fig. A.4B) employs a tiny transmitter and a receiver. The transmitting and receiving coils are separated by a plastic sheet so close together such that antennas are not necessary and no appreciable radio wave signal escapes to the environment. For these patient protection devices to function, the patient side of the device must be powered either by batteries or by a mains supply that has equally well-insulated components (see SELV below). Even within the category "medical," instruments on the market differ in the degree of protection against electric shock and excessive leakage currents.

Therefore, IEC standards segregate medical instruments in classes and types, the most important of which are described briefly in a less formal way than in the corresponding IEC standards books below.

Class I has, in addition to basic insulation, its metal parts connected to a protective earth conductor. In short, grounded apparatus. In our homes, washing machines, refrigerators, etc. belong to this category.

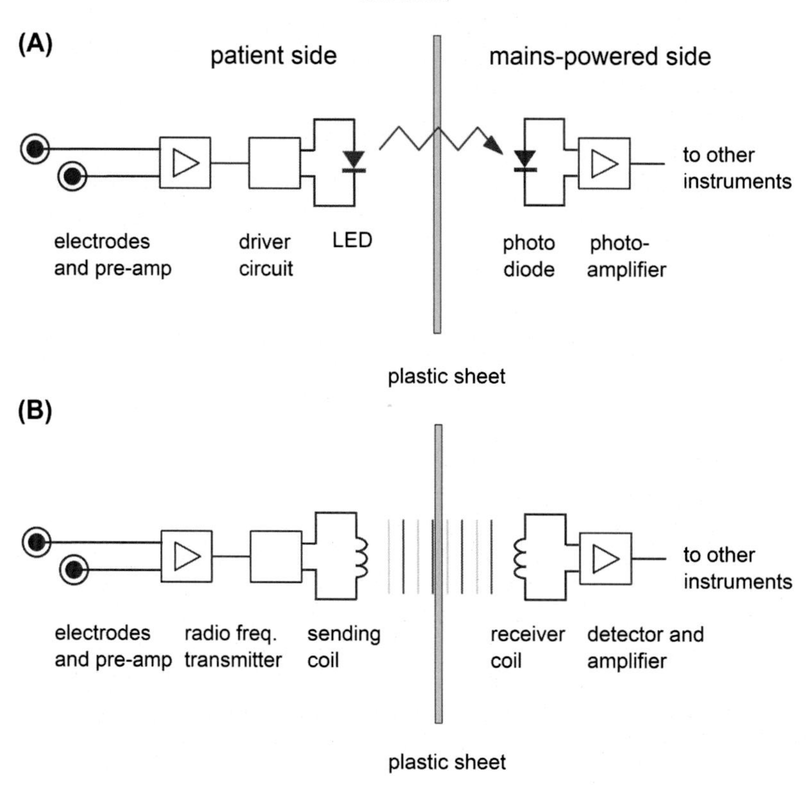

FIGURE A.4 Opto-isolator (A) and radio frequency (RF) isolator (B).

Class II has an extra layer of insulation. In other words, these are double-insulated instruments. Many household appliances fall into this category: shavers, hair dryers, portable audio, etc.
Type B Medical Equipment specified as to maximum leakage current and reliability of earth connection. As a guideline, the patient leakage current in this type of apparatus under normal circumstances will be lower than 100 μA.
Type BF is the same as B but having an attached part of type F (floating; see below). The floating part consists usually of the preamplifier in contact with the patient electrodes.
Type C meets higher demands as to leakage and safety. It is fitted usually with a floating part (type CF). As a guideline, the patient leakage current in this type of apparatus under normal circumstances will be lower than 10 μA.
Type F refers a part of the equipment that is floating, i.e., so well insulated from ground that no appreciable leakage current will flow even when that part should come into contact with the mains voltage. It is used as an addition to types B and C.

Instruments with the highest degree of protection are designated as "safety extra low voltage," or SELV equipment. This is defined as being lower than 25 V AC or 60 V DC. These instruments are powered by low voltages, usually derived from the mains voltage by a special, so-called SELV transformer. Such transformers meet very high insulation standards to avoid electrical shocks and excessive leakage currents in the patient circuit. Obviously, instruments used in heart surgery are built to the highest safety standards.

Instruments that do not meet the appropriate standard from the list above are not safe enough to be used for the respective purposes. It is the responsibility of the laboratory staff to verify whether their instruments meet the standards required to work safely.

COMPLEX NUMBERS AND COMPLEX FREQUENCY

Complex numbers are denoted in general by $z = a + jb$, where a and b are the real and imaginary part of the complex number, z. The imaginary unit vector, j, has the property: $j^2 = -1$. Thus, j is identical to the imaginary unit, notated i in mathematical texts. Since in electricity theory the letter i is used to represent electrical current, the letter j is chosen for the imaginary unit (the convention in technical texts).

Basic operations of two complex numbers are straightforward:

Addition:

$$(a + jb) + (c + jd) = (a + c) + j(b + d)$$

Multiplication:

$$(a + jb)\cdot(c + jd) = (ac - bd) + j(ad + bc) \ \left(\text{here we used } j^2 = -1\right)$$

Division:

$$\begin{aligned}(a + jb)/(c + jd) &= (a + jb)\cdot(c + jd)/((c + jd)\cdot(c + jd)) \\ &= ((ac + bd) + j(bc - ad))/\left(c^2 + d^2\right)\end{aligned}$$

The so-called complex conjugate is very useful to simplify complex formulae, e.g., to segregate real and imaginary parts. The conjugate of a complex number, z, has j replaced by −j and is denoted here by z*. Hence, if $z = a + jb$, then $z^* = a - jb$. Of course the conjugate of the conjugate of z is z itself: $(z^*)^* = z$. It can be easily verified that if x and y are complex numbers:

$$x + y = x + y \quad \text{and} \quad xy = yx$$

The real part of $x = a + jb$, Re(x), can be reformulated using the conjugate:

$$\text{Re}(x) = (x + x^*)/2, \text{since}((a + jb) + (a - jb))/2 = a \tag{A.1}$$

and similarly,

$$\text{Im}(x) = (x - x^*)/2j, \text{because } ((a + jb) - (a - jb))/2j = b \tag{A.2}$$

If $x = a + jb$, then $xx^* = (a + jb)\cdot(a - jb) = a^2 + b^2$. The value $a^2 + b^2$ is always positive or zero. Its positive square root is called the modulus or the absolute value of the complex number x and is denoted by |x|. Hence $|x|^2 = xx$.

Since complex numbers are actually pairs of numbers, they can be depicted as vectors in a two-dimensional space, where the vector projections on the two orthogonal axes represent the real and imaginary part of the complex numbers (Fig. A.5).

Suppose a is the angle between a vector ($\mathbf{v}$) of length r_1 and the real axis, then:

$$\mathbf{v} = x + jy = r_1 \cdot (\cos a + j \sin a),$$

where x and y are the coordinates of $\mathbf{v}$ in complex space.

Now, if we have another vector $\mathbf{w} = r_2 \cdot (\cos b + j \sin b)$, then

$$\mathbf{vw} = r_1 r_2 \cdot ((\cos a \cos b - \sin a \sin b) + j(\sin a \cos b + \sin b \cos a))$$

which simplifies to:

$$\mathbf{vw} = r_1 r_2 \cdot ((\cos(a+b) + j \sin(a+b))$$

Hence multiplication of two complex numbers involves the multiplication of their moduli and the summation of their angles.

It has been shown by Euler (using an argument based on Taylor series expansion, which we will not repeat here) that:

$$\cos a + j \sin a = e^{ja}$$

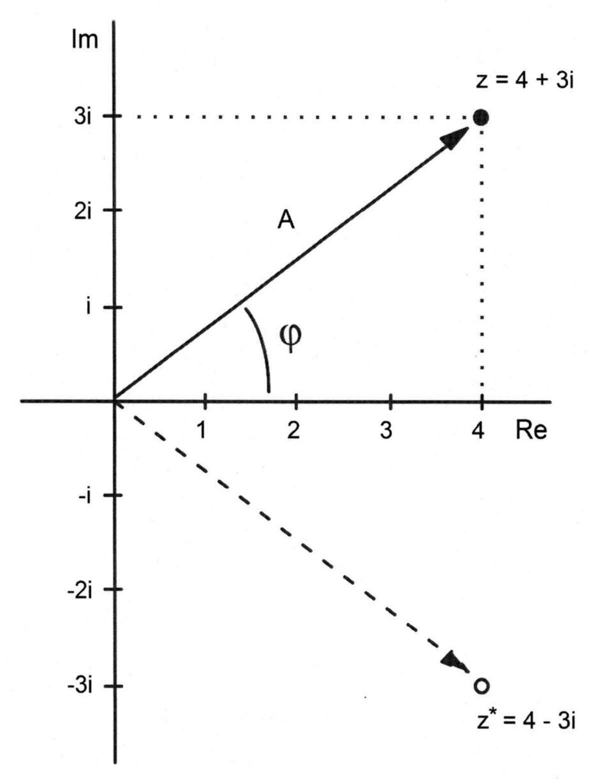

FIGURE A.5 A complex number and its conjugate as vectors on the real (Re) and imaginary (Im) plane.

when replacing a by −a it follows that:

$$\cos a - j\sin a = e^{-ja}$$

Combining this result with Eq. (A.1) and Eq. (A.2) gives the useful equations:

$$\cos a = \left(e^{ja} + e^{-ja}\right)/2 \quad \text{and} \quad \sin a = \left(e^{ja} - e^{-ja}\right)/2j$$

Rewriting with $v = r_1 \cdot e^{ja}$ and $w = r_2 \cdot e^{jb}$ gives: $vw = r_1 r_2 \cdot e^{j(a+b)}$

THE MEANING OF COMPLEX FREQUENCY

It was stated earlier that the Laplace transform can be used for any signal, being either periodical (sinusoid, square, and so on) or transient (pulse, exponential decay, etc.). This follows from the variable s, which is called complex frequency. This variable is complex and so is in fact a pair of variables:

$$s = \sigma + j\omega$$

The quantity behind the j, the imaginary part, ω, represents the conventional frequency ($\omega = 2\,\pi\,f$ rad/s). In addition, the real part of s, σ, stands for nonperiodic (transient) functions, such as growth or decay. This can be illustrated best by plotting in a real (σ) versus imaginary (jω) graph again. Fig. A.6 shows how all sorts of signals fit into this graph: sine waves of increasing frequency are lying on the y-axis. Exponential growth and decay are laid out on the x-axis, with faster decay farther left, faster growth farther to the right. The origin represents a direct current (frequency zero; no growth and no decay).

THE MATHEMATICS OF MARKOV CHAINS

The presentation of the problem in Chapter 9 is a special case of the Markov chain approach, as in addition, Markov chains may have both forward and backward rate constants and the linear chain as shown in Eq. (9.4) may have side chains. In that general case and supposing we have 3 states, S_1 through S_3,

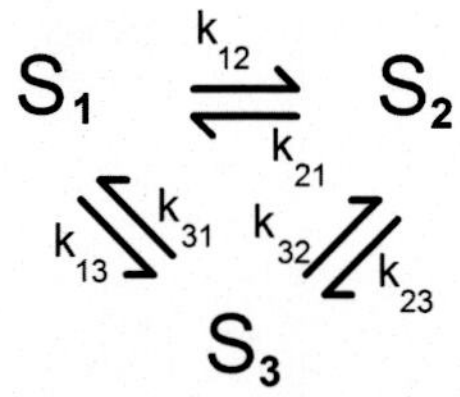

$S_1 \underset{k_{21}}{\overset{k_{12}}{\rightleftharpoons}} S_2$; $S_1 \rightleftharpoons S_3$ (k_{13}, k_{31}); $S_2 \rightleftharpoons S_3$ (k_{23}, k_{32})

then the system may be represented by a transition matrix, A:

$$A = \begin{bmatrix} . & k_{21} & k_{31} \\ k_{12} & . & k_{32} \\ k_{13} & k_{23} & . \end{bmatrix}$$

where the rate constants k_{ft} indicate the rate from state f to state t. Note that zero matrix elements are left out for clarity.

A Markov chain resumes in fact to a problem of solving a set of differential equations. With probabilities p_1, p_2, and p_3 to be in each of the three states S_1 through S_3, and the rate constants k_{ft} as shown above, the set of differential equations is:

$$\begin{aligned} dp_1/dt &= p_2k_{21} + p_3k_{31} - p_1(k_{12} + k_{13}) \\ dp_2/dt &= p_1k_{12} + p_3k_{32} - p_2(k_{21} + k_{23}) \\ dp_3/dt &= p_2k_{13} + p_2k_{23} - p_3(k_{31} + k_{32}) \end{aligned} \tag{A.3}$$

or in matrix form:

$$\frac{dp}{dt} = \begin{bmatrix} -k_{12} - k_{13} & k_{21} & k_{31} \\ k_{12} & -k_{21} - k_{23} & k_{32} \\ k_{13} & k_{23} & -k_{31} - k_{32} \end{bmatrix}$$

Hence it suffices to complement the Markov transition matrix at the diagonal entries with the negative of the sum over each column to obtain the corresponding set of differential equations. Equations like Eq. (A.3) are written in matrix shorthand as:

$$X' = AX \tag{A.4}$$

where X is a vector, X′ is the derivative of X, and A is a square matrix.

This equation is suspected to have solutions of the form:

$X = Ce^{\lambda t}$, i.e., an exponential function, with C, a constant vector.

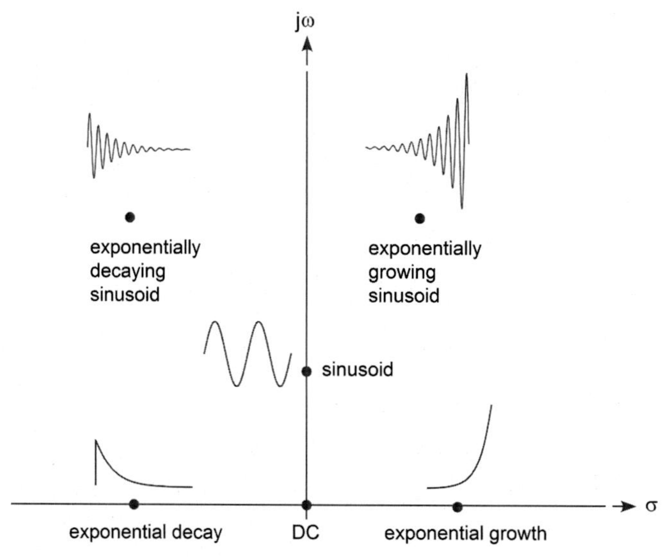

FIGURE A.6 Explanation of the meaning of the complex frequency Laplace variable s. The real part is rendered along the x-axis, and the imaginary part, along the y-axis. The fat dots are examples of the different signal categories. Except at DC, a stylized figure shows what the signal at each point would look like.

Substitution into Eq. (A.4) gives:

$$C\lambda e^{\lambda t} = ACe^{\lambda t}$$

or

$$(AC - \lambda C)e^{\lambda t} = 0 \qquad (A.5)$$

This can only be true if:

$$AC - \lambda C = 0$$

Because C = IC, where I is the identity matrix (a matrix with 1s on the main diagonal and 0s elsewhere), Eq. (A.5) becomes:

$$(A - \lambda I)C = 0$$

It has been proven that Eq. (A.5) only has solutions if the determinant of $(A - \lambda I)$ is 0,[1] hence:

$$|A - \lambda I| = 0 \qquad (A.6)$$

The determinant of a matrix is rather difficult to calculate by hand but becomes much easier if the matrix is triangular, since the determinant of a triangular matrix is simply the product of its diagonal elements. Therefore most computer programs try to reduce a matrix as in Eq. (A.3) to triangular form by so-called similarity transforms:

$$A_2 = \begin{bmatrix} a_{11} & a_{21} & a_{31} \\ . & a_{22} & a_{32} \\ . & . & a_{33} \end{bmatrix}$$

See Ref. Watkins[2] for a more in-depth treatment of the problem. A_2 indicates the transformed matrix A and a_{ij} the, a priori, nonzero elements of A_2. Then:

$$A_2 - \lambda I = \begin{bmatrix} a_{11} - \lambda & a_{21} & a_{31} \\ . & a_{22} - \lambda & a_{32} \\ . & . & a_{33} - \lambda \end{bmatrix}$$

and therefore:

$$|A_2 - \lambda I| = (a_{11} - \lambda)(a_{22} - \lambda)(a_{33} - \lambda) = 0$$

This is a polynomial of the third degree having three solutions:

$$\lambda = a_{11}, \lambda = a_{22} \quad \text{and} \quad \lambda = a_{33}$$

The three lambdas are called the eigenvalues of the square matrix A. The next thing to do is to find the eigenvectors, C, of Eq. (A.6). For $\lambda = a_{11}$ this will be:

$$(A_2 - \lambda I)C = \begin{bmatrix} 0 & a_{21} & a_{31} \\ 0 & a_{22} - a_{11} & a_2 \\ 0 & 0 & a_{33} - a_{11} \end{bmatrix} \cdot \begin{pmatrix} c_1 \\ c_2 \\ c_3 \end{pmatrix} = 0 \tag{A.7}$$

where the vector C has three elements: c_1, c_2, and c_3.

It follows that:

$$c_2 \cdot a_{21} + c_3 \cdot a_{31} = 0, \quad c_2 \cdot (a_{22} - a_{11}) + c_3 \cdot a_{32} = 0 \quad \text{and} \quad c_3 \cdot (a_{33} - a_{11}) = 0$$

This gives $c_2 = c_3 = 0$ and c_1 is free to choose. Hence the first eigenvector corresponding to the eigenvalue $\lambda = a_{11}$ is:

$$C_1 = \begin{pmatrix} 1 \\ 0 \\ 0 \end{pmatrix}$$

The other two eigenvectors are obtained similarly. Since the three vectors correspond to the transformed system A_2, the last thing that has to be done is to carry out the inverse similarity transform on the three vectors to obtain the final result which is of the form:

$$p = z_1 C_1 \exp(a_{11} t) + z_2 C_2 \exp(a_{22} t) + z_3 C_3 \exp(a_{33} t)$$

with p, C_1, C_2, and C_3 vectors. The parameters z_1 through z_3 are scalars that are determined by the boundary conditions.

Now that we know how to proceed in general, let us return to the initial problem of example 9.4. In the case of example 9.4, with $n = 4$ the transition matrix would have been:

$$\begin{bmatrix} -\mu_1 & 0 & 0 & 0 \\ \mu_1 & -\mu_2 & 0 & 0 \\ 0 & \mu_2 & -\mu_3 & 0 \\ 0 & 0 & \mu_3 & 0 \end{bmatrix}$$

Actually, we are not interested in the final observable fourth state, but only in the time spent in the pathway leading to it, so the system to solve reduces to:

$$A = \begin{bmatrix} -\mu_1 & 0 & 0 \\ \mu_1 & -\mu_2 & 0 \\ 0 & \mu_2 & -\mu_3 \end{bmatrix}$$

Because A is already triangular, no transformations are required to determine the eigenvalues, which obviously are $-\mu_1$, $-\mu_2$, and $-\mu_3$. Solving $(A - \lambda I)C$ as in Eq. (A.7) results in:

$$C_1 = \begin{pmatrix} (\mu_3 - \mu_1)(\mu_2 - \mu_1) \\ \mu_1(\mu_3 - \mu_1) \\ \mu_1\mu_2 \end{pmatrix} \quad C_2 = \begin{pmatrix} 0 \\ \mu_3 - \mu_2 \\ \mu_2 \end{pmatrix} \quad C_3 = \begin{pmatrix} 0 \\ 0 \\ 1 \end{pmatrix}$$

and

$$p = z_1C_1\exp(-\mu_1 t) + z_2C_2\exp(-\mu_2 t) + z_3C_3\exp(-\mu_3 t) \tag{A.8}$$

Initially, at $t = 0$, only the first state is occupied or:

$$p_{t=0} = \begin{pmatrix} 1 \\ 0 \\ 0 \end{pmatrix} = z_1C_1 + z_2C_2 + z_3C_3$$

hence

$$z_1(\mu_3 - \mu_1)(\mu_2 - \mu_1) = 1$$

$$z_1\mu_1(\mu_3 - \mu_1) + z_2(\mu_3 - \mu_2) = 0$$

$$z_1\mu_1\mu_2 + z_2\mu_2 + z_3 = 0$$

Solving the three equations gives:

$$z_1 = \frac{1}{(\mu_3 - \mu_1)(\mu_2 - \mu_1)}$$

$$z_2 = \frac{-\mu_1}{(\mu_3 - \mu_2)(\mu_2 - \mu_1)}$$

$$z_3 = \frac{\mu_1\mu_2}{(\mu_3 - \mu_2)(\mu_3 - \mu_1)}$$

Back substitution of this result into Eq. (A.8) gives the time-dependent occupation for each of the three states:

$$p_1(t) = \exp(-\mu_1 t)$$

$$p_2(t) = \frac{\mu_1}{\mu_2 - \mu_1}\cdot\{\exp(-\mu_1 t) - \exp(-\mu_2 t)\}$$

$$p_3(t) = \frac{\mu_1\mu_2\{(\mu_3 - \mu_2)\exp(-\mu_1 t) - (\mu_3 - \mu_1)\exp(-\mu_2 t) + (\mu_2 - \mu_1)\exp(-\mu_3 t)\}}{(\mu_2 - \mu_1)(\mu_3 - \mu_2)(\mu_3 - \mu_1)}$$

Before exiting to state 4, the system remains in one of the three lower states with probability $p(t) = p_1(t) + p_2(t) + p_3(t)$. The waiting time distribution, W, is a probability density function and therefore $dp(t)/dt$ needs to be calculated. After summation and differentiation:

$$W(3) = \mu_1\mu_2\mu_3\left\{\frac{\exp(-\mu_1 t)}{(\mu_2 - \mu_1)(\mu_3 - \mu_1)} - \frac{\exp(-\mu_2 t)}{(\mu_2 - \mu_1)(\mu_3 - \mu_2)} + \frac{\exp(-\mu_3 t)}{(\mu_3 - \mu_2)(\mu_3 - \mu_1)}\right\} \tag{A.9}$$

Fig. A.7 shows an example for $\mu_1 = 1$, $\mu_2 = 2$, $\mu_3 = 3$.

One might hope that Eq. (A.9) would reduce to Eq. (9.3) for $\mu_1 = \mu_2 = \mu_3 = \mu$ and $n = 3$. Clearly this is not the case, as the denominators of the quotients in Eq. (A.9) become 0. In fact, a problem already occurs when trying to solve $(A - \lambda I)C = 0$ to obtain three (independent) eigenvectors:

$$A = \begin{bmatrix} -\mu & 0 & 0 \\ \mu & -\mu & 0 \\ 0 & \mu & -\mu \end{bmatrix}$$

Clearly, A has three repeated eigenvalues, $-\mu$.

$$(A - \lambda I)C = \begin{bmatrix} 0 & 0 & 0 \\ \mu & 0 & 0 \\ 0 & \mu & 0 \end{bmatrix} \cdot \begin{pmatrix} c_1 \\ c_2 \\ c_3 \end{pmatrix} = 0$$

Apart from the identity $0 = 0$, this gives two equations:

$$\mu c_1 = 0$$

$$\mu c_2 = 0$$

and therefore $c_1 = c_2 = 0$ and c_3 may be chosen freely. Hence, one solution is:

$$\begin{pmatrix} 0 \\ 0 \\ 1 \end{pmatrix} \cdot e^{-\mu t} \tag{A.10}$$

and because that is the only one we will get by trying $e^{-\mu t}$ as a solution, other solutions have to be looked for. Suppose that the solution X has the form:

$X = C(t) \cdot e^{-\mu t}$ with C(t) a vector with time-dependent elements: $\begin{pmatrix} c_1(t) \\ c_2(t) \\ c_3(t) \end{pmatrix}$

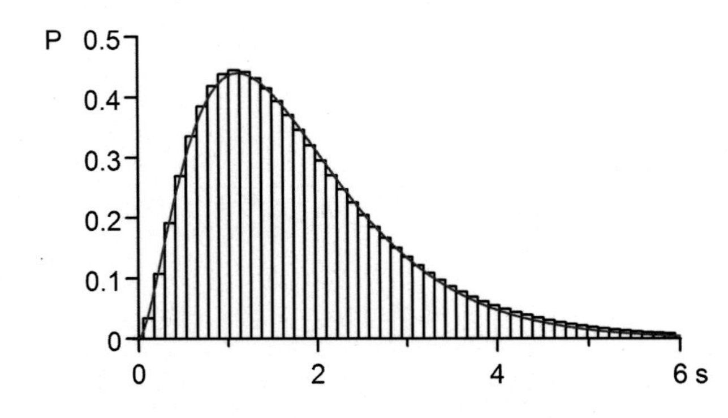

FIGURE A.7 Waiting time distribution for four states with transition rate constants $\mu_1 = 1$, $\mu_2 = 2$, $\mu_3 = 3/s$. The distribution was fitted with the Gamma distribution of Eq. (4.9), yielding $\mu = 1.535$ and $n = 2.724$. Despite the fact that the model, assuming $\mu_1 = \mu_2 = \mu_3$, is incorrect, the fit is rather good.

Substitution into Eq. (A.4) gives:

$$\begin{pmatrix} c_1(t) \\ c_2(t) \\ c_3(t) \end{pmatrix} \cdot -\mu e^{-\mu t} + \begin{pmatrix} c_1'(t) \\ c_2'(t) \\ c_3'(t) \end{pmatrix} \cdot e^{-\mu t} = \begin{bmatrix} -\mu & 0 & 0 \\ \mu & -\mu & 0 \\ 0 & \mu & -\mu \end{bmatrix} \cdot \begin{pmatrix} c_1(t) \\ c_2(t) \\ c_3(t) \end{pmatrix} \cdot e^{-\mu t}$$

After elimination of $e^{-\mu t}$ and rewriting, the result is: $\begin{pmatrix} c_1'(t) \\ c_2'(t) \\ c_3'(t) \end{pmatrix} = \begin{bmatrix} 0 & 0 & 0 \\ \mu & 0 & 0 \\ 0 & \mu & 0 \end{bmatrix} \cdot \begin{pmatrix} c_1(t) \\ c_2(t) \\ c_3(t) \end{pmatrix}$

and thus:

$$\begin{aligned} c_1'(t) &= 0 \\ c_2'(t) &= \mu c_1(t) \\ c_1'(t) &= \mu c_2(t) \end{aligned}$$

After integration of each of the three equations:
$c_1(t) = a$ and so $c_2'(t) = \mu a$
$c_2(t) = \mu at + b$ and so $c_3'(t) = \mu^2 at + \mu b$

$c_3(t) = 0.5\mu^2 at^2 + \mu bt + c$

where a, b, and c are arbitrary (integration) constants. Writing out this result shows that the solutions for X are of the form:

$$\begin{pmatrix} a \\ b \\ c \end{pmatrix} \cdot e^{-\mu t} + \begin{pmatrix} 0 \\ \mu a \\ \mu b \end{pmatrix} \cdot te^{-\mu t} + \begin{pmatrix} 0 \\ 0 \\ a\mu^2/2 \end{pmatrix} \cdot t^2 e^{-\mu t}$$

Choosing $a = b = 0$ and $c = 1$ yields the eigenvector found in Eq. (A.10). The two other solutions are found similarly, by setting $a = c = 0$, $b = 1$ and by setting $a = 1$, $b = c = 0$. The result is a linear combination of the three solutions:

$$p = z_1 \begin{pmatrix} 0 \\ 0 \\ 1 \end{pmatrix} \cdot e^{-\mu t} + z_2 \left\{ \begin{pmatrix} 0 \\ 0 \\ 1 \end{pmatrix} \cdot t + \begin{pmatrix} 0 \\ 1/\mu \\ 0 \end{pmatrix} \right\} \cdot e^{-\mu t}$$

$$+ z_3 \left\{ \begin{pmatrix} 0 \\ 0 \\ 1 \end{pmatrix} \cdot t^2 + \begin{pmatrix} 0 \\ 2/\mu \\ 0 \end{pmatrix} \cdot t + \begin{pmatrix} 2/\mu^2 \\ 0 \\ 0 \end{pmatrix} \right\} \cdot e^{-\mu t}$$

At $t = 0$ only the first state is occupied, which yields $z_1 = z_2 = 0$ and $z_3 = \mu^2/2$.

Before exiting to state 4, the system remains in one of the three lower states with probability $p(t) = p_1(t) + p_2(t) + p_3(t)$. The waiting time distribution, W(3), is obtained as previously by differentiating p(t):

$$W(3) = \mu^3 t^2 e^{-\mu t}/2$$

which is identical to Eqs. (9.2) and Chapter 10.13 for $n = 3$.

This little exercise shows first the essential principles for the manipulation of Markov chains and second that Eqs. (9.3) constitutes a special case of the Markov chain approach, indicating that the latter is the more general of the two. It may appear that the eigen decomposition as shown above is somewhat complicated and rather cumbersome to apply each time we wish to test or study a kinetic model. Actually, it is quite the contrary as the analysis that was carried out here by hand is always the same, no matter the model at hand. It can therefore be well performed by a computer routine, which makes the analysis almost as easy as a mouse click.

PSEUDOCODE TO CALCULATE THE MACROSCOPIC CURRENT AND DWELL-TIME DISTRIBUTIONS FROM A TRANSITION MATRIX

```
Routine FirstLatency (matrix A, array pnul, array S, int n, array evals, array amps)
{   # matrix A is the n*n transition matrix
    # array pnul of length n contains the probabilities to be in each state at t=0
    # array S indicates whether the state is closed (S[i]=0) or open (S[i]=1)
    # n is the number of states
    # The time constant of each exponential is returned in array evals
    # The amplitude of each exponential is returned in array amps
    # The routines Eigenvalues() and Invertmatrix() are standard routines
    call Routine FillDiagonal(A)
    # First modify matrix A to have transitions between closed states in the upper left
    # corner
    call Routine SortMatrix(A, n, S, pnul)
    set m to the number of closed states
    declare matrix evecs[m*m]
    # get the eigenvalues and eigenvectors of the upper left corner of A and return
    # the result in evals and evecs
    call Routine Eigenvalues(A, evals, evecs, m)
    # find scaling factors for eigenvectors
    declare array factors[m]
    call Routine Invertmatrix(evecs, pnul, m, factors)
    # Differentiate from infinity to 0, scale and sum the elements of the vectors
    # For symplicity, only isolated real roots are assumed. Differentiation of
    # repeated roots or complex roots is somewhat more elaborate.
    for (i=0 to i=m-1)
    { cum=0
      for (j=0 to j=m-1) cum=cum+evecs[j,i]
      amps[i]=-cum*factors[i]*evals[i]
    } # next i
} # end of routine
```

Routine **Closedtimes**(matrix A, array pnul, array S, int n, float t, array evals, array amps)

```
{     # matrix A is the n*n transition matrix
      # array pnul of length n contains the probabilities to be in each state at t=0
      # array S indicates whether the state is closed (S[i]=0) or open (S[i]=1)
      # n is the number of states
      # t is the integration time span
      # The time constant of each exponential is returned in array evals
      # The amplitude of each exponential is returned in array amps
      # The routines Eigenvalues() and Invertmatrix() are standard routines
      call Routine FillDiagonal(A)
      # First modify matrix A to have transitions between closed states in the upper left
      # corner
      call Routine SortMatrix(A, n, S, pnul)
      declare matrix evecs[n*n]
      # get the eigenvalues and eigenvectors of the upper left corner of A and return
      # the result in evals and evecs
      call Routine Eigenvalues(A, evals, evecs, n)
      # find scaling factors for eigenvectors
      declare array factors[n]
      call Routine Invertmatrix(evecs, pnul, n, factors)
      set m to the number of closed states
      # Integrate the open states over t seconds and store result in pnul
      # For symplicity, only isolated real roots are assumed. Integration of
      # repeated roots or complex roots is somewhat more elaborate.
      for (i=m to i=n-1)
      { cum=0
        for (j=0 to j=n-1) cum=cum+factors[j]*evecs[i,j]*(1-exp(evals[j]*t))/evals[j]
        pnul[i]=cum
      } # next i
      # get transition probabilities from open to closed states and store result in pnul
      for (i=0 to i=m-1)
      { cum=0
        for (j=m to j=n-1) cum=cum+A[j,i]*pnul[j]
        pnul[i]=cum
      } # next i
      call Routine FirstLatency(A, pnul, S, n, evals, amps)
} # end of routine
```

Routine **Opentimes**(matrix A, array pnul, array S, int n, float t, array evals, array amps)

```
{     # The meaning of the input variables is as for routine Closedtimes()
      for (i=0 to i=n-1)
      { if S[i]=1 then S[i]=0
        else S[i]=1
      } # next i
```

```
    call Routine Closedtimes(A, pnul, S, n, t, evals, amps)
} # end of routine

Routine Macrocurrent(matrix A, array pnul, array S, int n, array evals, array amps)
{   # matrix A is the n*n transition matrix
    # array pnul of length n contains the probabilities to be in each state at t=0
    # array S contains channel conductances for each state
    # n is the number of states
    # The time constant of each exponential is returned in array evals
    # The amplitude of each exponential is returned in array amps
    # The routines Eigenvalues() and Invertmatrix() are standard routines
    call Routine FillDiagonal(A)
    # Get the eigenvalues and eigenvectors of A and return the result in evals and evecs
    call Routine Eigenvalues(A, evals, evecs, n)
    # find scaling factors for eigenvectors
    declare array factors[n]
    call Routine Invertmatrix(evecs, pnul, n, factors)
    # scale vectors, sum their elements weighed by S
    for (i=0 to i=n-1)
    { cum=0
      for (j=0 to j=n-1) cum=cum+evecs[j,i]*S[j]
      amps[i]=cum*factors[i]
    } # next i
} # end of routine

Routine FillDiagonal(matrix A, int n)
{ # matrix A is the n*n transition matrix
    # n is the number of states
    for (i=0 to i=n-1)
    { cum=0
      for (j=0 to j=n-1)
      { if i is not j then cum=cum+A[j,i]
      } # next j
      A[i,i]=cum
    } # next i
} # end of routine

Routine SortMatrix(matrix A, int n, array S, array pnul)
{   # matrix A is an n*n matrix
    # pnul is an array of length n
    # array S indicates whether the state is closed (S[i]=0) or open (S[i]=1)
    s1=0,s2=0
    while s2<n
    {   while (s1<n) and (S[s1] is not 1) s1=s1+1
```

```
        s2=s1+1
        while (s2<n) and (S[s2] is not 1) s2=s2+1
        if (s1<n-1) and (s2<n)
           {   swap columns s1 and s2 of A
               swap rows s1 and s2 of A
               swap pnul[s1] and pnul[s2]
               swap S[s1] and S[s2]
           } # end if
     } # end while
} # end of routine
```

THE RECEIVER OPERATING CHARACTERISTIC

The statistical theory behind the receiver operating characteristic (ROC) originally comes from the problem of detecting an enemy aircraft by reflected radar pulses against background noise. Part of the problem can be solved by data processing techniques such as averaging, correlation, and filtering, but at the end a person or the machine has to decide whether a blip on an oscilloscope screen indicates an incoming aircraft or not.

Suppose the background noise is described by a Gaussian distribution with mean 0 and the signal the same distribution with a mean different from 0 (signal mean is 35 in Fig. A.8, left panels). Now, suppose also that the detector sets off an alarm as soon as the signal is 25 or above (the red vertical lines in the left and middle panels of the figure). The probability that a real signal is present corresponds to the part of the surface under the Gaussian distribution, labeled "signal," at the right of the vertical line. The middle panel traces this surface as a function of signal strength. In Fig. A.8A, the probability to correctly identify a signal (true positive or "hit"), given a threshold of 25, is 0.63. The probability of a false positive is 0.18. Plotting the probability of true positives as a function of the probability false positives obtained with different thresholds gives the ROC curve (rightmost panels). The variance of the distributions for signal and noise are not necessarily identical. Fig A.8B shows that the ROC curve skews somewhat if the variances are different. Moreover, the underlying probability density distributions may not be Gaussian.

What is the utility of an ROC curve? Let us consider two extremes. If the distributions for signal and noise are identical, then the cumulative distributions are identical and the true positive rate equals the false-negative rate (the gray diagonal in the ROC curves). The other extreme occurs when the noise and signal curves do not overlap at all, for example, if the variance is 0. The detector never makes errors in that case. The ROC curve then corresponds to the horizontal line going from [0, 1] to [1, 1]. The area under the ROC curve (AUC) may be taken as a token of detector quality. The AUC is exactly 1 for a flawless detector. In the case of completely overlapping distributions, the AUC is 0.5, meaning the detector does not do better than someone flipping a coin. "Accuracy," defined as the sum of true positives and true negatives divided by the total number of test samples, is another estimate of quality (it is 0.83 for the curve in Fig. A.8B). Drawing a ROC curve can be applied to questions such as "To what extent is method X well adapted to decide whether drugs cause *torsade de pointes*?" and "Is method Y better than method X?".

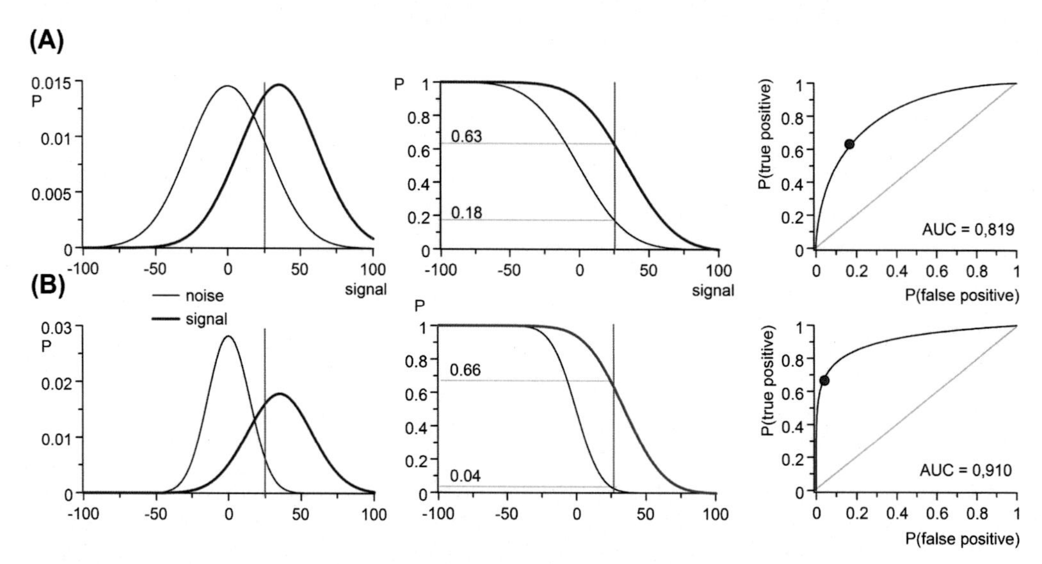

FIGURE A.8 From left to right: probability density functions, cumulative probability functions, and ROC curves for noise (*thin grey lines*) and signal + noise (*thick blue lines*). The *vertical red lines* in the graph represent the decision threshold (25 in this example). The dots in the ROC curves correspond to the coordinates [0.18, 0.63] and [0.04, 0.66], respectively. The variances of Gaussian signal and noise curves are identical in (A) and different in (B).

References

1. Rainville ED, Bedient PE. *Elementary differential equations*. 6th ed. Macmillan Publishing Co, Inc.; 1981.
2. Watkins DS. *Fundamentals of matrix computations*. 2nd ed. New York: Wiley Interscience; 2002.

Recommended Literature

Electricity and Electronics

Horowitz P, Hill W. *The art of electronics*. Cambridge: Cambridge University Press; 1989.
Stanley WD. *Electronic devices: circuits & applications*. Englewood Cliffs (NJ): Prentice Hall; 1989.
Boyes W. *Instrumentation reference book*. 4th ed. Burlington (MA): Butterworth-Heinemann; 2009.

Electrochemistry

Bard AJ, Faulkner LR. *Electrochemical methods: fundamentals and applications*. New York: Wiley; 2001.

Neurophysiology and Recording Methods

Hille B. *Ionic channels of excitable membranes*. 3 ed. USA: Oxford University Press; 2001. ISBN: 978−0878933211.
Neher, E., Sakmann, B. Single-channel recording. Plenum Press, New York and London; London, 1995. ISBN: 978−1441912305.
Standen NB, Gray PTA, Whitaker MJ, editors. *Microelectrode techniques*. Cambridge: The Company of Biologists Limited; 1987, ISBN 0 948601 09 4.

Signal Analysis and Mathematics

Dempster J. *Computer analysis of electrophysiological signals*. London: Academic Press; 1993.

Hamming RW. *Digital filters*. 2nd ed. Englewood Cliffs (NJ): Edt. Oppenheim, A.V. Prentice Hall; 1983, ISBN 0-13-212506-4.

Rieke F, Warland D, Ruyter van Steveninck R, de Bialek W. *Spikes — exploring the neural code*. Cambridge (MA): MIT Press; 1997, ISBN 0-262-18174-6.

Wilson HR. *Spikes — decisions and actions*. Oxford: Oxford University Press; 1999, ISBN 0-19-852430-7.

Marmarelis PZ, Marmarelis VZ. *Analysis of physiological systems, the white noise approach*. New York: Plenum Press; 1978, ISBN 0-306-31066-X.

Rainville ED, Bedient PE. *Elementary differential equations*. 8th ed. USA: Pearson; 1996, ISBN 9780135927755.

Watkins DS. *Fundamentals of matrix computations*. 2nd ed. New York: John Wiley & Sons Inc.; 2002, ISBN 0-471-21394-2.

Elementary Microscopy

Inoué S. *Video microscopy*. 2nd ed. Berlin: Springer Verlag; 1997. ISBN: 978–0306455315.

Programming Language Websites

Java: https://www.java.com.
R: https://www.r-project.org.
LabView: http://www.ni.com/en-us/shop.html.
MATLAB®: https://www.mathworks.com/products/matlab.html.
Mathematica: https://www.wolfram.com/mathematica.

Equipment Websites

Multichannel systems: http://www.multichannelsystems.com.
Flyion: https://www.digitimer.com/pdf/flyion/flyscreen8500.pdf.
Molecular devices: https://www.moleculardevices.com.
Cytocentrix: http://cytobioscience.com.
Nanion: http://www.nanion.de.
Bionas: http://www.bionas-discovery.com/prodservices/instruments/system2500.

Index

'*Note:* Page numbers followed by "f" indicate figures, "t" indicate tables.'

B

C

N

Printed in the United States
By Bookmasters